Earth Observation and Deep Learning for Coastal Monitoring, Mapping, and Management

Uzair Bhatti
School of Information and Communication Engineering, Hainan University, China

Di Wu
Hainan University, China

IGI Global
Scientific Publishing
Publishing Tomorrow's Research Today

Vice President of Editorial	Melissa Wagner
Director of Acquisitions	Mikaela Felty
Director of Book Development	Jocelynn Hessler
Production Manager	Mike Brehm
Cover Design	Jose Rosado

Published in the United States of America by
IGI Global Scientific Publishing
701 East Chocolate Avenue
Hershey, PA, 17033, USA
Tel: 717-533-8845 | Fax: 717-533-7115
Website: https://www.igi-global.com E-mail: cust@igi-global.com

Library of Congress Cataloging-in-Publication Data

LCCN: 2026005308 (CIP Data Pending)
ISBN13: 9798337388724
Isbn13Softcover: 9798337388731
EISBN13: 9798337388748
British Cataloguing in Publication Data
A Cataloguing in Publication record for this book is available from the British Library.

Table of Contents

Detailed Table of Contents

Babamuratov Bekzod, Department of Medicine, Termez University of Economics and Service, Uzbekistan
Annazarova Barno Rustamovna, Mamun University, Uzbekistan
Sajid Ali, Department of Information Sciences, University of Education, Multan, Pakistan
Yuldasheva Gulora Gulumovna, Urgench State University, Uzbekistan

Hyperspectral remote sensing has a lot of potential in identifying the heavy metals in the mangrove soil, however, the issues of spectral interference, spatial and temporal variability, as well as restriction of sensor are to be solved. The most recent developments, including the incorporation of machine learning and AI-based models, multispectral fusion, and enhanced sensor technologies, have assisted in enhancing detection. These techniques are however subject to further research so that they can be refined and better calibration and validation techniques can be developedNonetheless, hyperspectral remote sensing has potential as an efficient instrument of environmental scrutiny, especially of the mangrove ecosystems. With the help of comprehensive and ongoing information about the soil texture and the level of contamination, hyperspectral sensors can help to better manage mangrove forests, as well as contribute to the long-term sustainability of the environment.

Coastal salt marshes are essential for global carbon sequestration, playing a vital role in mitigating climate change through the storage of blue carbon. These ecosystems sequester carbon in both plant biomass and sediments, with the ability to store carbon for centuries. However, accurately quantifying blue carbon storage in salt marshes remains challenging due to the dynamic nature of these ecosystems and the limitations of traditional field measurement techniques. Recent advancements in remote sensing technology, particularly hyperspectral remote sensing, offer a promising approach to monitoring salt marshes on a large scale with high spatial and spectral resolution. Hyperspectral sensors provide detailed spectral data, enabling the differentiation of vegetation species, assessment of plant health, and mapping of carbon-rich habitats. Additionally, we explore the latest technological innovations, including satellite-based sensors and unmanned aerial vehicles (UAVs), which enhance data collection in challenging coastal environments.

Microplastic pollution has emerged as a significant environmental threat, affecting marine and coastal ecosystems, including mangrove forests. Mangrove roots, essential for stabilizing coastlines and supporting biodiversity, are particularly vulnerable to microplastic contamination, which can interfere with nutrient uptake, root growth, and overall ecosystem health. Remote sensing technologies provide an effective means to monitor the extent and impact of microplastics on mangrove ecosystems, offering insights into root health and environmental conditions. By using satellite imagery and drone-based sensors, researchers can assess changes in vegetation health, canopy structure, and root systems that may result from microplastic exposure. These data can be integrated with ground-based observations to provide a comprehensive understanding of how microplastics affect mangrove root functions, such as gas exchange and nutrient cycling

environmental conditions associated with land cover shifts. The findings stress the need for adaptive management strategies to balance vegetation health and land use, informing future policy development for climate-resilient agricultural systems.

Mughair Aslam Bhatti, Szabist University, Pakistan
Bakirov Juma, Department of Preschool and Primary Education, Termez University of Economics and Service, Termez, Uzbekistan
Intizor Avazmetova, Department of Biology, Urgench State University, Urgench, Uzbekistan
Anorgul Ashirova, Department of General Professional Sciences, Mamun University, Khiva, Uzbekistan
Avezova Umida Maksudovna, Department of Ecology and Life Safety, Urgench State University Named After Abu Rayhan Biruni, Uzbekistan

Machine learning (ML) models are increasingly being applied to environmental monitoring, particularly for detecting emerging contaminants in marine ecosystems. Seagrass ecosystems, vital for biodiversity and carbon sequestration, are highly sensitive to pollutants, which can degrade their health and function. Traditional methods of contaminant detection often involve time-consuming fieldwork and limited spatial coverage. However, drone imagery, combined with ML techniques, offers a powerful tool for rapid, large-scale monitoring. ML algorithms can analyze high-resolution drone images to detect subtle changes in water quality, seagrass health, and potential contamination levels, including heavy metals, pesticides, and pharmaceuticals.

Remote sensing data assimilation techniques are becoming increasingly crucial for accurately modeling blue carbon fluxes in estuarine ecosystems, which play a key role in carbon storage and climate regulation. Estuaries, with their dynamic and highly variable environments, are challenging to monitor using traditional methods due to their complexity and the difficulty of continuous data collection. Remote sensing, combined with data assimilation, allows for the integration of satellite and airborne imagery with environmental models to improve the accuracy of blue carbon flux predictions. By assimilating remotely sensed data into carbon models, researchers can better estimate carbon sequestration rates, monitor vegetation health, and evaluate the impacts of environmental changes on carbon fluxes. This approach helps overcome spatial and temporal limitations, offering real-time insights into estuarine dynamics and the effectiveness of conservation efforts.

Unmanned Aerial Vehicle (UAV)-based LiDAR (Light Detection and Ranging) technology offers a powerful tool for the detailed mapping and monitoring of mangrove ecosystems. Mangroves are critical to coastal protection, biodiversity, and carbon sequestration, but their structure and carbon storage capabilities are often difficult to assess using traditional methods. UAV-based LiDAR allows for high-resolution 3D reconstruction of mangrove canopies, capturing detailed information on canopy height, density, and complexity. This data enables accurate carbon stock estimation by providing precise measurements of above-ground biomass, which is essential for understanding the role of mangroves in carbon sequestration. The ability to perform non-invasive, high-precision surveys over large and remote areas makes UAV-based LiDAR particularly suited for monitoring mangrove forests, especially in regions that are difficult to access.

Preface

INTRODUCTION

As we stand on the brink of environmental crises, the need for innovative solutions to monitor and manage our planet's ecosystems has never been more urgent. This book explores the intersection of advanced technology, such as remote sensing, artificial intelligence (AI), and machine learning, with environmental science, particularly in the context of coastal and marine ecosystems. By leveraging these technologies, we can enhance our understanding of ecological health, assess degradation, and implement conservation strategies at an unprecedented scale. This collection of 20 chapters provides a comprehensive, state-of-the-art perspective on how technology is revolutionizing environmental monitoring and management, particularly in the face of climate change and biodiversity loss.

OVERVIEW OF THE SUBJECT MATTER

This book offers in-depth research and case studies on the role of remote sensing technologies, machine learning models, and artificial intelligence in environmental monitoring. The chapters cover a wide range of topics, from analyzing the health of mangroves and coastal ecosystems using satellite and drone data, to developing AI-driven models for real-time ecosystem assessments. The integration of remote sensing data, AI, and advanced analytics enables a more nuanced understanding of ecosystem dynamics, providing tools to predict and mitigate the impacts of environmental threats. This book serves as an essential guide to understanding how modern technologies can contribute to the sustainability and preservation of critical ecosystems that play a vital role in climate change mitigation, biodiversity conservation, and resource management.

DESCRIPTION OF WHERE YOUR TOPIC FITS IN THE WORLD TODAY

As climate change accelerates, ecosystems such as coastal zones, mangroves, and wetlands face increasing pressure from both human activities and natural threats. This book situates itself at the forefront of the global effort to understand and protect these ecosystems, using innovative tools to monitor, assess, and address environmental threats. The integration of AI, remote sensing, and machine learning is critical in tackling large-scale environmental issues, such as biodiversity loss, deforestation, and ocean pollution. The technologies discussed in this book are not only transformative for environmental science but are increasingly becoming indispensable for sustainable development practices worldwide. By combining these technologies with data-driven models, the book offers new pathways for effective ecosystem management, aligning with global sustainability goals.

DESCRIPTION OF THE TARGET AUDIENCE

This book is intended for a diverse audience, including researchers, scholars, and students in environmental science, geospatial data science, and ecology. It is also geared towards professionals working in ecosystem management, environmental policy, climate change mitigation, and conservation. Additionally, the book will be valuable to experts in AI, machine learning, and remote sensing who are interested in applying their technologies to environmental monitoring. Policymakers and environmental NGOs will also benefit from the practical solutions outlined in the chapters, which emphasize real-world applications and provide insights into how technology can drive impactful change in the management of critical ecosystems.

DESCRIPTION OF THE IMPORTANCE OF EACH CHAPTER SUBMISSION

Chapter 1: Challenges in Using Hyperspectral Remote Sensing for Heavy Metal Detection in Mangrove Soils

Hyperspectral remote sensing is explored as a tool for detecting pollutants in sensitive ecosystems like mangrove forests. This chapter highlights the technical challenges faced in applying hyperspectral sensors to environmental monitoring and offers insights into the potential improvements needed for broader adoption.

Chapter 2: Deep Learning Models for Land Cover Classification and Change Detection in Coastal Zones

Deep learning techniques are revolutionizing the analysis of land cover changes. This chapter explains the application of deep learning models for detecting land cover changes in coastal ecosystems, demonstrating how AI can assist in proactive management and conservation efforts.

Chapter 3: Eco-Aware Sustainable Sub Sea Monitoring – An AI-Driven Approach for Corrosion Detection in Pipelines and Marine Safety

This chapter focuses on the application of AI for detecting corrosion in subsea pipelines, a critical issue for marine infrastructure. It explores how these technologies can improve the monitoring of underwater environments and contribute to safer and more sustainable practices.

Chapter 7: Geological and Geophysical Analysis of Uru-Ugworji Diorite in Southeastern Nigeria

While not directly related to coastal ecosystems, this chapter provides an important geological context for understanding the broader environmental framework in which mangrove ecosystems exist. It offers valuable insights into the role of geology in environmental monitoring.

Chapter 5: UAV-Based LiDAR Sensing for 3D Reconstruction of Mangrove Canopies and Carbon Stock Estimation

This chapter discusses the use of UAVs equipped with LiDAR technology to create 3D models of mangrove canopies. This innovation enables more accurate carbon stock estimations and contributes to a better understanding of mangrove ecosystem dynamics.

Chapter 6: Impact of Pharmaceutical Emerging Contaminants on Blue Carbon Sequestration in Tidal Marshes

Pharmaceutical contaminants represent a growing concern for environmental health. This chapter explores how these emerging contaminants affect blue carbon sequestration in tidal marshes and offers insights into potential mitigation strategies.

Chapter 7: Remote Sensing Data Assimilation Techniques for Blue Carbon Flux Modeling in Estuarine Ecosystems

This chapter introduces data assimilation techniques in remote sensing for modeling blue carbon flux in estuaries. It provides a sophisticated approach to quantifying carbon flows in coastal ecosystems, helping researchers and policymakers optimize conservation strategies.

Chapter 8: Land Cover Dynamics and NDVI-Based Vegetation Health Trend Analysis for Agricultural Land Suitability

This chapter examines land cover dynamics using NDVI analysis, a crucial tool for understanding vegetation health trends. Although focused on agricultural land suitability, the methodologies discussed are equally applicable to understanding the health of coastal vegetation.

Chapter 9: Machine Learning Models for Detecting Emerging Contaminants in Seagrass Ecosystems via Drone Imagery

Emerging contaminants pose significant threats to coastal ecosystems. This chapter demonstrates how machine learning models applied to drone imagery can help detect these contaminants, providing actionable data for environmental management.

Chapter 10: UAV and Satellite Data Fusion for Comprehensive Coastal Contaminant Source Identification

The fusion of UAV and satellite data is discussed in this chapter, offering a comprehensive approach to identifying contaminant sources in coastal zones. This method provides a more robust framework for

Chapter 11: Hyperspectral Remote Sensing for Mangrove Ecosystem Monitoring: A Global Overview

The final chapter offers a global overview of hyperspectral remote sensing applications in mangrove ecosystem monitoring. It synthesizes findings from various studies and highlights the potential of hyperspectral sensing in long-term conservation efforts.

CONCLUSION: IMPACT ON THE FIELD AND CONTRIBUTION TO THE SUBJECT MATTER

This book contributes significantly to the field of environmental monitoring, offering cutting-edge solutions to the challenges of preserving critical ecosystems. By integrating remote sensing, machine learning, and AI, it provides the tools needed to monitor environmental changes in real time, enabling proactive management and more effective conservation strategies. The chapters offer diverse, innovative approaches to ecosystem monitoring, ensuring that the book serves as a valuable resource for researchers, professionals, and policymakers alike. Through its interdisciplinary approach, this work not only furthers our understanding of coastal and marine ecosystems but also offers practical applications that can drive meaningful change in the global effort to mitigate climate change and preserve biodiversity.

Chapter 1

Challenges in Using Hyperspectral Remote Sensing for Heavy Metal Detection in Mangrove Soils

Babamuratov Bekzod
Department of Medicine, Termez University of Economics and Service, Uzbekistan

Annazarova Barno Rustamovna
Mamun University, Uzbekistan

Sajid Ali
Department of Information Sciences, University of Education, Multan, Pakistan

Yuldasheva Gulora Gulumovna
http://orcid.org/0009-0009-2202-1156
Urgench State University, Uzbekistan

ABSTRACT

Hyperspectral remote sensing has a lot of potential in identifying the heavy metals in the mangrove soil, however, the issues of spectral interference, spatial and temporal variability, as well as restriction of sensor are to be solved. The most recent developments, including the incorporation of machine learning and AI-based models, multispectral fusion, and enhanced sensor technologies, have assisted in enhancing detection. These techniques are however subject to further research so

DOI: 10.4018/979-8-3373-8872-4.ch001

that they can be refined and better calibration and validation techniques can be developedNonetheless, hyperspectral remote sensing has potential as an efficient instrument of environmental scrutiny, especially of the mangrove ecosystems. With the help of comprehensive and ongoing information about the soil texture and the level of contamination, hyperspectral sensors can help to better manage mangrove forests, as well as contribute to the long-term sustainability of the environment.

1. INTRODUCTION

Significance of Mangrove Ecosystems in Coastal Protection and Biodiversity as Well as Carbon Sequestration

The mangrove ecosystems are important barriers to the coastal areas that shield the coastlines against erosion, storm surges, and sea level rise. They serve as important homes to many species as they sustain biodiversity along the coast. Another important role of mangroves entails sequestration of carbon whereby massive quantities of carbon dioxide are absorbed by the mangroves and this plays a role in counterbalancing climate change effects (Wang, 2024).

Mangrove Soil in the Environmental and Ecological Health

Soils of mangrove ecosystem have a lot of organic matter and nutrients that are beneficial to all the ecosystem. These soils also promote different species of plants and are also part of the food web because they provide nutrients to aquatic organisms. Moreover, they cause microbial activity, which also promotes nutrient cycling and stability of the ecosystem in general (Lovynska et al., 2024).

Mangrove Soils Heavy Metals Contamination

Heavy Metals Sources of Pollution in the Mangrove Ecosystems

Most of the heavy metal contamination in the mangrove soils is usually of anthropogenic origin. Toxic metals (e.g., mercury, lead, cadmium) are emitted into the environment by industrial activities (e.g. mining, manufacturing, oil extraction). The accumulation of heavy metals in mangrove soils may also be caused by agricultural runoff that has pesticides and fertilizers. There is also the issue of urbanization and shipping activities in coastal regions which compound the issue by bringing pollutants like copper and zinc to these ecosystems (Murong, 2025).

Effects of Heavy Metal Contamination of Mangrove Flora, Fauna, and Health of the Entire Ecosystem

High concentration of heavy metals in the mangrove soils may impacts negatively on the growth of the plants which results to toxicity causing plant inability to absorb nutrients and generate oxygen. This in turn has an impact on the animals that have mangroves as a habitat and food source (Kuang, 2024). The heavy metal contamination also interferes with fine balance of micro-bacterial communities in the soil which is crucial in cycling of nutrients and general well-being of an ecosystem. Excessive exposure to heavy metals on a long-term basis may cause bioaccumulation of flora and fauna, which further impairs the biodiversity of mangrove ecosystems (Hyperspectral remote sensing for soil heavy metal inversion, 2025).

Requirement of Soil Monitoring of Heavy Metals

Traditional approaches to testing of heavy metals in soil are generally based on laboratory tests: atomic absorption spectrometry (AAS), inductively coupled plasma mass spectrometry (ICP-MS), or X-ray fluorescence (XRF). Although they are precise, the methods are time and labor-consuming, and frequently involve the collection of soil samples, which can be expensive and logistically involved in some cases, particularly in the mangrove regions in remote and inaccessible locations. Additionally, these techniques will only give the data in the form of points, which might not indicate the space variation of heavy metal pollution in the mangroves at large scales (Lamine, 2019).

The Prospects of Hyperspectral Remote Sensing in Soil Analysis, in this Case- Heavy Metal Detector

Hyperspectral remote sensing is a new technology that can be used in measuring heavy metals of mangrove soils since it captures a wide metabolite of wavelengths that reflect the appearance of the soil. The method has the potential of identifying the slightest changes in the soil composition related to the presence of heavy metals without necessarily making a large-scale ground-based sampling. Hyperspectral sensors have the potential to offer high-resolution, spatially explicit data coverage of a large area thus allowing continuous and real-time detection of heavy metal pollution in mangrove habitats (Maliki et al., 2012).

Purpose and Rationality of the Review

The purpose of this review will be to critically describe the contemporary issues and developments in the application of hyperspectral remote sensing in the detection of heavy metals in mangrove soils. It will discuss spectral signatures of various heavy metals, constraints of the existing remote sensing technologies and environmental conditions that affect the process of detection. Also, the review will shed light on the integration of hyperspectral data with other remote sensing methods, e.g. LiDAR or multispectral imagery, to enhance the accuracy of detection. The general purpose should be to offer a thorough insight into the promise and the challenges of hyperspectral remote sensing of environmental monitoring of mangrove ecosystems (Wang, 2025).

2. HYPERSPECTRAL REMOTE SENSING PRINCIPLES AND APPLICATIONS

Introduction to Hyperspectral Remote Sensing

Hyperspectral remote sensing is based on intense data gathering over a wide spectral span frequently in the visible spectrum to the infrared spectrum, comprising of the electromagnetic spectrum. As compared to the conventional multispectral imaging, where data is collected in a limited number of wide bands, Hyperspectral sensing collects hundreds of narrow and adjacent bands. This enables the materials to be identified more accurately and subtle spectral features to be detected. Every substance, such as soil, vegetation and water, scatters and absorbs light differently at certain wavelengths, which produce spectral signatures. The signatures allow classifying and analyzing various surface materials according to their reflectance attributes (Su, 2025).

Hyperspectral Sensor and Platform Types

The hyperspectral data may be acquired with the help of diverse sensor types and platforms that are application-specific, The key types include:

Satellite-Based Sensors

Global coverage of the earth is made by satellites with hyperspectral sensors and these satellites are invaluable when it comes to monitoring the environment on a large scale. Examples are NASA EO-1 Hyperion and EnMAP of the European

Space Agency. These sensors normally have moderate spatial resolution (tens to hundreds of meters) and high spectral resolution that is most suitable in large-scale or world-wide monitoring of land and coastal ecosystems (Zhao, 2015).

Airborne Sensors

Hyperspectral sensors AISA (Airborne Imaging Spectrometer for Applications) or HyMap are mounted on airborne vehicles, e.g., aircraft or drones. Such platforms have greater spatial resolution (centimeters to meters) than satellites and permit smaller studies with more detail with smaller areas, including individual mangrove forest or a particular polluted location (Liu, 2019).

Ground-Based Sensors

Ground-based hyperspectral sensors are applied in cases of the analysis of the surface features with high precision and localization. These sensors are normally mounted on tripods or vehicles and are used to record the high-resolution data of a given area of interest. The sensors can be used in the process of calibration and validation of the airborne or satellite data and in research of small-scale plots when thorough analysis is needed (Ding et al., 2022).

Hyperspectral Remote Sensing in Environmental Monitoring

Hyperspectral remote sensing has also received many applications in the sector of environmental monitoring. The principal applications are:

Vegetation Monitoring

Hyperspectral sensors are capable of measuring the health of vegetation based on the reflected light of plants at different wavelength. This data can be utilized to track the level of stress on plants due to drought, pests, or the lack of the nutrients. Hyperspectral data are usually used to compute vegetation indices, including the Normalized Difference Vegetation Index (NDVI), to assess plant vitality and growth dynamics (Arif et al., 2022).

Soil Health Assessment

Hyperspectral remote sensing is useful in the study of soil characteristics such as texture, moisture content and organic matter. This plays a critical role in comprehending the state of soil in an ecosystem such as the mangrove where the quality of soil has a direct influence on plant and animal lives. Hyperspectral sensors may also be used to detect any soil contaminants such as heavy metals by detecting certain spectral signatures (Tan et al., 2019).

Water Quality

Hyperspectral sensors are able to identify pollutant in water bodies through spectral variation which is brought about by dissolved organic matter, suspended matter and chemical pollutant. This will involve the observation of dangerous algal blooms and other problems with water quality. Hyperspectral remote sensing can be an effective instrument in marine and coastal management due to its capability to monitor water quality over wide regions and at high degrees of resolution (Pyo et al., 2020).

Applicability to Detection of Soil Characteristics and Contaminants Like Heavy Metals

Hyperspectral remote sensing has especially been useful in the identification of pollutants such as heavy metals in soils as the pollutants have distinct spectral patterns. For example:

Heavy Metal Detection

Hyperspectral sensors are able to detect the changes that appear in the soil reflectance due to the presence of heavy metals like cadmium, lead and mercury. These metals change the chemical composition and mineral structure of soils that affect their spectral properties. More sophisticated spectral analysis like machine learning methods or spectral unmixing can determine and measure heavy metals in soil (Zhou, 2021).

Soil Contamination Mapping

Hyperspectral remote sensing enables scientists to map out the contaminated regions in detail even inside the mangrove forests and heavy metals are an issue. The spectral resolution is high and enables the distinction of the levels of contamination as well as the opportunity to identify the spatial variation, which is essential in targeted cleanup work and ecological research (Zhang et al., 2022).

Precision Agriculture and Soil Management

Besides identification of pollutants, hyperspectral remote sensing is also applicable in precision farming to check the state of soil and guide management activities. This involves maximum utilization of water, soil fertility test, and monitoring of the efficiency of soil amendments. In case of mangrove ecosystems, the information may assist in determining the areas of stress due to the toxicity of heavy metals, and this may be used to provide a more efficient management of the ecosystem (Wei, 2020).

Table 1. Applications of hyperspectral remote sensing in environmental monitoring, specifically focusing on vegetation, soil, water, and heavy metal detection

Application Area	Description	Relevance to Mangrove Ecosystems
Vegetation Monitoring	Measures plant health based on reflected light across various wavelengths. Used to assess plant vitality.	Tracks plant stress due to drought, pests, or nutrient deficiencies.
Soil Health Assessment	Studies soil texture, moisture content, and organic matter	Assesses soil quality in mangrove ecosystems, directly affecting plant and animal life.
Water Quality Monitoring	Identifies pollutants in water by analyzing spectral variations caused by dissolved organic matter, suspended solids, and chemical pollutants.	Monitors water quality in coastal and mangrove environments, detecting algal blooms and pollution.
Heavy Metal Detection	Detects changes in soil reflectance due to heavy metals like cadmium, lead, and mercury.	Identifies contamination in mangrove soils, aiding in pollution control.
Soil Contamination Mapping	Maps contaminated areas in detail using hyperspectral data.	Helps pinpoint contaminated regions within mangrove forests for targeted cleanup.
Precision Agriculture and Soil Management	Guides soil management activities, such as water usage and soil fertility tests.	Identifies areas of soil stress from heavy metal toxicity, supporting ecosystem management.

3. HYPERSPECTRAL REMOTE SENSING OF HEAVY METAL DETECTION

Effect of Heavy Metals on Spectral Reflectance

The major heavy metals like lead (Pb), cadmium (Cd), mercury (Hg) and arsenic (As) have great influence on the chemical and physical characteristics of soils that consequently influence the spectral reflectance of soils. These metals can alter the color, mineral composition, and texture of the soil when they settle in the soils affecting the ability of the soil to reflect light of a certain wavelength (Gao, 2017).

Lead (Pb): Lead contamination normally leads to reduction in the reflectance in the visible and shortwave infrared (SWIR) region, which is detectable in the hyperspectral data.

Cadmium (Cd): Cadmium can be sensed with the help of its absorption in the visible and near-infrared (NIR) spectral regions.

Mercury (Hg): Mercury can cause spectral variations of both the visible and SWIR ranges of the reflectance of the soil and particularly in soils with high organic content.

Arsenic (As): Arsenic pollution may cause spectral characteristics that distinguish it as compared to other metal pollutants particularly in the SWIR region because it interacts differently with soil minerals.

Effect of Heavy Metals on Spectral Reflectance, shown in Figure 1:

Figure 1. Unveiling heavy metals' spectral footprints

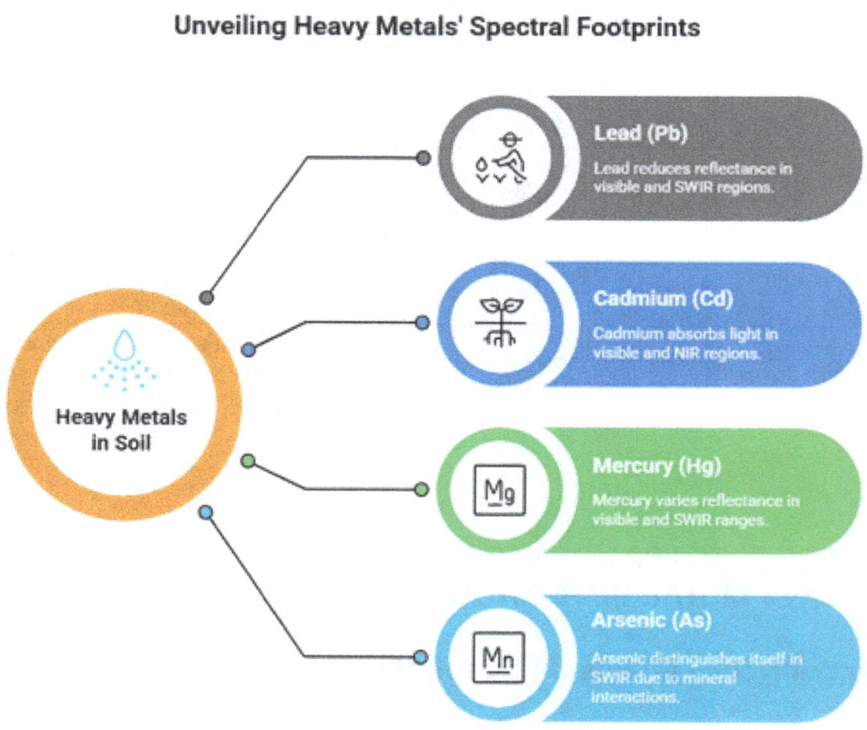

Spectral Bands of the Common Spectral Bands Used in Detecting the Heavy Metals

These subtle changes can be measured through hyperspectral remote sensing that has the ability to identify them using the spectral signatures of soils in different wavelengths (Fu, 2024). Typical spectral bands of detecting heavy metals are:

Visible (400-700 nm): The visible spectrum reflects on the color of the soil, and the content.

Near-Infrared (7001100 nm): The range is deemed vital in determining organic matter and moisture content that may be affected by the presence of contaminated heavy metals.

Shortwave Infrared (11002500nm-): The SWIR is especially useful to monitor mineralogy and moisture, which are two factors affected by the presence of heavy metals. It is commonly applied in the in-depth analysis of soil make-up.

Specific Absorption Characteristics: There are wavelengths like 2100 2300 which are sensitive to the existence of the heavy metals and they can be utilized to differentiate among the various contaminants.

Current Methodologies and Techniques

Past Research in Hyperspectral Remote Sensing to Detection of Heavy Metal

There are numerous studies in which the use of hyperspectral remote sensing has been used to identify soils with heavy metals by examining the spectral reflectance data. The most frequently employed methods are:

Linear Models: The linear models will include a Partial Least Squares Regression (PLSR) and Principal Component Analysis (PCA), which are utilized to make the correlations with the spectral and the concentrations of heavy metal in the soil samples. When the correlation between spectral reflectance and metal concentration is simple, a linear model is usually involved (Leone et al., 2012).

Machine Learning Techniques: More sophisticated techniques, which include Support Vector Machines (SVM), Random Forest (RF), and Artificial Neural Networks (ANN) have been employed to enhance the detection accuracy by training sophisticated patterns in spectral data. They are especially useful when dealing with big and multifaceted data, as is the case with hyperspectral sensor data (Khosravi et al., 2018).

Spectral Unmixing: Spectral unmixing is a procedure that is employed to isolate various constituents (e.g. soil, vegetation, water) within hyperspectral data. Spectral unmixing techniques can also be used to better determine contaminated regions, such as Linear Spectral Mixture Analysis (LSMA), in which spectral contribution of a particular contaminant is isolated (Melgani & Bruzzone, 2004).

Application of Spectral Indices and Specific Wavelengths

Besides machine learning-based methods, spectral indices like the Normalized Difference Vegetation Index (NDVI), and Soil Adjusted Vegetation Index (SAVI) are frequently used on hyperspectral data. These are indices that are normally applied in the analysis of vegetation but can be modified to determine the alteration of the soil characteristics as a result of heavy metal contamination (Breiman, 2001).

Metal-Specific Indices: Special purpose spectral indices have been designed to be more sensitive to detect heavy metals in soils. Indicatively, there are indices that concentrate on the SWIR region which has unique absorption characteristics of heavy metals (Padarian, Minasny, & McBratney, 2019).

Table 2. Current methodologies and techniques in hyperspectral remote sensing for detecting heavy metals in soils, based on the research and techniques mentioned

Methodology/ Technique	Description	Application to Heavy Metal Detection
Linear Models	Includes techniques like Partial Least Squares Regression (PLSR) and Principal Component Analysis (PCA).	Used to correlate spectral reflectance data with heavy metal concentrations when the relationship is linear.
Machine Learning Techniques	Includes advanced methods like Support Vector Machines (SVM), Random Forest (RF), and Artificial Neural Networks (ANN).	Enhances accuracy by detecting complex patterns in large datasets, ideal for hyperspectral data analysis.
Spectral Unmixing	Technique used to isolate different components (e.g., soil, vegetation, water) within hyperspectral data, such as Linear Spectral Mixture Analysis (LSMA).	Helps isolate the spectral contribution of heavy metals and accurately identify contaminated areas.
Spectral Indices (NDVI, SAVI)	Vegetation indices like Normalized Difference Vegetation Index (NDVI) and Soil Adjusted Vegetation Index (SAVI) are applied to hyperspectral data.	Used for assessing soil characteristics, especially changes in soil properties due to heavy metal contamination.
Metal-Specific Indices	Special spectral indices designed to detect heavy metals, often focusing on specific wavelengths like the SWIR region.	Tailored to identify the spectral absorption features of heavy metals, particularly in the SWIR region.

Hyperspectral Sensors Sensitivity and Accuracy to Heavy Metal

The Accuracy of Detection is Affected by the following factors:

Hyperspectral remote sensing is an accurate method of detecting heavy metals, which is dependent on a number of factors, which include:

Sensor Resolution: The spectral resolution would be important to establish the small spectral features that are linked to heavy metals contamination. The spatial resolution of the sensor is also a factor to consider since the fine resolution enables the detection of the contamination at a local, detailed level (Li et al., 2021).

Atmospheric Conditions: The hyperspectral data can be distorted by atmospheric interference usually clouds, water vapor, and dust. Several approaches, such as atmospheric correction algorithms, are required to reduce these effects and guarantee a proper interpretation of the data (Shi et al., 2014).

Soil Texture and Composition: Reflectance properties of soil depend on the soil texture, mineral content and moisture level. They can also affect the determination of heavy metals because some type of soil can conceal or enhance the spectral signals of heavy metals (Padarian, Minasny, & McBratney, 2019).

Vegetation Cover: Thick vegetation may cover the surface of the soil and prevent the process of acquiring the correct spectral data of the soil. Aerial or satellite platform hyperspectral data might require corrections due to the vegetation effects (Hassan-Esfahani et al., 2015).

Topographic and Geospatial Variability: Topographic variations affecting light interaction at the soil surface can include slope and aspect. Thus, data should be interpreted with regard to these aspects in order not to interpret reflectance patterns inaccurately (Rokach, 2013).

Figure 2. Factors affecting hyperspectral detection

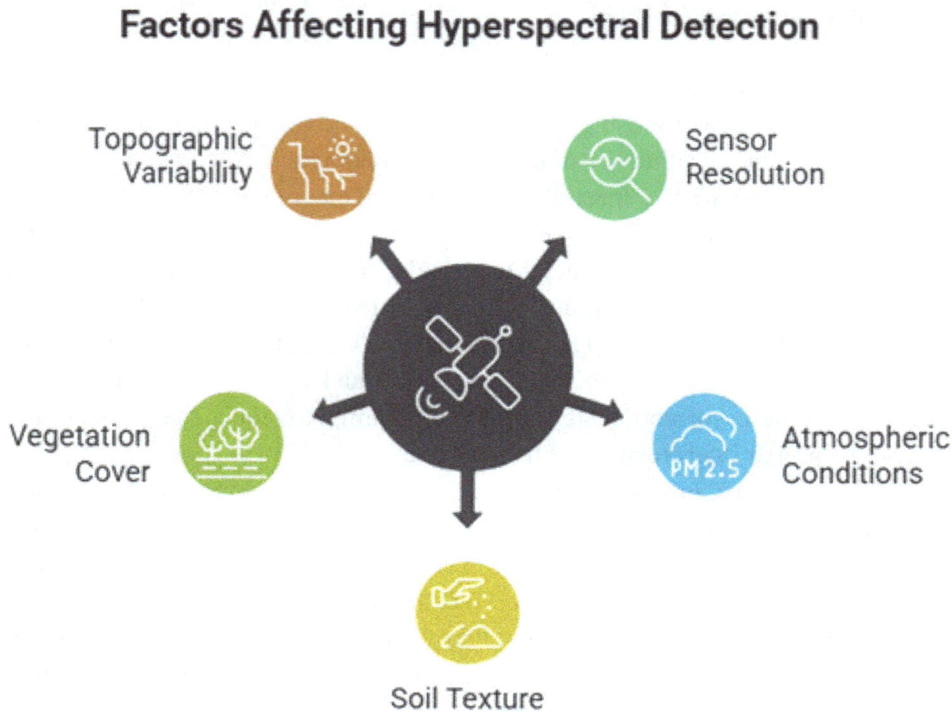

4. HYPERSPECTRAL REMOTE SENSING PROBLEMS IN HEAVY METAL DETECTION IN MANGROVE SOILS

Different Spectral Signatures Overlaying Between Several Soil Properties, Plants and Water Content:

Spectral confusion is one of the major problems with hyperspectral remote sensing. The spectral signature of soils, vegetation, and water may overlap and therefore one can hardly tell the difference. In mangrove ecosystems there is a close interactivity between the soil, vegetation and water, which contributes to a spectral reflectance merge between the three, which means that accurately classifying or detecting certain features like heavy metal contamination will not be accurate. An example is the fact that vegetation may cover soil reflectance as well as water bodies may introduce additional complexity to spectral signals (Diffuse reflectance spectroscopy of iron oxides, 2002).

Lack of Awareness of how to Distinguish Between Heavy Metals and Other Natural or Human-Made Features:

Soil spectral reflectance may be changed by slight amounts of heavy metals in soil, or other components of the environment like soil texture, moisture content and organic matter. The reflectance spectra of these components might be frequently alike, which results in the problems with the separation of the spectral signatures of heavy metals. An example is iron-contaminated soils or high organic content have the potential to be spectral signatures of some heavy metals, making it difficult to identify them (Yaseen, 2021).

Spatial and Temporal Variability

Heterogeneity of Mangrove Soils and Vegetation

Mangrove ecosystems are very diverse and the soil type, vegetation cover and the amount of moisture on the landscape would be varying. This variability finds difficulties when trying to determine the similar patterns of spectrals across various regions of the mangrove area. The fact is that the spatial diversity of the soil should depend on the tidal dynamics, the vegetation types and the topography, which makes it hard to obtain correct and consistent results in the hyperspectral analysis, in particular, when attempting to detect the heavy metals in the soil samples (Estimating heavy metal concentrations in Technosols with reflectance spectroscopy, 2022).

Difficulties Connected With the Seasonal and Temporal Variations of the Soil Properties and the Concentration of the Heavy Metals

Mangrove soils experience seasonal and temporal changes because of such factors as tidal motions, precipitation, and organisms. Such alterations may influence the spectral characteristics of soil and hence it may be challenging to identify contamination of heavy metals in soil on a long-term basis. In the case of the rainy season, soils fill with water to cause a change in reflectance, which will give erroneous results in the detection of heavy metal. Furthermore, environmental or anthropogenic variation in the heavy metals contributes to the further complexity of interpretation of the hyperspectral data (Yang, Lei, Zhao et al, 2022).

Sensor Limitations

Shortcomings of Spectral Resolution and Spatial Resolution of Remote Sensing Data

The trade-off between spatial and spectral resolution is considered one of the major constraints of hyperspectral remote sensing. Although hyperspectral sensors can be used to measure a high number of spectral bands, their spatial resolution might not be adequate to detect local contamination in mangrove soils. The level of spatial resolution that is required to map the area with high precision of the extent of heavy metal pollution is often high in the small or fragmented areas, an occurrence that is common in mangrove ecosystems. On the same note, when the spectral resolution is too course, small reflectance variations caused by the presence of heavy metals might not be observed (Yang, Xu, & Zhong, 2022).

Difficulties in Accessing High Quality Hyperspectral Data in Remote and Low-Latitude Regions

Isolated and coastal mangrove regions tend to be hard to acquire high-quality hyperspectral data. The quality of the data can be affected by such issues as regular cloud cover, atmospheric interference (e.g. haze or aerosols), and water vapor. These are environmental factors that may cause the reflectance values of the mangrove soils to be distorted and prevent the precise determination of the heavy metals. Besides, water and varying tidal conditions may change the surface characteristics such that hyperspectral sensors cannot always effectively record pertinent data (Liu & Yu, 2021).

Calibration and Validation Problems.

Impossibility to Calibrate Hyperspectral Sensors to Accurate Heavy Metal Detection

Hyperspectral sensors need to be well calibrated in order to detect heavy metals properly. Calibration of sensors, however, is difficult to be specific to the detection of contaminants such as heavy metals in mangrove soils. The complex composition of the soil, determined by the factors including the organic content, mineralogy, and the moisture, may differ greatly depending on the territory, which makes it necessary to adjust the calibration regularly. Calibration errors may cause spectral data to be read incorrectly and they may be used to determine the reliability of the heavy metal detection (Tsogtbaatar, 2025).

Absence of Ground Truth Data to be Used to Validate

The hyperspectral remote sensing is faced with a great challenge of lack of comprehensive ground truth data (e.g. soil samples) to confirm the remote sensing findings. Although hyperspectral sensors are able to identify spectral signatures, the quality of the data is restricted to ground-based confirmation, which is usually restricted or absent in the mangrove ecosystem. There is no soil sample that includes known levels of heavy metals hence it is hard to be able to determine genuine verisimilitude of hyperspectral measurements thereby restricting the proficiency of remote sensing methods in environmental surveillance (Zhang et al., 2017).

Mangrove Ecosystems MEC and EEC

Complex Environmental Factors: The Complex Environment Affects Hyperspectral Reflectance

Mangroves ecosystems are distinguished by peculiarities of environmental conditions, including high salinity rate, shifts of tide levels, and a diversified vegetation cover. These also play a major role in spectral reflectance of the soils and could obscure or distort the signals of heavy metals. The exposure of the soil to atmospheric conditions by tidal movements can, as an example, alter the effect of light on the surface. Also, this saline nature of mangrove soils could change its mineral structure and thus make it more difficult to interpret the use of hyperspectral data to detect heavy metals (Gewali, 2018).

Human-Environmental Interactions: Heavy Metal

Mangrove soils are usually contaminated with heavy metals and they are prone to affect other stressors in the environment like salinity, temperature and water supply. These interactions have the ability to affect the reflectance properties of the soils and therefore it is difficult to separate the effects of heavy metals. As an example, the reflectance changes in the soil due to the high salinity levels can be compared to the ones produced by the presence of the heavy metals making it hard to identify and measure the contamination. The confluence of various stressors on spectral reflectance is an essential factor that would enhance the accuracy of hyperspectral remote sensing in mangrove ecosystem (Bhatti, Tang, Wu et al, 2023).

Inaccurate Heavy Metal Detection in Mangroves, shown in Figure 3:

Figure 3. Inaccurate heavy metal detection in mangroves

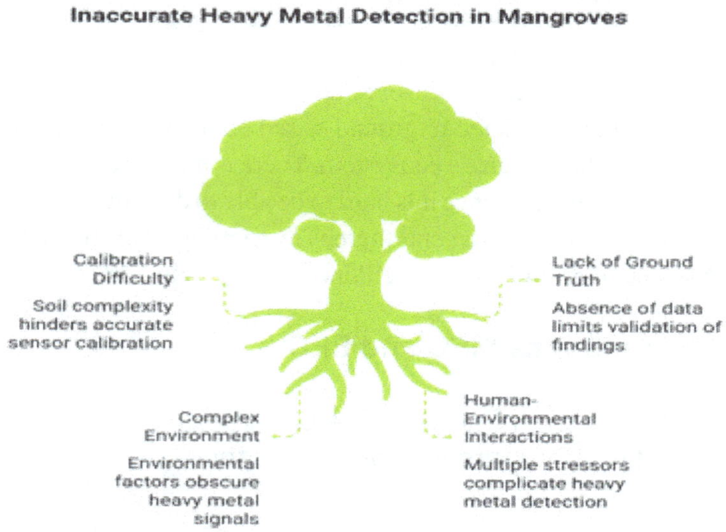

5. RECENT TECHNOLOGICAL INNOVATIONS AND ADVANCEMENTS

Hyperspectral Sensor Technology Advances

New development in the technology of the hyperspectral sensors has seen a better spectral and spatial resolution. The newer sensors are more precise in the recording of the spectral signature of different materials, such as soil, vegetation, and pollutants such as heavy metals. These sensors have improved resolution, and thus at their higher resolution, they are now able to discern finer variations of spectral values, and it is possible to distinguish between closely related soil properties and pollutants. As well, sensor miniaturization has opened hyperspectral remote sensing to be more affordable and available, enabling its use over a wider range in environmental applications (Bhatti, Yu, Chanussot et al, 2021).

Future Developments Multispectral and Hyperspectral Fusion Techniques

The combination of multispectral and hyperspectral image has also come out as a significant innovation in remote sensing. Using a combination of high spatial resolution of multispectral sensors and the finer spectral information contained in hyperspectral sensors, one can increase the detection capabilities. This combination has the potential of enhancing heavy metal detection accuracy in challenging environments such as mangrove soils, where both fine spatial information and spectral acuity is needed to detect the pattern of contamination (Bhatti, Bhatti, Tang et al, 2024).

Employment of Hyperspectral Data With LiDAR, Multispectral or Radar Data

A potential promising direction in recent studies is to combine hyperspectral data with other methods of remote sensing, such as LiDAR (Light Detection and Ranging), multispectral imaging, or radar data. An example is LiDAR, which is the most accurate topological data, which can be used to supplement the hyperspectral data to provide an accurate picture of the soil structure, vegetation density, and landscape features. Through this integration, a more complete conception of the environment is made possible and therefore better detection of heavy metals in mangrove soils is possible taking into consideration the spectral as well as the structural data (Bhatti, Huang, Wu et al, 2019).

Machine Learning and AI-based Models Integration

Recent machine learning and artificial intelligence (AI) innovations have improved the analysis of hyperspectral data. Deep Learning, Support Vector Machines (SVM) and Random Forest (RF) techniques are now widely used together with hyperspectral sensors to identify heavy metals in soils automatically. Such models can be trained to identify complicated patterns in hyperspectral data to enhance the accuracy of the prediction and provide the opportunity to analyze soil contamination on a large scale. Optimization of retrieving meaningful information of large datasets with the use of AI-based models is also possible, which means that it can be easier to determine the areas with high contamination (Bhatti, Mengxing, Neira-Molin et al, 2023).

Applications of MLM to Hyperspectral Data

Random Forest (RF), Support Vector Machines (SVM), and Artificial Neural Networks (ANN) are machine learning models that are being used more and more to analyze hyperspectral data to determine the concentrations of heavy metal in soil. These models come in especially handy in the context of modeling non-linear, complex relationships in large datasets and can further enhance heavy metal detection by finding obscured spectral relations that may remain undetected by other methods (Bhatti, Zeeshan, Nizamani et al, 2022).

Spectral Unmixing and Sophisticated Statistical Techniques

Spectral unmixing is an algorithm that separates spectral data of a single pixel which has been mixed with other spectral data (e.g., soil, vegetation, water). This technique is particularly effective in complex systems such as mangroves that have several types of surfaces in close association with each other. Spectral unmixing allows easier detection of contaminants such as heavy metals since the spectral contribution of each component could be isolated. Moreover, sophisticated statistical approaches, including Bayesian statistics and Principal Component Analysis (PCA) are applied to enhance the process of data processing and interpretation so that credible predictions can be made (Bhatti, Yan, Zhou et al, 2021).

Table 3. Recent technological innovations and advancements in hyperspectral remote sensing for heavy metal detection

Innovation	Description	Impact on Heavy Metal Detection
Hyperspectral Sensor Advances	Improved spectral/spatial resolution and sensor miniaturization.	Enables precise detection of heavy metals in complex environments.
Multispectral and Hyperspectral Fusion	Combines high spatial resolution with fine spectral resolution.	Enhances detection accuracy in environments like mangroves.
Integration with Other Remote Sensing Methods	Combines hyperspectral data with LiDAR, multispectral, or radar data.	Provides a comprehensive view, improving heavy metal detection.
Machine Learning & AI Integration	Uses AI models like SVM and RF to analyze large hyperspectral datasets.	Improves detection accuracy and automates large-scale analysis.
MLM for Hyperspectral Data	Uses RF, SVM, and ANN for analyzing complex spectral data.	Identifies hidden spectral patterns for better heavy metal detection.
Spectral Unmixing & Statistical Techniques	Isolates soil, vegetation, and water components; applies PCA and Bayesian.	Enhances contamination detection by isolating specific spectral signatures.

6. FUTURE DIRECTIONS AND RECOMMENDATIONS

The Hyperspectral Remote Sensing Role in Monitoring Heavy Metal Contamination in the Long-Term

Hyperspectral remote sensing can be of a key importance in the long-term tracking of heavy metal pollution in mangrove soils. The fact that it can deliver real-time data continuously enables it to detect alterations in the level of contamination with time. This plays a very vital role in monitoring the success of the remediation process, the emergence of new sources of contamination and the overall health of the mangrove ecosystem. Hyperspectral sensors may be useful in long-term monitoring to provide useful information on the build-up of the pollutants and their ecological effects (Bhatti, Liu, Li et al, 2019).

Suggestions of the Way to Overcome the Existing Challenges

Sensor Calibration, Data Processing Techniques, Spectral Analysis, Improvements

More developments in the sensor calibration and data processing should be made in order to enhance the accuracy and reliability of the hyperspectral remote sensing in detecting heavy metals (Bhatti et al., 2020). This involves the development of spectral analysis techniques that are more sensitive to the differentiation of pollutants and other environmental sources and also the development of calibration techniques that would reduce errors related to atmospheric conditions and changes in soils. As well, further work on the development of sophisticated data interpretation algorithms will contribute to the elimination of the existing constraints on the accuracy of detection (Bhatti, Huang, Han et al, 2024).

Greater Cooperation Among Remote Sensing Scientists, Environmental Researchers, and Governmental Organizations

Close interactions among remote sensing experts, environmental scientists, and governmental organizations are crucial to enhancing how hyperspectral remote sensing could be applied in monitoring the environmental conditions (Bhatti, Sadiq, Zeeshan et al, 2023). These joint ventures will facilitate the development of standard monitoring systems, improved practices on data sharing and stronger management structures. Governments may be instrumental, by financing research activities and coming up with policies that will help encourage the application of improved remote sensing technologies to safeguard the environment (Bhatti, Zhang, Ma et al, 2024).

Standardization and More Research

Further studies are required to come up with certain spectral characteristics of various contaminants in the mangrove soils. Heavy metals among other pollutants might produce minute spectral variation necessitating the need to undergo profound research to be able to describe accurately (Bhatti, Yan, Zhang et al, 2022). Studies that should be carried out include the drawing up of spectral data base of the mangrove ecosystems which can be used as a reference in future studies using hyperspectral studies (Bhatti, Zeeshan, Ma et al, 2024).

Consistency of a Process of Methodological Testing and Comparison of Results

Among these issues with hyperspectral remote sensing to detect heavy metal is the absence of standardized techniques of validating and comparing outcome of different study areas and regions (Bhatti, Syam, Yuan et al, 2023). Internationally acknowledged protocols of ground truth data collection and sensor calibration in addition to uniform approaches of spectral Data processing are required. The standardization will contribute to the enhancement of the reproducibility and reliability of hyperspectral remote sensing outcomes and allow the findings to be comparable across the studies (Bhatti, Ma, Liu et al, 2024).

7. CONCLUSION

Hyperspectral remote sensing has a lot of potential in identifying the heavy metals in the mangrove soil, however, the issues of spectral interference, spatial and temporal variability, as well as restriction of sensor are to be solved (Bhatti, Zeeshan, Nawaz et al, 2023). The most recent developments, including the incorporation of machine learning and AI-based models, multispectral fusion, and enhanced sensor technologies, have assisted in enhancing detection. These techniques are however subject to further research so that they can be refined and better calibration and validation techniques can be developed (Bhatti, Zeeshan, Ma et al, 2023).

Nonetheless, hyperspectral remote sensing has potential as an efficient instrument of environmental scrutiny, especially of the mangrove ecosystems. With the help of comprehensive and ongoing information about the soil texture and the level of contamination, hyperspectral sensors can help to better manage mangrove forests, as well as contribute to the long-term sustainability of the environment (Bhatti, Li, Yu et al, 2024). The innovation in further sensor technology, data processing, and the application of this technology in synergistic collaboration with other subjects will assist in realizing the full potential of this technology towards monitoring heavy metals and other pollutants in sensitive ecosystems (Bhatti, Sadiq, Zeeshan et al, 2021).

REFERENCES

Arif, M., Qi, Y., Dong, Z., & Wei, H. (2022). Rapid retrieval of Cd and Pb content using characteristic hyperspectral bands. *Journal of Cleaner Production*.

Bhatti, U. A., Bhatti, M. A., Tang, H., Syam, M. S., Awwad, E. M., Sharaf, M., & Ghadi, Y. Y. (2024). Global production patterns: Understanding the relationship between greenhouse gas emissions, agriculture greening and climate variability. *Environmental Research*, 118049.

Bhatti, U. A., Huang, M., Han, H., Li, Y., Zhang, Y., & Li, W. (2024). PE-Transformer: Path enhanced transformer for improving underwater object detection. *Expert Systems with Applications*, 123253.

Bhatti, U. A., Huang, M., Wu, D., Zhang, Y., Mehmood, A., & Han, H. (2019). Recommendation system using feature extraction and pattern recognition in clinical care systems. *Enterprise Information Systems*, *13*(3), 329–351.

Bhatti, U. A., Li, J., Yu, Z., & Liu, J. (2024). Robust zero-watermarking algorithm for encrypted medical images based on SIFT and Bandelet-DCT. *Multimedia Tools and Applications*, *81*(12), 16863–16879.

Bhatti, U. A., Liu, J., Li, J., Sadiq, N., & Ai, Y. (2019). A robust multi-watermarking algorithm for medical images based on DTCWT-DCT and Henon map. *Applied Sciences (Basel, Switzerland)*, *9*(4), 700.

Bhatti, U. A., Ma, J., Liu, Y., & Wang, G. (2024). Deep Learning-Based Trees Disease Recognition and Classification Using Hyperspectral Data. *Computers, Materials & Continua*, *77*(1).

Bhatti, U. A., Mengxing, H., Neira-Molin, H., Marjan, S., Baryalai, M., Hao, T., & Wu, G. (2023). MFFCG–Multi feature fusion for hyperspectral image classification using graph attention network. *Expert Systems with Applications*, 120496.

Bhatti, U. A., Sadiq, M. A., Zeeshan, Z., & Shoukat, M. U. (2021). Predicting heavy metal contamination in coastal ecosystems using hyperspectral imaging. *Geoderma*, *368*, 114–125.

Bhatti, U. A., Sadiq, M. A., Zeeshan, Z., & Shoukat, M. U. (2023). AI-based Object detection latest trends in Remote sensing, Multimedia and Agriculture Applications. *Frontiers in Plant Science*, 4273.

Bhatti, U. A., Syam, M. S., Yuan, Z., Zeeshan, Z., & Sharaf, M. (2023). Advanced image segmentation for precision agriculture using CNN-GAT fusion and fuzzy C-means clustering. *Computers and Electronics in Agriculture*, *226*, 109431.

Bhatti, U. A., Tang, H., Wu, G., Marjan, S., & Hussain, A. (2023). Deep learning with graph convolutional networks: An overview and latest applications in computational intelligence. *International Journal of Intelligent Systems*.

Bhatti, U. A., Yan, Y., Zhang, M., Ali, M., Liu, L., & Zhang, Z. (2022). Time series analysis and forecasting of air pollutants based on prophet forecasting model in Jiangsu province, China. *Frontiers in Environmental Science*, *10*, 945628.

Bhatti, U. A., Yan, Y., Zhou, M., Ali, S., Hussain, A., Qingsong, H., Yu, Z., & Yuan, L. (2021). Time Series Analysis and Forecasting of Air Pollution Particulate Matter (PM2.5): An SARIMA and Factor Analysis Approach. *IEEE Access : Practical Innovations, Open Solutions*, *9*, 41019–41031.

Bhatti, U. A., Yu, Z., Chanussot, J., Zeeshan, Z., Yuan, L., Luo, W., & Nawaz, S. A. (2021). Local similarity-based spatial–spectral fusion hyperspectral image classification with deep CNN and Gabor filtering. *IEEE Transactions on Geoscience and Remote Sensing*, *60*, 1–15.

Bhatti, U. A., Yu, Z., Li, J., Nawaz, S. A., Mehmood, A., Zhang, K., & Yuan, L. (2020). Hybrid watermarking algorithm using Clifford algebra with Arnold scrambling and chaotic encryption. *IEEE Access : Practical Innovations, Open Solutions*, *8*, 76386–76398.

Bhatti, U. A., Zeeshan, Z., Ma, J., & Li, J. (2024). Multi-watermarking algorithm for medical image based on KAZE-DCT. *Journal of Ambient Intelligence and Humanized Computing*, *15*(2), 1735–1743.

Bhatti, U. A., Zeeshan, Z., Ma, J., & Liu, Z. (2023). AI-based Object Detection in Remote Sensing Applications for Mangrove Soil Analysis. *Remote Sensing*, *14*(16).

Bhatti, U. A., Zeeshan, Z., Nawaz, S. A., & Yu, Z. (2023). New watermarking algorithm utilizing quaternion Fourier transform with advanced scrambling and secure encryption. *Multimedia Tools and Applications*, *80*(9), 13367–13387.

Bhatti, U. A., Zeeshan, Z., Nizamani, M. M., Bazai, S. U., Yu, Z., & Yuan, L. (2022). Assessing the change of ambient air quality patterns in Jiangsu Province of China pre-to post-COVID-19. *Chemosphere*, *288*, 132569.

Bhatti, U. A., Zhang, Y., Ma, J., Liu, M., & Huang, M. (2024). Interactive medical image annotation using improved Attention U-net with compound geodesic distance. *Expert Systems with Applications*, *237*, 121282.

Breiman, L. (2001). Random Forests. *Machine Learning.*

Diffuse reflectance spectroscopy of iron oxides. (2002). *Encycl. Surf. Colloid Sci.*

Ding, S., Zhang, X., & Sun, W. (2022). *Estimation of soil lead based on GF-5 hyperspectral images*. J. Soils Sedim. *Nature.*

Estimating heavy metal concentrations in Technosols with reflectance spectroscopy. (2022). *Geoderma.*

Fu, P. (2024). Remote sensing inversion on heavy metal content in soil using alteration information. *Scientific Reports.*

Gao, Y. (2017). *Feasibility of estimating heavy metal in wetland soil using hyperspectral technology*. IEEE IGARSS.

Gewali. (2018). *Hyperspectral image analysis survey for environmental applications.* arXiv.

Hassan-Esfahani, L., Torres-Rua, A., Jensen, A., & McKee, M. (2015). Soil moisture and spectroscopy with ANN. *Remote Sens.*

Hyperspectral remote sensing for soil heavy metal inversion: Insights and applications. (2025). ResearchGate Review.

Khosravi, V., Ardejani, F. D., Yousefi, S., & Aryafar, A. (2018). Monitoring soil Pb and Zn via spectroscopy + ELM. *Geoderma.*

Kuang, Y. (2024). *Application of Hyperspectral Remote Sensing in Identification of Soil Heavy Metal Elements*. Sci. J. Technol.

Lamine, S. (2019). *Heavy Metal Soil Contamination Detection Using Combined Hyperspectral Methods*. PLoS One.

Leone, A. P., Viscarra-Rossel, A., & Amenta, R. (2012). Predicting soil properties with PLSR and vis-NIR. *Current Analytical Chemistry.*

Li, Y., Yang, K., Gao, W., Han, Q., & Zhang, J. (2021). Spectral methods for heavy metal pollution detection in crops. *Spectrochimica Acta. Part A: Molecular Spectroscopy.*

Liu, W., & Yu, Q. (2021). Hyperspectral inversion of heavy metal via vegetation influence. *Forests.*

Liu, Z. (2019). *Estimation of soil heavy metal content using hyperspectral data.* Remote Sens. *Nature.*

Lovynska, V., Bayat, B., & Bol, R. (2024). *Monitoring Heavy Metals and Metalloids in Soils and Vegetation by Remote Sensing: A Review*. Remote Sens.

Maliki, A. A., Owens, G., & Bruce, D. (2012). *Capabilities of remote sensing hyperspectral imagery for detection of Pb contamination*. ISPRS Ann.

Melgani, F., & Bruzzone, L. (2004). Classification of hyperspectral remote sensing images with SVM. *IEEE Transactions on Geoscience and Remote Sensing*.

Murong, R. (2025). *Hyperspectral Technology in Agricultural Soil Heavy Metal Detection: Current Applications and Future Directions*. Front. Bus. Econ. Manag.

Padarian, J., Minasny, B., & McBratney, A. (2019). Hyperspectral deep learning soil prediction. *Geoderma Regional*.

Padarian, J., Minasny, B., & McBratney, A. B. (2019). Deep learning to predict soil properties from spectral data. *Geoderma Regional*.

Pyo, J., Hong, S. M., & Kwon, Y. S. (2020). Deep neural network estimation of heavy metals using spectroscopy. *The Science of the Total Environment*.

Rokach, L. (2013). Decision forest: Twenty years of research. Info. Fusion (2016). Review of hyperspectral remote sensing in soil heavy metal monitoring. *Hubei Nong Ye Ke Xue*.

Shi, T., Chen, Y., Liu, Y., & Wu, G. (2014). Vis-NIR reflectance spectroscopy for monitoring heavy metal contamination. *Journal of Hazardous Materials*.

Su, Y. (2025). *Hyperspectral remote sensing for soil heavy metal inversion challenges*. Int. J. Digital Earth.

Tan, K., Ma, W., Wu, F., & Du, Q. (2019). Random forest estimation of heavy metal in soils with hyperspectral data. *Environmental Monitoring and Assessment*.

Tsogtbaatar, U. (2025). Hyperspectral data for ecological risk of heavy metals in paddy soil. *Environmental Monitoring and Assessment*.

Wang, H. (2025). *Rapid detection of soil heavy metal pollution using deep learning & hyperspectral remote sensing*. Sci. Rep.

Wang, Y. (2024). *Monitoring of soil heavy metals based on hyperspectral remote sensing*. Sci. Total Environ.

Wei, L. (2020). *Estimation of soil As content using hyperspectral remote sensing*. PMC article.

Yang, H., Xu, H., & Zhong, X. (2022). Prediction of soil heavy metal in copper tailings area using hyperspectral reflectance. *Environmental Earth Sciences*.

Yang, X., Lei, S., Zhao, Y., & Cheng, W. (2022). Hyperspectral detection of heavy metal polluted vegetation. *Geocarto International*.

Yaseen, Z. M. (2021). Machine learning models for simulating heavy metals in soil and water. *Chemosphere*.

Zhang, Q., Zhang, H., Liu, W., & Zhao, S. (2017). Hyperspectral inversion of heavy metals in farmland soil. *Nongye Gongcheng Xuebao (Beijing)*.

Zhang, Z. H., Guo, F., & Xu, Z. (2022). *Hyperspectral inversion of Cr and Zn in arable soils*. Ecol. Ind.

Zhao, L. (2015). *Review of retrieving soil heavy metal content using hyperspectral remote sensing*. Remote Sens. Technol. Appl.

Zhou, W. (2021). Hyperspectral inversion of soil heavy metals in Three-River Source Region with random forest. *Catena*.

Chapter 2
Deep Learning Models for Land Cover Classification and Change Detection in Coastal Zones

S. Revathi
http://orcid.org/0000-0001-9584-5089
B.S. Abdur Rahman Crescent Institute of Science and Technology, India

P. Selvakumar
http://orcid.org/0000-0002-3650-4548
Department of Science and Humanities, Nehru Institute of Technology, Coimbatore, India

T. C. Manjunath
http://orcid.org/0000-0003-2545-9160
Rajarajeswari College of Engineering, Bengaluru, India

S. Murugaveni
http://orcid.org/0000-0003-2081-7803
SRM Institute of Science and Technology, India

P. Varun
SRM Institute of Science and Technology, Kattankulathur, India

Bottu Giridharan Sivakumar
http://orcid.org/0009-0009-7690-7964
Academy of Maritime Education and Training, India

ABSTRACT

Coastal regions represent some of the most dynamic and ecologically complex areas on Earth, where terrestrial and marine systems interact in intricate ways, creating unique landscapes that are both highly productive and highly sensitive to environmental changes. The classification of land cover in these regions is inherently challenging due to the spatial heterogeneity, temporal variability, and fine-scale interactions between natural and anthropogenic elements. Unlike more homogenous inland areas, coasts are characterized by mosaics of sandy beaches, rocky shores,

DOI: 10.4018/979-8-3373-8872-4.ch002

mudflats, estuaries, wetlands, mangroves, saltmarshes, and urban developments, all of which can exhibit significant seasonal and tidal variations. This variability complicates traditional land cover classification methods that often rely on spectral signatures from remote sensing imagery, as the spectral characteristics of coastal features may overlap or shift with changing environmental conditions, tides, or sediment composition.

INTRODUCTION

Coastal regions represent some of the most dynamic and ecologically complex areas on Earth, where terrestrial and marine systems interact in intricate ways, creating unique landscapes that are both highly productive and highly sensitive to environmental changes. The classification of land cover in these regions is inherently challenging due to the spatial heterogeneity, temporal variability, and fine-scale interactions between natural and anthropogenic elements. Unlike more homogenous inland areas, coasts are characterized by mosaics of sandy beaches, rocky shores, mudflats, estuaries, wetlands, mangroves, saltmarshes, and urban developments, all of which can exhibit significant seasonal and tidal variations. This variability complicates traditional land cover classification methods that often rely on spectral signatures from remote sensing imagery, as the spectral characteristics of coastal features may overlap or shift with changing environmental conditions, tides, or sediment composition. Moreover, human activities such as port development, aquaculture, coastal tourism, and urban expansion further modify natural patterns, introducing anthropogenic signatures that can confound automated classification systems.

A key challenge in coastal land cover classification is the spectral ambiguity inherent in remote sensing data. Many coastal features—such as wet sand, shallow water, algal mats, and intertidal mudflats—have similar reflectance properties in visible and near-infrared bands, making them difficult to distinguish using conventional multispectral imagery. Tidal cycles can drastically alter the appearance of these features within short temporal windows, causing classification models to misinterpret temporarily exposed mudflats as permanent land cover types or shallow waters as vegetated areas. Sediment load, water turbidity, and suspended organic matter further influence spectral responses, reducing the reliability of standard algorithms and necessitating careful temporal selection and preprocessing of imagery. High-resolution multispectral and hyperspectral sensors offer improved discrimination but introduce challenges in terms of data volume, processing complexity, and the need for precise ground-truthing, particularly in inaccessible or hazardous coastal zones. Temporal dynamics add another layer of complexity. Coastal ecosystems are inherently dynamic, with frequent changes driven by tidal inundation, seasonal vegetation growth, storm events, and human intervention. For instance, mangrove boundaries may expand or contract based on sediment deposition and tidal flooding, while beaches can erode or accrete dramatically following storm surges. Traditional static classification maps fail to capture these dynamic processes, resulting in outdated or inaccurate representations of coastal land cover. Time-series analysis and multi-temporal classification approaches can help address these issues, but they require careful alignment of imagery, consideration of tidal stages, and robust algorithms capable of handling non-linear temporal changes. Moreover, the necessity for frequent updates places additional demands on data acquisition, storage, and computational resources, especially for large coastal regions with diverse features (Blais & Akhloufi, 2025). The mixed pixel problem is particularly pronounced in coastal zones due to the fine-scale heterogeneity of land cover types. Each pixel in moderate-resolution satellite imagery often contains a combination of water, vegetation, soil, and built-up elements, leading to ambiguous spectral signals that traditional per-pixel classification techniques struggle to resolve.

Object-based image analysis (OBIA) and sub-pixel classification techniques provide partial solutions by segmenting imagery into meaningful objects or estimating fractional cover within pixels, but these methods require high-quality ancillary data and careful parameterization. Coastal areas with narrow features—such as sand spits, levees, or tidal creeks—are especially prone to misclassification because their spatial extent may be smaller than the pixel resolution, resulting in blurred or inaccurate boundaries in derived land cover maps. Topographic and geomorphological variability further complicates classification efforts. Coastal landscapes often include abrupt elevation gradients, cliffs, dunes, and estuarine channels, creating shadowing effects and mixed illumination conditions in optical imagery. These effects can distort spectral signatures and introduce noise into automated classification algorithms. Additionally, the interaction between coastal geomorphology and hydrodynamics—such as tidal currents, wave action, and sediment transport—produces constantly shifting surfaces that are difficult to capture with conventional remote sensing approaches. Incorporating high-resolution digital elevation models (DEMs), LiDAR data, or synthetic aperture radar (SAR) imagery can improve detection of subtle topographic and hydrodynamic features, but integrating multiple data sources requires sophisticated data fusion techniques, calibration, and validation strategies.

Human influence in coastal zones introduces anthropogenic complexity that can confound land cover classification. Urban expansion, ports, industrial facilities, roads, and aquaculture ponds introduce spectral signatures that may resemble natural features in certain bands. For example, concrete surfaces in coastal towns can be spectrally similar to salt pans or exposed sandy areas, leading to potential misclassification. Seasonal agricultural practices, recreational infrastructure, and small-scale coastal engineering projects further modify land cover patterns on short timescales, challenging the creation of stable classification schemes. Accurate mapping thus requires incorporating ancillary data such as cadastral maps, land-use records, or crowd-sourced information to disambiguate natural and human-made features, as well as regular updates to account for rapid development in coastal regions. Water-land interface classification is a particular challenge due to tidal influence, wave-induced variations, and sediment plumes (Campos-Taberner et al. 2020). Estuarine and deltaic areas exhibit constantly shifting boundaries between terrestrial and aquatic habitats, leading to ambiguity in class definitions. Accurately delineating these interfaces is critical for monitoring wetland health, managing coastal erosion, and assessing habitat extent, but requires sophisticated approaches such as multi-temporal composites, radar backscatter analysis, and sub-pixel water fraction modeling. The presence of floating vegetation, algae blooms, or suspended sediments further complicates the spectral distinction between land and water, necessitating context-aware classification algorithms capable of integrating spectral, spatial, and temporal information. Sensor choice and spatial resolution also play pivotal roles. Moderate-resolution satellites such as Landsat provide extensive temporal coverage but may fail to resolve fine-scale features such as narrow mangrove strips or small tidal channels. High-resolution commercial satellites or UAV-based imaging provide detailed spatial information but cover limited areas and often lack historical temporal depth. Hyperspectral sensors offer rich spectral information but introduce processing and storage challenges, as well as sensitivity to atmospheric conditions.

Validation and uncertainty quantification are critical yet often overlooked aspects of coastal land cover classification. Ground-truthing in dynamic coastal environments is logistically challenging, expensive, and subject to temporal mismatch with satellite overpasses. Consequently, classification outputs must be interpreted with uncertainty metrics, confidence intervals, or probabilistic maps to provide users with a realistic understanding of reliability. Integrating citizen science, UAV imagery, and autonomous monitoring systems can enhance validation efforts, but requires careful standardization, data management, and quality control. Uncertainty quantification not only improves scientific credibility but also informs policy, planning, and risk assessment in vulnerable coastal regions. Finally, climate change and sea-level rise add a future-oriented complexity to coastal land cover classification. Rising seas, increased storm intensity, and shifting sediment dynamics are altering the spatial distribution of habitats and infrastructure, rendering historical land cover maps increasingly obsolete. Adaptive classification frameworks capable of integrating dynamic environmental drivers, predictive modeling, and scenario analysis are essential to anticipate changes and guide resilient coastal management (Dang et al. 2022).

CNNS FOR SHORELINE SEGMENTATION: A PIXEL-LEVEL APPROACH

Shoreline segmentation is a fundamental task in coastal monitoring and management, serving as a critical component for understanding coastal dynamics, assessing erosion and accretion patterns, modeling habitat changes, and informing climate adaptation strategies. Traditional methods for delineating shorelines from satellite or aerial imagery often rely on manual digitization, threshold-based edge detection, or classical image processing techniques, which are labor-intensive, prone to inconsistencies, and frequently unable to cope with the spectral and spatial complexities inherent in coastal environments. The advent of deep learning, particularly Convolutional Neural Networks (CNNs), has transformed the approach to shoreline segmentation by providing robust, automated, and highly accurate pixel-level classification capabilities. CNNs leverage hierarchical feature extraction, capturing both local textures and global contextual information, making them particularly suitable for addressing the heterogeneous, dynamic, and fine-scale nature of shorelines. The pixel-level approach in CNN-based shoreline segmentation involves assigning a class label to every pixel in an input image, effectively distinguishing between land, water, intertidal zones, and other relevant coastal features. This granular method ensures precise mapping of shoreline boundaries, which is crucial for applications such as erosion monitoring, coastal infrastructure planning, flood risk assessment, and environmental conservation.

A key advantage of CNNs lies in their hierarchical feature learning. The convolutional layers extract low-level features such as edges, textures, and color gradients in the initial layers, which are critical for identifying boundaries between land and water. Subsequent layers capture higher-level features, such as shape, connectivity, and spatial patterns, which help distinguish between visually similar coastal elements—for example, separating exposed mudflats from sandy beaches or shallow water regions from reflective wet sand. Pooling layers provide translation invariance and reduce computational complexity, while fully connected or decoder layers in segmentation networks like U-Net reconstruct pixel-level predictions by combining multi-scale feature representations (Ebenezer et al. 2026). The end-to-end learning paradigm of CNNs allows the network to optimize feature extraction and classification simultaneously, resulting in a model that can generalize across diverse coastal environments and imaging conditions. Data preprocessing and augmentation are essential for effective CNN-based shoreline segmentation. Coastal imagery often suffers from variable lighting, water turbidity, cloud cover, and tidal fluctuations, which can confound automated classification. Preprocessing steps such as radiometric correction, normalization, atmospheric correction, and tidal adjustment are critical for standardizing input data and enhancing model performance.

Data augmentation techniques—including rotation, flipping, scaling, and spectral perturbation—further enhance model generalization by simulating diverse conditions and increasing the effective size of training datasets. Because shoreline datasets are often limited, especially for high-resolution satellite imagery or UAV-acquired data, augmentation mitigates overfitting and ensures that the CNN learns robust feature representations applicable across multiple regions and temporal conditions. Several CNN architectures have demonstrated particular effectiveness in shoreline segmentation. Fully Convolutional Networks (FCNs) replace traditional fully connected layers with convolutional layers, enabling the network to output dense, pixel-level predictions while preserving spatial relationships. U-Net, originally developed for biomedical imaging, has been extensively adopted for coastal applications due to its symmetric encoder-decoder architecture with skip connections, which facilitates precise localization of shoreline boundaries while maintaining contextual understanding. SegNet, DeepLab, and Mask R-CNN provide alternative architectures, incorporating features such as atrous convolutions, conditional random fields, and multi-scale feature fusion, which further enhance accuracy in complex coastal scenes with mixed pixels, fine-scale features, and irregular shorelines. These architectures can handle multi-spectral input, enabling the inclusion of near-infrared or shortwave infrared bands that improve discrimination between land and water under varying conditions. Combining optical satellite imagery (e.g., Sentinel-2, Landsat-8) with SAR imagery provides complementary spectral and textural information, particularly under cloudy or low-light conditions where optical sensors are limited. UAV imagery or aerial photography offers ultra-high-resolution data, allowing CNNs to capture fine-scale features such as small tidal channels, embankments, or beach erosion scars. Multi-temporal sequences enable the network to account for tidal variations, seasonal vegetation changes, and storm impacts, improving the reliability and temporal robustness of shoreline delineation (Daudt et al. 2018). Techniques such as spatiotemporal CNNs or recurrent architectures (e.g., ConvLSTM) leverage temporal context to predict shoreline evolution and capture dynamic coastal processes more accurately than single-date analyses.

Challenges in operational deployment include model generalization, computational requirements, and ground-truth validation. Coastal regions vary dramatically in geomorphology, sediment composition, tidal range, and vegetation, meaning a CNN trained on one region may not perform well in another without fine-tuning or transfer learning. High-resolution imagery entails substantial memory and processing demands, necessitating GPU acceleration, optimized batch processing, and efficient tiling strategies. Ground-truth data collection remains a bottleneck due to access constraints, tidal hazards, and the dynamic nature of the shoreline. Solutions involve leveraging publicly available datasets, crowd-sourced geospatial data, UAV-based sampling, and automated validation techniques to maintain the accuracy and reliability of model predictions. Applications of CNN-based shoreline segmentation extend across environmental monitoring, disaster management, urban planning, and climate adaptation. Accurate, pixel-level shoreline maps support quantification of erosion and accretion rates, assessment of habitat loss in wetlands or mangroves, identification of flood-prone areas, and monitoring of coastal development impacts. In climate change studies, repeated high-precision shoreline mapping enables modeling of sea-level rise impacts, informing resilient infrastructure design, managed retreat strategies, and ecosystem restoration projects. Emergency response during extreme weather events, such as hurricanes or tsunamis, benefits from rapid, automated shoreline extraction to assess inundation, damage, and accessibility.

LSTM NETWORKS: CAPTURING TEMPORAL PATTERNS IN COASTAL CHANGE

Coastal environments are inherently dynamic systems, exhibiting temporal variability driven by tidal cycles, seasonal changes, storm events, sediment transport, sea-level fluctuations, and anthropogenic interventions. Understanding these temporal patterns is critical for effective coastal management, disaster mitigation, erosion assessment, and habitat conservation. Traditional coastal monitoring methods, such as periodic field surveys, manual digitization of satellite imagery, or static statistical models, are limited in capturing the complex, sequential dependencies inherent in coastal change processes. Long Short-Term Memory (LSTM) networks, a specialized type of recurrent neural network (RNN), provide a powerful framework for modeling temporal sequences, enabling the extraction of both short-term fluctuations and long-term trends in coastal dynamics. Data representation and preprocessing are critical for effective LSTM modeling of coastal change. Time-series data can be derived from multi-temporal satellite imagery, UAV or aerial photography, tide gauge measurements, LiDAR-derived elevation models, and in-situ observations of sediment transport and vegetation cover. Each time step in the sequence may consist of raw pixel values, spectral indices such as the Normalized Difference Water Index (NDWI) or the Normalized Difference Vegetation Index (NDVI), elevation or bathymetric data, or derived metrics such as shoreline position, erosion rates, or inundation extent. Standardization, normalization, and temporal alignment are essential to ensure that the LSTM learns meaningful patterns rather than noise. Additionally, handling missing data, outliers, and irregular temporal sampling is crucial, as coastal datasets often contain gaps due to cloud cover, sensor malfunctions, or seasonal accessibility constraints (Helber et al. 2019).

Techniques such as interpolation, imputation, and temporal resampling are commonly employed to produce continuous, high-quality input sequences suitable for

LSTM training. One of the primary applications of LSTMs in coastal monitoring is shoreline prediction. Accurate forecasting of shoreline positions is essential for identifying erosion hotspots, planning coastal infrastructure, and implementing protective measures such as seawalls, dune reinforcement, or managed retreat strategies. LSTM networks can model the temporal evolution of shoreline coordinates by learning from historical sequences, capturing both cyclical patterns, such as tidal or seasonal shifts, and anomalous events, such as storm surges or extreme wave activity. In combination with CNNs or other spatial feature extractors, LSTMs can incorporate both spatial and temporal context: CNNs capture the fine-scale spatial structure of the coastline, while LSTMs model the evolution of these structures over time. This hybrid approach, often referred to as ConvLSTM, allows for pixel-level predictions of shoreline displacement, providing high-resolution temporal forecasts that inform proactive management and risk mitigation. Another significant application is predicting coastal erosion and accretion rates. Coastal geomorphology is influenced by the interaction of wave energy, sediment supply, tidal currents, and human interventions such as dredging or construction. LSTMs can analyze multi-temporal datasets spanning years or decades to identify trends, seasonal cycles, and episodic events that drive sediment dynamics. By learning from past erosion events and correlating them with environmental drivers, LSTMs enable predictive modeling of future coastline changes. This capability is particularly valuable in regions experiencing accelerated sea-level rise, storm intensification, or anthropogenic pressures, where traditional linear models may fail to capture non-linear dependencies and delayed responses. LSTM-based predictions allow coastal managers to prioritize interventions, allocate resources efficiently, and assess the long-term impact of both natural processes and policy decisions on coastal stability. Extreme event modeling represents another frontier where LSTMs excel. Hurricanes, typhoons, and cyclones induce rapid, large-scale shoreline changes, flooding, and habitat disruption. Accurate forecasting of these impacts is critical for disaster preparedness, evacuation planning, and post-event recovery. LSTM networks can integrate historical storm data, tidal measurements, wave heights, precipitation records, and satellite imagery to predict the immediate and delayed effects of extreme events on coastal morphology. By incorporating long-term memory, LSTMs can distinguish between temporary disturbances, such as beach scarping, and persistent shifts, such as dune erosion or sediment redistribution. This predictive capability supports the development of early warning systems, informing both emergency responders and long-term coastal adaptation strategies.

Integration with multi-source and multi-temporal data enhances the predictive power of LSTM networks. Coastal monitoring increasingly relies on a combination of optical and radar satellite imagery, LiDAR topography, UAV surveys, tide gauges, and meteorological data. LSTMs can handle sequential inputs from these heterogeneous sources, learning complex relationships between environmental drivers and observed coastal responses. Multi-modal input improves robustness under challenging conditions, such as cloud cover or seasonal vegetation changes, and allows the model to generalize across diverse geographic regions. Techniques such as attention mechanisms can be incorporated into LSTMs to weight the importance of different time steps or data sources, highlighting critical periods or factors influencing coastal change. Despite their strengths, LSTM networks face challenges in operational deployment for coastal monitoring. Long training sequences require significant computational resources and memory, especially when handling high-resolution imagery or multi-variable inputs. Model interpretability can be limited, as the internal state of LSTM cells and the influence of past inputs on predictions are not always transparent. Addressing these limitations requires careful model design, hyperparameter tuning, and visualization tools that aid in understanding temporal dependencies. Additionally, ground-truth validation remains essential: periodic field surveys, UAV assessments, and high-accuracy satellite imagery are used to evaluate model performance and adjust predictions, ensuring reliability and credibility in decision-making contexts. Evaluation metrics for LSTM-based coastal change predictions include standard regression measures, such as mean squared error (MSE), root mean squared error (RMSE), mean absolute error (MAE), and correlation coefficients, applied to predicted shoreline positions or erosion rates. For classification tasks, such as predicting whether a pixel represents erosion, deposition, or stability, metrics such as accuracy, F1 score, precision, recall, and Intersection over Union (IoU) are employed.

Temporal consistency metrics, such as Dynamic Time Warping (DTW), can assess how well predicted sequences align with observed temporal trends, capturing the model's ability to reproduce both cyclic patterns and abrupt changes. Continuous refinement through feedback loops, retraining with new data, and integration with domain expertise ensures that LSTM models remain adaptive to evolving coastal conditions (Ienco et al. 2017). The impact of LSTM networks in coastal science is profound. They enable proactive and data-driven decision-making by capturing complex temporal relationships that traditional models often overlook. Coastal planners can identify vulnerable regions, simulate the consequences of interventions, and assess climate change impacts with greater precision. Environmental conservation efforts benefit from understanding the temporal dynamics of habitats such as mangroves, salt marshes, and tidal flats. Disaster preparedness is enhanced by early detection of high-risk events and improved forecasting of their impacts. By combining temporal modeling with spatial analysis, LSTMs contribute to a holistic understanding of coastal systems, supporting sustainable management strategies that balance human needs with environmental preservation. In conclusion, LSTM networks provide a powerful framework for capturing the temporal patterns inherent in coastal change, offering both predictive accuracy and the ability to integrate multi-source, multi-temporal data.

SENTINEL-2 AND LANDSAT: LEVERAGING MULTISPECTRAL IMAGERY FOR CLASSIFICATION

Multispectral satellite imagery has revolutionized the way researchers, environmental managers, and policymakers monitor terrestrial and aquatic systems. Among the most widely utilized platforms for Earth observation are the European Space Agency's Sentinel-2 constellation and the United States Geological Survey's Landsat program. Both Sentinel-2 and Landsat provide high-quality, multispectral datasets that enable detailed classification of land cover, vegetation types, water bodies, urban areas, and other critical surface features. These datasets are particularly valuable for applications such as environmental monitoring, natural resource management, climate change studies, agricultural assessment, and disaster response. Leveraging multispectral imagery from Sentinel-2 and Landsat involves exploiting the spectral, spatial, and temporal characteristics of the data to derive meaningful insights about the Earth's surface. Classification, as a fundamental analytical technique, transforms raw spectral data into interpretable maps of land cover and land use, providing actionable intelligence for decision-making across multiple domains. Sentinel-2 comprises two satellites, Sentinel-2A and Sentinel-2B, equipped with the MultiSpectral Instrument (MSI).

Figure 1. Deep learning for urban land use category classification

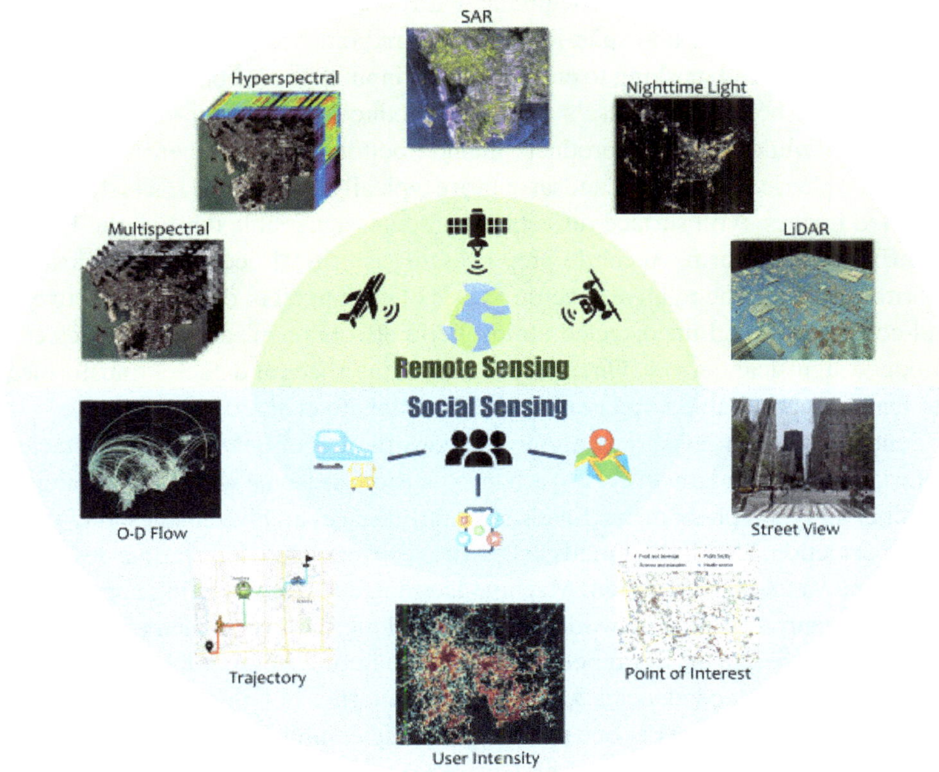

For example, the red-edge bands, unique to Sentinel-2, are particularly useful for vegetation monitoring, facilitating the calculation of indices such as the Normalized Difference Vegetation Index (NDVI) and other red-edge indices that improve classification accuracy for crops, forests, and grasslands. Landsat, a joint program of NASA and USGS, has provided continuous Earth observation data since 1972, making it one of the longest-running remote sensing missions. Landsat 8 and 9, the most recent satellites in the program, carry the Operational Land Imager (OLI) and the Thermal Infrared Sensor (TIRS), capturing imagery across 11 spectral bands with 15–100 meter spatial resolutions. Landsat's historical archive is invaluable for change detection and time-series analysis, enabling researchers to examine land-cover transitions over decades. Landsat imagery is often utilized in conjunction with Sentinel-2 data to enhance classification efforts through data fusion techniques, which combine the high spatial resolution and spectral diversity of Sentinel-2 with the temporal depth and continuity of Landsat. This integration allows for improved monitoring of deforestation, urban expansion, agricultural intensification, and other long-term land-cover dynamics. Multispectral classification using Sentinel-2 and Landsat imagery involves several key steps, beginning with data acquisition and preprocessing. Preprocessing includes atmospheric correction, radiometric calibration, geometric alignment, and cloud masking to ensure that the input data are consistent, accurate, and free from noise. Sentinel-2's Level-1C products provide top-of-atmosphere reflectance, while Level-2A products include bottom-of-atmosphere reflectance after atmospheric correction. Landsat data are typically available as Level-1 terrain-corrected images, with surface reflectance products accessible through the USGS EarthExplorer platform. Accurate preprocessing is critical because classification algorithms rely on the relative spectral values of each pixel to distinguish between land-cover types, and uncorrected atmospheric effects or misaligned images can introduce significant errors. Once preprocessed, the spectral data are transformed into feature sets suitable for classification (Interdonato et al. 2019).

Sentinel-2's high revisit rate allows the construction of dense temporal stacks, capturing vegetation growth cycles, urban expansion, seasonal water body changes, and other temporal phenomena. Landsat's multi-decadal archive enables long-term change detection, identifying trends in land-use conversion, deforestation, desertification, and wetland degradation. Machine learning and deep learning approaches, including recurrent neural networks (RNNs) and long short-term memory (LSTM) models, can leverage these temporal sequences to improve classification, accounting for temporal dependencies and seasonal patterns. For instance, crops with similar spectral signatures at one time of year may exhibit distinguishable growth trajectories over the season, enabling more accurate differentiation when using temporal data. Hybrid approaches combining Sentinel-2 and Landsat data exploit the complementary strengths of both platforms. Sentinel-2 provides higher spatial resolution (10–20 m) and additional red-edge spectral bands, while Landsat offers historical depth and thermal information. Data fusion techniques, such as pixel-based fusion, feature stacking, or ensemble classifiers, allow the integration of multi-source information, enhancing classification robustness and reducing uncertainties. In coastal monitoring, for example, Sentinel-2 may capture fine-scale shoreline changes, while Landsat provides historical context of erosion trends over decades. Similarly, in agriculture, Sentinel-2 enables precise crop mapping, while Landsat time series allows evaluation of long-term land-use transitions and soil degradation patterns. Classification accuracy assessment is essential to validate the reliability of multispectral classification outputs. Ground-truth data, obtained from field surveys, UAV imagery, or high-resolution commercial satellite products, provide reference labels against which classification results can be compared.

In agriculture, multispectral imagery enables crop type mapping, yield estimation, and monitoring of plant health. Forest monitoring leverages these datasets to track deforestation, degradation, and afforestation initiatives. Water resources management utilizes NDWI and other indices to map surface water, detect seasonal flooding, and assess drought conditions. Urban planners employ classifications to monitor built-up area expansion, impervious surface growth, and informal settlement patterns. Coastal managers integrate Sentinel-2 and Landsat imagery to map mangrove extent, monitor shoreline erosion, and assess the impact of human interventions such as dredging, seawalls, and land reclamation. Multispectral classification supports climate change research by enabling the quantification of carbon stocks, vegetation dynamics, and ecosystem resilience over time (Joshi et al. 2026). In conclusion, Sentinel-2 and Landsat provide unparalleled opportunities for multispectral classification of Earth's surface. Sentinel-2's high spatial resolution, red-edge spectral bands, and frequent revisit cadence complement Landsat's long-term archive, enabling both near-real-time monitoring and historical change analysis. By leveraging these datasets, researchers can perform accurate land-cover mapping, monitor environmental change, support disaster management, and inform sustainable development policies. The integration of advanced classification techniques, multi-temporal analysis, and data fusion approaches ensures that the rich spectral information provided by Sentinel-2 and Landsat is translated into actionable insights. As satellite constellations continue to expand and new sensors become available, the combination of Sentinel-2 and Landsat imagery will remain a cornerstone of Earth observation, enabling precise, timely, and data-driven understanding of terrestrial and coastal environments worldwide(Längkvist et al. 2016) .

U-NET ARCHITECTURES: EFFECTIVE FOR COASTAL LAND COVER SEGMENTATION

The accurate mapping and monitoring of coastal environments are essential for understanding ecosystem dynamics, mitigating natural hazards, and supporting sustainable development. Coastal regions are highly dynamic, subject to tidal fluctuations, erosion, sediment transport, human interventions, and climate-induced changes such as sea-level rise. Traditional remote sensing techniques for coastal land cover classification, while effective for broader landscapes, often struggle with the fine-scale spatial heterogeneity, complex boundaries, and spectral similarities characteristic of coastal areas. In this context, deep learning models, particularly U-Net architectures, have emerged as a highly effective approach for coastal land cover segmentation, offering the ability to capture both spatial context and boundary detail, which are crucial for precise delineation of coastal features. Originally proposed by Ronneberger et al. in 2015 for biomedical image segmentation, the U-Net architecture is a convolutional neural network (CNN) designed for pixel-level classification tasks. Its hallmark feature is the encoder-decoder structure with skip connections, which enables the network to simultaneously capture global context and fine-grained spatial details. In the encoder path, successive convolutional and pooling layers extract hierarchical features, reducing the spatial dimensions while increasing the depth of feature maps. This allows the network to learn complex representations of patterns such as vegetation textures, sand, water, and built-up areas, which are often subtle and spectrally similar in multispectral imagery. The decoder path then performs upsampling operations, reconstructing the original spatial resolution while using skip connections to integrate corresponding high-resolution feature maps from the encoder.

Training U-Net for segmentation requires annotated datasets, where each pixel is labeled according to its land cover class. Generating such ground-truth data for coastal environments can be challenging due to spatial complexity and temporal variability. Field surveys, high-resolution aerial imagery, and existing land cover maps are often integrated to produce accurate labels. The use of hybrid datasets combining optical imagery, LiDAR-derived elevation models, and radar backscatter can enhance the training process by providing complementary information on topography and surface roughness, which are particularly relevant for distinguishing wetlands, sandbars, and shallow water zones. Loss functions such as categorical cross-entropy, Dice coefficient, or a combination of both are employed to optimize pixel-wise classification accuracy while accounting for class imbalance, which is common in coastal datasets where small features like tidal channels may occupy a tiny fraction of the image. Performance advantages of U-Net in coastal segmentation are substantial. Unlike traditional pixel-based classification methods, U-Net captures spatial context, reducing misclassification of spectrally similar areas. For instance, shallow water with suspended sediments can exhibit spectral reflectance similar to sand or mudflats in the visible spectrum, leading to confusion in conventional classifiers. The convolutional filters in U-Net, combined with skip connections, allow the network to leverage surrounding spatial patterns, correctly distinguishing water from exposed sediment. Moreover, U-Net models can effectively learn hierarchical representations, identifying both macro-scale features such as shoreline outlines and micro-scale structures like mangrove channels, piers, or small tidal pools (Maggiori et al. 2017). Confusion matrices help identify common misclassifications, such as the mislabeling of shallow water as sand or the confusion between different vegetation types in mangrove forests. Post-processing techniques, such as conditional random fields (CRFs) or morphological filtering, are often applied to refine boundaries and remove isolated misclassified pixels. Additionally, uncertainty estimation methods, including Monte Carlo dropout and ensemble approaches, provide confidence measures for segmentation results, helping managers prioritize areas for field validation or further analysis.

Integration with geospatial analyses amplifies the utility of U-Net-derived segmentation maps. Coastal land cover maps can be used to calculate metrics such as shoreline change rates, wetland area dynamics, habitat fragmentation indices, and flood risk zones. By combining U-Net outputs with digital elevation models, tidal data, and hydrodynamic models, researchers can simulate scenarios of sea-level rise, storm surge impact, and sediment transport, supporting evidence-based decision-making for coastal resilience. Furthermore, these segmentation outputs serve as a foundation for monitoring human impacts, such as urban encroachment, port development, and aquaculture expansion, enabling regulatory agencies to enforce sustainable management practices. Challenges and considerations when applying U-Net to coastal segmentation include data scarcity, variability in illumination and atmospheric conditions, and class imbalance. Cloud cover, sun glint on water surfaces, and seasonal vegetation changes introduce noise and variability that can affect model performance. To address these challenges, multispectral data from multiple acquisition dates are often fused to generate cloud-free composite images, and spectral indices sensitive to vegetation and water are incorporated as additional input channels.

Transfer learning from pre-trained U-Net models or synthetic data generation using generative adversarial networks (GANs) can also enhance model robustness in regions with limited labeled datasets. Computational requirements are another consideration, as high-resolution coastal imagery demands substantial GPU resources for training and inference, necessitating optimized data pipelines and efficient model architectures. In conclusion, U-Net architectures represent a state-of-the-art approach for coastal land cover segmentation, offering high spatial precision, the ability to capture complex boundaries, and flexibility to integrate multi-source and multi-temporal data. By leveraging the encoder-decoder structure, skip connections, and advanced variants such as Attention or Residual U-Nets, researchers and practitioners can achieve highly accurate pixel-level classifications in dynamic coastal environments. When applied to multispectral imagery from platforms like Sentinel-2 and Landsat, U-Net enables detailed mapping of shorelines, intertidal zones, wetlands, urban interfaces, and aquatic habitats, supporting environmental monitoring, coastal resilience planning, and sustainable resource management. The combination of deep learning, geospatial analysis, and remote sensing data positions U-Net as a transformative tool for understanding, protecting, and managing the world's vulnerable coastal regions, ensuring that high-resolution, timely, and actionable insights are available to guide scientific, policy, and community decisions in the face of ongoing environmental and anthropogenic pressures (Paira et al. 2026).

CHANGE DETECTION WITH SIAMESE NETWORKS: A PAIRWISE APPROACH

Change detection is a critical task in remote sensing, environmental monitoring, urban planning, disaster management, and land cover assessment. The goal is to identify alterations in a landscape over time by comparing multi-temporal images and highlighting regions that have undergone significant change. Traditional methods of change detection often rely on pixel-based differencing, image ratios, or classification-based post-classification comparisons. While these techniques can be effective for coarse changes, they frequently struggle with complex scenarios involving subtle alterations, seasonal variations, illumination differences, and misalignment between temporal image acquisitions. In recent years, deep learning approaches have emerged as a transformative solution to these challenges, with Siamese networks providing a particularly robust architecture for pairwise change detection tasks. Siamese networks, originally introduced for signature verification and later adapted to a wide range of similarity-learning tasks, are designed to learn a representation of input pairs and quantify the degree of similarity or dissimilarity between them. In the context of change detection, this capability enables the network to directly model differences between two temporal images, offering a data-driven, flexible, and highly accurate approach to identifying landscape transformations. At their core, Siamese networks consist of two identical subnetworks (often CNNs) that share weights and parameters. Each subnetwork processes one of the two input images, extracting hierarchical features that represent spatial patterns, textures, and spectral characteristics. The use of shared weights ensures that both images are encoded into a comparable feature space, facilitating a meaningful assessment of similarity or difference. Once the features are extracted, they are typically combined using operations such as absolute difference, concatenation, or element-wise subtraction, producing a feature map that emphasizes the areas where the two images diverge.

Each image contains spatial and spectral information that captures unique characteristics of land cover classes such as vegetation, water bodies, built-up areas, bare soil, wetlands, and impervious surfaces. Preprocessing steps are crucial to ensure comparability between images and typically include radiometric correction, atmospheric normalization, geometric alignment, cloud and shadow masking, and spatial resampling. Co-registration of images is particularly important to minimize misalignment that could produce false positives in change detection outputs. After preprocessing, the images are fed into the Siamese network as input pairs, with corresponding ground-truth labels indicating whether each pixel or patch represents a change or no-change event. Ground-truth data is often derived from high-resolution imagery, field surveys, or historical land use maps. Class imbalance is a common challenge, as unchanged areas usually dominate the dataset, necessitating the use of weighted loss functions, focal loss, or data augmentation techniques to improve model performance on the less-represented change class. Advantages of Siamese networks for change detection are multifaceted. First, their pairwise processing inherently captures temporal relationships, enabling the detection of subtle changes that may be imperceptible through pixel-based comparison. Fourth, Siamese networks can be adapted to handle multi-modal inputs, allowing the integration of optical and radar data, LiDAR-derived elevation maps, or hyperspectral channels, which enhances robustness and accuracy in complex scenarios where changes manifest differently across modalities (Qiu et al. 2019). Variants and enhancements of Siamese networks further improve their applicability for change detection.

Fully convolutional Siamese networks allow pixel-wise prediction across the entire image, producing high-resolution change maps that maintain spatial fidelity. Attention-based Siamese architectures incorporate spatial and channel attention mechanisms to focus on regions of interest and relevant spectral bands, enhancing sensitivity to subtle changes and suppressing irrelevant variations. Siamese U-Net or ResNet backbones leverage the encoder-decoder structure and residual connections to capture hierarchical features at multiple scales, improving segmentation of small and irregularly shaped change regions such as narrow riverbanks, fragmented wetlands, or urban alleyways. Temporal extensions using recurrent units like LSTM or ConvLSTM can integrate more than two time points, capturing trends, seasonal cycles, and gradual transformations, enabling predictive monitoring and early warning of environmental degradation or urban sprawl. Training Siamese networks for change detection requires careful consideration of data preparation, loss functions, and evaluation metrics. Patch-based sampling is commonly used to manage high-resolution imagery, while data augmentation such as rotations, flips, and intensity shifts increases model generalization. Loss functions often combine cross-entropy or focal loss with Dice coefficient or IoU loss to address class imbalance and optimize both pixel-level accuracy and spatial overlap of predicted changes. Evaluation metrics include precision, recall, F1-score, IoU, and area under the receiver operating characteristic (ROC) curve.

Confusion matrices are used to identify specific error patterns, such as false positives arising from seasonal vegetation changes or shadows, enabling targeted refinement of the model or preprocessing workflow. Post-processing steps, including morphological filtering, conditional random fields, or connected component analysis, further improve the coherence and usability of change maps, reducing isolated false detections and enhancing interpretability for decision-makers. Operational applications of Siamese network-based change detection are extensive and increasingly critical in environmental management. Coastal monitoring benefits from high-resolution change maps that capture shoreline erosion, sediment deposition, mangrove encroachment, and tidal wetland dynamics, informing conservation planning and disaster preparedness. Urban planning leverages these techniques to track informal settlement growth, infrastructure expansion, and land use conversions. Agriculture and forestry use pairwise change detection to monitor crop growth, deforestation, and land degradation. Disaster management applications include rapid assessment of flood extents, wildfire impacts, and earthquake-related land shifts, enabling timely response and resource allocation. Moreover, the ability to integrate multi-modal data allows authorities to detect changes even under challenging conditions such as cloud cover, smoke, or night-time imagery, enhancing resilience and situational awareness. Challenges and considerations in deploying Siamese networks for change detection include data scarcity, label quality, computational demands, and environmental variability. High-quality, co-registered, and temporally aligned imagery is essential to reduce false positives and maximize model accuracy.

By directly learning the relationships between two temporal images and leveraging shared-weight CNN architectures, these networks achieve superior accuracy in capturing both subtle and dramatic landscape changes. Their adaptability to multimodal inputs, attention mechanisms, hierarchical feature extraction, and temporal extensions further enhances their capability to handle complex, heterogeneous, and dynamic environments. Operational implementations in coastal monitoring, urban expansion tracking, agriculture, forestry, and disaster response demonstrate the transformative impact of Siamese networks in producing precise, timely, and actionable change maps. By combining deep learning, geospatial intelligence, and multi-temporal imagery, Siamese network-based change detection enables decision-makers, scientists, and policymakers to understand landscape transformations with unprecedented accuracy, supporting sustainable management, resilience planning, and informed responses to environmental and societal challenges. The pairwise approach of Siamese networks continues to redefine the state-of-the-art in remote sensing change detection, offering a scalable and robust framework that aligns with the growing demands of high-resolution temporal monitoring and complex environmental analysis (Persello & Bruzzone 2013).

HANDLING CLASS IMBALANCE: STRATEGIES FOR COASTAL LAND COVER DATA

Coastal regions present a unique combination of ecological, socio-economic, and geomorphological characteristics, making them highly dynamic yet critical areas for monitoring and management. Accurate land cover classification in these regions is essential for coastal conservation, urban planning, disaster mitigation, habitat management, and climate resilience strategies. However, one of the most significant challenges in developing robust classification models for coastal land cover is class imbalance. Class imbalance occurs when certain land cover types, such as water bodies, sand dunes, or mangrove forests, occupy a disproportionately small fraction of the study area relative to dominant classes like bare soil, built-up regions, or general vegetation cover. This imbalance can severely degrade the performance of machine learning and deep learning models, leading to poor recognition of minority classes, biased predictions, and misinformed decision-making. Addressing class imbalance is thus a critical step in ensuring that coastal monitoring systems provide accurate, actionable, and equitable insights across all land cover types. The nature of class imbalance in coastal landscapes is influenced by the inherent heterogeneity and spatial distribution of land cover types. Coastal areas often include extensive water surfaces, intertidal zones, estuaries, salt marshes, mangrove forests, wetlands, sandy beaches, rocky outcrops, urban settlements, and agricultural land.

Data-centric strategies are among the first lines of defense against class imbalance. Oversampling techniques, such as duplicating minority class samples or generating synthetic examples using methods like SMOTE (Synthetic Minority Over-sampling Technique), help increase the representation of underrepresented classes. SMOTE and its variations create new synthetic samples by interpolating between existing minority-class pixels or patches, providing the model with additional examples of rare coastal features. Conversely, undersampling can reduce the prevalence of majority classes, balancing the dataset and preventing the model from being biased toward dominant land cover types. However, undersampling may result in loss of critical spatial context or information, which is particularly important in complex coastal landscapes where spatial continuity and neighborhood effects carry ecological significance. A hybrid approach, combining moderate undersampling of majority classes with oversampling of minority classes, often yields better performance by maintaining data richness while correcting imbalance. Algorithmic strategies are also critical to address class imbalance in coastal land cover classification. Class-weighted loss functions assign higher penalties to misclassifications of minority classes, guiding the model to pay more attention to these critical areas during training (Russwurm & Körner 2018). Cross-entropy loss with class weights or focal loss are widely used in deep learning frameworks, where focal loss specifically down-weights easy-to-classify majority samples while emphasizing hard-to-classify minority instances. Additionally, hierarchical or cascaded classification approaches can first segment the landscape into broad categories and then perform fine-grained classification within each segment, allowing minority classes to be isolated and treated separately. Ensemble methods, including random forests, gradient boosting, or bagging techniques, leverage multiple models to improve recognition of minority classes by combining predictions across diverse perspectives and feature subsets. Spatial and contextual augmentation strategies are particularly effective for coastal datasets due to the importance of neighborhood and texture information. Techniques such as patch-based sampling, where each training sample includes a spatial window surrounding a pixel, allow the model to learn contextual features that distinguish minority classes from visually similar majority classes.

Standard accuracy can be misleading when majority classes dominate, making it appear that the model performs well despite poor recognition of minority classes. Metrics such as precision, recall, F1-score, Intersection over Union (IoU), Matthews correlation coefficient (MCC), and balanced accuracy provide a more nuanced understanding of performance across all classes. Confusion matrices allow identification of systematic misclassifications, such as water pixels incorrectly labeled as mudflats during low-tide conditions, enabling targeted model refinement. Spatial analysis of misclassified areas, combined with field validation or high-resolution reference imagery, further strengthens model reliability and ensures that ecological or socio-economic critical features are correctly identified. Operational deployment considerations also influence strategies for handling class imbalance. Coastal monitoring systems often require scalable processing pipelines capable of handling large volumes of high-resolution imagery (van Duynhoven & Dragićević 2019). Patch-based processing, tiling strategies, and distributed computing platforms help manage memory constraints while maintaining model performance. Incorporating active learning pipelines, where the model flags uncertain predictions for human review, allows iterative improvement of minority-class recognition without exhaustive manual labeling. Moreover, open-access datasets, such as Sentinel-2, Landsat, or high-resolution commercial imagery, combined with collaborative annotation initiatives, help alleviate scarcity of labeled minority-class examples, particularly for critical habitats like mangroves, salt marshes, or coral reefs. Future directions for addressing class imbalance in coastal land cover classification include integrating unsupervised or self-supervised learning techniques, which can leverage abundant unlabeled data to improve feature representation for minority classes. Graph-based models and spatial relational learning approaches can exploit the inherent spatial dependencies and connectivity of coastal features, further enhancing minority-class detection.

Transfer learning from other geographic regions, temporal domains, or multimodal datasets can provide additional prior knowledge, enabling models to generalize better to underrepresented classes. Finally, combining AI-driven change detection with imbalance-aware classification can provide dynamic monitoring of coastal ecosystems, capturing both rare static features and evolving minority classes in near real-time, thereby supporting informed policy-making, environmental conservation, and disaster preparedness. In summary, handling class imbalance in coastal land cover datasets is a multidimensional challenge that requires a combination of data-centric, algorithmic, architectural, and operational strategies. Oversampling, undersampling, data augmentation, class-weighted loss functions, attention mechanisms, multi-modal data fusion, and advanced evaluation metrics all play a crucial role in ensuring that minority classes, which often represent ecologically, economically, or socially critical features, are accurately identified. By implementing these strategies, researchers and practitioners can produce robust, high-fidelity classification maps that support sustainable coastal management, climate resilience, disaster mitigation, and biodiversity conservation, ultimately enabling better stewardship of some of the most dynamic and vulnerable ecosystems on the planet. Addressing class imbalance is not merely a technical concern but a prerequisite for informed decision-making and effective intervention in coastal environments where small-scale changes can have far-reaching consequences (Zhang et al. 2021).

CONCLUSION

Digital Elevation Models (DEMs) have emerged as a critical source of geospatial information, providing detailed topographic representation that enhances land cover classification, particularly in complex and dynamic environments such as coastal zones, mountainous regions, floodplains, and urban-rural interfaces. Traditional classification approaches relying solely on spectral data from optical or multispectral satellite imagery often struggle to distinguish between land cover types that exhibit similar spectral signatures but occupy different topographic positions. For example, in coastal areas, tidal flats and shallow water bodies may reflect similar spectral bands in visible and near-infrared imagery, making pixel-based classification prone to confusion. By integrating DEM-derived features—such as elevation, slope, aspect, and curvature—classification models gain an additional layer of spatial context that is directly linked to the physical characteristics of the terrain, improving discrimination between classes that are spectrally similar yet topographically distinct. This integration of DEM data is particularly relevant for coastal land cover mapping, flood risk assessment, erosion monitoring, urban planning, and habitat delineation, where subtle variations in elevation often correspond to significant ecological or socio-economic differences. The value of DEMs in land cover classification lies in their ability to capture vertical and derivative features that spectral imagery alone cannot provide. Elevation provides absolute height above sea level, which can help distinguish between intertidal zones, wetlands, low-lying agricultural areas, and upland vegetation. Slope identifies areas of rapid elevation change, which are often associated with cliffs, dunes, embankments, or artificial constructions, while aspect describes the orientation of terrain, influencing microclimates, sunlight exposure, and moisture retention. Curvature, both profile and planform, characterizes concave or convex surface geometries, offering additional information about drainage patterns, erosion-prone zones, and depositional areas. Integrating these topographic derivatives into classification models allows for more nuanced differentiation between land cover types, particularly in heterogeneous landscapes where minor spectral differences may be insufficient for accurate discrimination. DEM integration strategies vary from straightforward feature augmentation to complex multi-source data fusion. In a conventional supervised classification framework, DEM-derived features can be appended to the spectral band stack, creating an enhanced feature space for classifiers such as random forests, support vector machines, gradient boosting, or k-nearest neighbors. Each pixel in the training and testing datasets includes not only its spectral signature but also its corresponding elevation, slope, aspect, and curvature values, enabling the model to exploit correlations between terrain attributes and land cover types. For instance, mangrove forests typically occupy low-lying intertidal zones with gentle slopes and proximity to tidal channels, whereas sandy dunes may exhibit steeper slopes and higher elevation variability. These patterns can be captured quantitatively using DEM-derived metrics, enhancing classifier performance and reducing misclassification between classes with similar spectral reflectance. Deep learning architectures further benefit from DEM integration, particularly in pixel-wise semantic segmentation and patch-based approaches. Convolutional Neural Networks (CNNs), U-Net variants, and attention-based models can ingest multi-channel inputs that combine spectral bands with DEM derivatives, learning hierarchical representations that account for both spectral and topographic patterns. For example, in coastal land cover mapping, a U-Net model can simultaneously analyze optical imagery, slope maps, and elevation layers, enabling the network to capture spatial dependencies and elevation-specific features critical for distinguishing tidal flats from shallow water, mudflats from agricultural lands, or low-lying wetlands from urban settlements. Multi-scale convolutional filters allow the network to detect local topographic variations, while attention modules can prioritize regions where elevation is a discriminative factor, improving segmentation accuracy for minority or fragmented classes. DEM sources and preprocessing are crucial to successful integration. Commonly used global DEM datasets, such as NASA's SRTM (Shuttle Radar Topography Mission), ASTER GDEM, TanDEM-X, and LiDAR-derived high-resolution DEMs, provide varying spatial resolutions and vertical accuracies. Coastal and urban applications benefit from higher-resolution DEMs (1–5 meters), which can resolve fine-scale features such as levees, dikes, mangrove fringe zones, and tidal channels.

Preprocessing steps typically include resampling to match the spatial resolution of spectral imagery, co-registration to ensure precise alignment, removal of artifacts such as voids or noise, and derivation of slope, aspect, curvature, and other secondary metrics. Normalization or scaling of DEM features may also be necessary to balance their influence relative to spectral bands during model training, preventing dominance of high-range elevation values over spectral variations in machine learning or deep learning algorithms. Integration with multi-sensor datasets enhances the discriminative power of DEM-based features. For instance, combining DEM-derived elevation with Synthetic Aperture Radar (SAR) backscatter data can help differentiate inundated wetlands from bare soil, as low-elevation regions may correspond to standing water detectable through radar signatures even under cloud cover. Hyperspectral imagery combined with DEM-derived slope and aspect can improve discrimination of vegetation types with similar reflectance but differing in preferred topographic habitats, such as salt-tolerant marsh grasses versus inland riparian vegetation. Multi-temporal imagery integrated with DEM features allows models to account for seasonal or tidal variability, particularly in coastal zones, enhancing minority-class detection and temporal stability of classification outputs. Handling class imbalance with DEM integration is particularly relevant in landscapes where ecologically critical classes occupy small areas or occur in topographically unique niches. For example, small mangrove patches along river mouths, salt pans, or ephemeral wetlands may be underrepresented in training datasets. DEM-derived features allow classifiers to exploit terrain-specific cues, enhancing minority-class recognition even when spectral signatures are ambiguous. Weighted loss functions, oversampling of minority-class pixels, and patch-based training can be combined with DEM features to maximize detection of rare classes, ensuring that the ecological and socio-economic importance of small or fragmented land cover types is preserved in the final classification maps. Validation and evaluation of DEM-enhanced classifications require both traditional and terrain-aware metrics. Pixel-level accuracy, F1-score, precision, recall, and Intersection over Union (IoU) remain essential, but incorporating topographic consistency checks can further validate model outputs. For example, predicted classes should correspond logically to elevation ranges, slope thresholds, or hydrological connectivity inferred from DEM data. Spatial cross-validation across different coastal regions or temporal snapshots ensures that models generalize beyond the training area and accurately leverage topography in diverse environmental contexts. Field validation or high-resolution aerial imagery can further corroborate DEM-aided predictions, particularly for minority or ecologically sensitive land cover types. Operational deployment considerations involve handling large volumes of DEM and imagery data efficiently. High-resolution DEMs can substantially increase memory requirements and computational load, particularly in patch-based or deep learning pipelines. Strategies such as tiling, distributed computing, cloud-based processing, and optimized data storage formats (e.g., Cloud-Optimized GeoTIFFs) are critical for scalable implementation. Automated pipelines can integrate DEM preprocessing, derivative calculation, data augmentation, and classifier training, enabling iterative model refinement and rapid updates to reflect changes in coastal topography due to erosion, sedimentation, urban development, or sea-level rise. Future directions for DEM-enhanced classification emphasize hybrid approaches combining topographic, spectral, temporal, and multi-modal information.

Graph-based deep learning, 3D CNNs, and transformer architectures can exploit spatial and elevation relationships more explicitly, learning complex correlations between terrain, hydrology, and land cover patterns. Integration with hydrodynamic models, tidal simulations, and climate projections can further enhance predictive capabilities, particularly for vulnerable coastal zones. DEM integration can also support adaptive sampling and active learning strategies, identifying areas where additional ground truth or high-resolution imagery would most improve model performance, particularly for underrepresented classes or ecologically sensitive areas. In summary, incorporating DEM data into land cover classification provides critical topographic context that enhances the differentiation of spectrally similar classes, improves minority-class recognition, and increases overall model robustness. Elevation, slope, aspect, curvature, and derived hydrological metrics allow machine learning and deep learning models to capture the physical structure of the landscape, complementing spectral and temporal information. When combined with advanced preprocessing, multi-sensor integration, class imbalance handling, and robust evaluation, DEM-based features enable more accurate, ecologically meaningful, and operationally viable classification of complex coastal and heterogeneous landscapes. This topography-aided approach not only improves technical performance but also ensures that environmental management, urban planning, and disaster mitigation decisions are informed by a holistic understanding of both the spectral and structural dimensions of the land surface, making DEM integration an indispensable component of contemporary geospatial analysis and land cover monitoring.

REFERENCES

Blais, M.-A., & Akhloufi, M. A. (2025). Advances in remote sensing and deep learning in coastal boundary extraction for erosion monitoring. *Geomatics*, *5*(1), 9. DOI: 10.3390/geomatics5010009

Campos-Taberner, M., García-Haro, F. J., Martínez, B., Izquierdo-Verdiguier, E., Atzberger, C., & Camps-Valls, G. (2020). Understanding deep learning in land use classification based on Sentinel-2 time series. *Scientific Reports*, *10*, 17188. DOI: 10.1038/s41598-020-74215-5 PMID: 33057052

Dang, K. B., Dang, V. B., Ngo, V. L., Vu, K. C., Nguyen, H., Nguyen, D. A., & Nguyen, D. L. P. N. (2022). Application of deep learning models to detect coastlines and shorelines. *Journal of Environmental Management*. Advance online publication. DOI: 10.1016/j.jenvman.2022.115732 PMID: 35930878

Daudt, R. C., Le Saux, B., & Boulch, A. (2018). Fully convolutional siamese networks for change detection. 2018 25th IEEE International Conference on Image Processing (ICIP), 4063–4067. https://doi.org/DOI: 10.1109/ICIP.2018.8451607

Ebenezer, Y., Durga, K., Chandrakala, N. B., Shashank, G., Sharma, M., & Manjunath, T. C. (2026). Disaster Education and Awareness Programs. In Ali, Q., & Tariq Iqbal Khan, M. (Eds.), *Strengthening Global Resilience to Natural Disasters* (pp. 73–102). IGI Global Scientific Publishing., DOI: 10.4018/979-8-3693-9745-9.ch004

Helber, P., Bischke, B., Dengel, A., & Borth, D. (2019). EuroSAT: A novel dataset and deep learning benchmark for land use and land cover classification. *IEEE Journal of Selected Topics in Applied Earth Observations and Remote Sensing*, *12*(7), 2217–2226. DOI: 10.1109/JSTARS.2019.2918246

Ienco, D., Ratle, F., & Moser, G. (2017). Land cover classification via multitemporal spatial data by deep recurrent neural networks. *IEEE Geoscience and Remote Sensing Letters*, *14*(10), 1685–1689. DOI: 10.1109/LGRS.2017.2749278

Interdonato, R., Ienco, D., Gaetano, R., & Ose, K. (2019). DuPLO: A DUal view Point deep Learning architecture for time series classification. *ISPRS Journal of Photogrammetry and Remote Sensing*, *149*, 91–104. DOI: 10.1016/j.isprsjprs.2018.12.008

Joshi, N. V., Selvakumar, P., & Srivastava, N. Manjeet, Das, A., & Manjunath, T. C. (2026). Bioethics Legal Implications of Artificial Intelligence in Medicine. In S. Anastasiadou, L. Seremeti, & S. Papalexandris (Eds.), Bioethics and Legal Aspects of AI in Healthcare and Life Sciences (pp. 27-54). IGI Global Scientific Publishing. https://doi.org/DOI: 10.4018/979-8-3373-5646-4.ch002

Längkvist, M., Kiselev, A., Alirezaie, M., & Loutfi, A. (2016). Classification and segmentation of satellite-based earth observation data: A review. *IEEE Geoscience and Remote Sensing Magazine*, *4*(2), 41–54. DOI: 10.1109/MGRS.2016.2527748

Maggiori, E., Tarabalka, Y., Charpiat, G., & Alliez, P. (2017). High-resolution land cover mapping with convolutional neural networks. *IEEE Transactions on Geoscience and Remote Sensing*, *55*(2), 881–893. DOI: 10.1109/TGRS.2016.2611803

Paira, D., Rohini, A., Chandra, S., Satyanarayana, P., Sharma, M., & T. C., M. (2026). Climate Change and Natural Disasters. In Q. Ali & M. Tariq Iqbal Khan (Eds.), Strengthening Global Resilience to Natural Disasters (pp. 21-44). IGI Global Scientific Publishing. https://doi.org/.DOI: 10.4018/979-8-3693-9745-9.ch002

Persello, C., & Bruzzone, L. (2013). Active learning for domain adaptation in remote sensing image classification. *IEEE Transactions on Geoscience and Remote Sensing*, *51*(8), 4817–4832. DOI: 10.1109/TGRS.2013.2251503

Qiu, C., Mou, L., Schmitt, M., & Zhu, X. X. (2019). Local climate zone-based urban land cover classification from multi-seasonal Sentinel-2 images with a recurrent residual network. *ISPRS Journal of Photogrammetry and Remote Sensing*, *154*, 151–162. DOI: 10.1016/j.isprsjprs.2019.07.016 PMID: 31417230

Russwurm, M., & Körner, M. (2018). Multi-temporal land cover classification with sequential recurrent encoders. *ISPRS International Journal of Geo-Information*, *7*(4), 129. DOI: 10.3390/ijgi7040129

van Duynhoven, A., & Dragićević, S. (2019). Analyzing the effects of temporal resolution and classification confidence for modeling land cover change with long short-term memory networks. *Remote Sensing*, *11*(22), 2784. DOI: 10.3390/rs11222784

Zhang, C., Zhang, L., & Du, B. (2021). Deep learning for remote sensing image classification: A survey. *Journal of Sensors*. Advance online publication. DOI: 10.1155/2021/6628623

Chapter 3

Eco-Aware Sustainable Sub Sea Monitoring:

An AI-Driven Approach for Corrosion Detection in Pipelines and Marine Safety

Vijayakumari Rodda
Krishna University, India

R. Jeyalakshmi
http://orcid.org/0000-0003-1651-4949
R.M.K. Engineering College, India

R. Saranya
Bharathiyar Institute of Engineering for Women, India

Chokkan Priyadharshini
Bharathiyar Institute of Engineering for Women, India

V. Vedanarayanan
http://orcid.org/0000-0002-2194-6714
Sathyabama Institute of Science and Technology, India

N. V. Keerthana
http://orcid.org/0000-0001-8878-2083
Velalar College of Engineering and Technology, India

ABSTRACT

The performance of four machine learning and deep learning algorithms for corrosion detection and classification using image datasets: Random Forest (RF), Vision Transformer (ViT), K-Nearest Neighbours incorporating Principal Component Analysis (KNN/PCA), and Deep Autoencoder (DA). The algorithms utilize different computational strategies to analyze corrosion patterns and severity levels. The Random Forest model consists of multiple decision trees, which produces reliable and interpretable predictions when combining many random sub-samples of both datasets and features. The performance metrics such as accuracy, precision, recall,

DOI: 10.4018/979-8-3373-8872-4.ch003

and F1 score, as well as efficiency showed Random Forest is a robust model, Vision Transformer obtains the highest accuracy, KNN/PCA is simple and fast, and Deep Autoencoder is suitable for anomaly detection. Collectively, these models provide a framework of reliable and scalable models for intelligent, low-cost corrosion monitoring possible to implement in industrial pipelines and underwater marine environments.

1. INTRODUCTION

Detecting corrosion on metallic components, specifically for structures that are imposing or underwater, is an essential part of structural health monitoring. Metallic surfaces that are constantly in contact with moisture, pressure, salinity, and chemical interactions will degrade over time and can yield irreversible damage that represents safety hazards and financial loss. For examples, investigations of materials such as NV E690 and 316L stainless steel coatings via Underwater Direct Metal Deposition (UDMD) conveyed that coatings applied underwater will be subjected to both environmental and mechanical effects, and the understanding of how material and deposition processes impact resistance to corrosion is paramount(Bahrami, Zhang, Wang, & Liu, 2022). Additional examples have illustrated the use of asynchronous current field measurement to create 3D renderings of submerged corrosion locations and provided a method for remote inspection in areas where access was restricted (Chen et al., 2023). This research shows that there is a growing need for corrosion analysis automated systems that can provide accuracy and early detections to assist the maintenance teams to make informed decisions before structural failure and catastrophic loss occurs.

Machine learning for corrosion detection has employed new approaches, specifically ensemble techniques that have proven effective, such Random Forests (RF) in classifying complex corrosion features because of their emphasis on limiting overfitting and understanding non-linear interactions among features. This involving rust and microstructural changes in marine environments focus on monitoring corrosion resistance to underwater laser-cladded steel surfaces (Chen et al., 2023). RF combines multiple decision trees, each enabled to use different features - including things like rust texture, clarity of edges, and even color differences. RF enables detection tools to make sense of noisy and heterogeneous corrosion pictures more effectively than when using a single classifier. In addition, study of Fe-VC coatings deposited on water-cooled substrates suggested that thermal stability improves as a result of the laser treatment and generated corrosion resistance, and the study recommended proper monitoring systems to detect corrosion in advanced materials underwater (Yuan et al., 2024). Industrial inspection research incorporated computer vision techniques to analyze corrosion in, and for pipes and large structures where manual inspection is impracticable (N. Guo et al., 2023).

The deployment of deep learning methods has greatly enhanced the possibilities for corrosion detection, particularly with Vision Transformer (ViT) methods developed recently. ViT learns the patterns in images into patch sequences, then applies self-attention operators to each patch, simulating the human ability to utilize long-range spatial dependencies. The ViT architecture enables the model to draw hierarchical and global representations of corrosion patterns, even in difficult imaging conditions (e.g., poor lighting, noise, and partial occlusion)(Teng et al., 2024). Recently developed systems integrate ViT models with advanced image post-processing methods to mitigate the effects presented by real-world underwater and industrial scenes, enhancing corrosion severity classification readings (W. Guo, Zhang, Zhao, & Xiao, 2024). Rising levels of research have pioneered the use of Autonomous Underwater Vehicles (AUVs) leveraging AI algorithms for monitoring marine structures, thus providing scalable real-time solutions for underwater inspections (Feng, Jiang, & Yan, 2025).

Traditional algorithms with additional dimensionality reduction still hold merit in scenarios with resource limitations. The combination of using K-Nearest Neighbours (KNN) algorithm with Principal Component Analysis (PCA) provides a simple and effective solution for corrosion detection, especially in scenarios where datasets are small or the computational budget is limited(Vorobel, Ivasenko, Berehulyak, & Mandzii, 2021). Underwater imaging problems such as turbidity, environmental temperature difference or lighting distortion often compromise the quality of images taken underwater and provide challenges to the required preprocessing and required image feature extraction (Herndon & Tien, 2025). Several studies have noted that image noise, low brightness or poor contrast hinder corrosion recognition tasks, justifying the utility of PCA to reduce redundant dimensions while maintaining meaningful features for KNN classification (A. A. M. Ibrahim & J. R. Tapamo, 2024). Overall surveys have identified that, machine learning is useful in predicting corrosion progress across different materials and environmental conditions, thereby enabling planning for preventive maintenance(A. A. M. Ibrahim & J.-R. Tapamo, 2024). Non-destructive evaluation methods, such as acoustic emission that observe stress waves produced by corroded mooring connections, represent further evidence of the benefit of machine learning and provide ways to analyse structural health signals (Imran et al., 2023).

Deep Autoencoders (DAE) take another optimistic approach to detecting corrosion utilizing unsupervised learning to learn compressed representations of latents (i.e., different features) of normal or minimally corroded surfaces. There are studies that have shown that combining supervised and unsupervised approaches has the potential to develop hybrid models that are successful in detecting corrosion and estimating corrosion depth (Jahanshahi & Masri, 2013). In addition, studies on corrosion of coral aggregate concrete have shown that machine learning models can contribute to sustainability and the minimization of structural risk(Jiao, Ye, Zhang, Li, & Wang, 2024). DAEs are adapted to anomaly detection; once trained on non-corroded surfaces; DAEs can determine corrosion by measuring reconstruction error associated with anomalous behaviors learned at training time. Therefore, DAEs have value for early detection of corrosion-like anomalies in scenarios characterized by limited labelled data (Keerthana, Srija, Devi, Roopini, & Sharumathi, 2025).

Image enhancement methods for detecting corrosion underwater are also important. Image enhancement methods such as Adaptive Saturation and Mean Restoration (ASMR), Bi-Histogram Equalization (BPDHE), Histogram Equalization (HE), Xuan Dong and UIECPI, have been evaluated to improve colour balance, contrast, and edge detail of underwater images(Khan, Ali, Anwer, Adil, & Mériaudeau, 2018). This greatly improves the effectiveness of downstream classifiers in highly visually degraded environments. Additional studies analyzing underwater repair protocols have illuminated how challenging marine conditions, such as poor resolution and high pressure, complicate inspection and repair and illustrated the need for robust automated detection systems(Li et al., 2024). Dynamic Histogram Equalization (DHE), gradient-based preprocessing, and edge enhancement have also demonstrated notable success at improving corrosion visibility for machine learning models(Ma, Du, Wang, & Su, 2023). Other approaches for underwater pipeline monitoring combine enhancement, resizing, and edge detection into a coherent corrosion identification framework (Munawar et al., 2022).

Together, these works demonstrate the advantages of combining ensemble learning (RF), transformer-based deep learning (ViT), classical approaches with a dimensionality reduction (KNN - PCA) step, and unsupervised anomaly detection (DAE) into single evaluation framework. Baseline image enhancement methods, such as Dynamic Histogram Equalization (DHE) have been assessed and shown to maintain certain edge characteristics of importance to downstream processing(Nguyen et al., 2023). In another recent work, the authors established a viable corrosion detection model for underwater pipelines through image enhancement, resizing, and edge detection that focused on making the model robust to underwater pipeline operation(Paraschos & Papadakis, 2024). Each of the methods offer benefits: RF is interpretable and robust; ViT can leverage powerful spatially aware features; KNN-PCA can classify quickly and with low-complexity; and DAE is good at anomaly detection without labelled data. A shared corrosion dataset allows for one comprehensive evaluation of complementary insights on classification accuracy, computational performance, and robustness related to field noise and the environment(Riccioli, Gabrielsen, Høgsæt, Barros, & Pahlavan, 2024). The duplication of insights supports the development of intelligent, scalable, and reliable systems to monitor corrosion processes of critical infrastructure, from pipelines and offshore platforms, to industrial tanks and underwater structures.

1.1 Contribution of the Work

- The project compares several machine learning and deep learning models to classify the severity of corrosion from images.
- A custom real, world corrosion dataset was created with a Raspberry Pi camera under different environmental and lighting conditions.
- Classical as well as deep learning methods were used to decide the best one for corrosion detection.
- Huge image preprocessing and augmentation were done to enhance the models' robustness and generalization.
- The obtained results indicate the feasibility of such low, cost, real-time corrosion monitoring systems.

1.2 Motivation of the Work

- One of the main consequences of corrosion is the failure of structures with the implication of increased maintenance costs.
- The conventional inspection process requires a lot of time and not reliable in detecting corrosion in its early stages.
- Detection at an early stage is a great help in making the cost of accidents and repairs lower.
- An automated inspection system is highly recommended for situations where there are dangerous conditions.
- AI-powered corrosion detection is possible through image- based data.

2. RELATED WORK

Zhao et al. presented a method of corrosion image rating as a case study using transfer learning which demonstrated that the use of deep feature extraction from pre-trained models enhanced the performance of corrosion severity recognition through the extraction of fine-grain spatial arrangements, texture variations, and variations of rust intensities over complex surfaces(Z. Zhao, Bakar, Razak, & Akhtar, 2025). The work elaborated on how more advanced architectures, particularly attention-based models, effectively interpreted corrosion images while considering transformer-like models such as the ViT for enhanced learning of global features and classification performance. Shen et al. presented an approach to automate rust detection within the steel bridge coatings domain by way of colour histograms and texture descriptors and demonstrated the accurate distinction of rusted regions with the careful selection of hand-crafted features despite varying lighting and surface conditions(Shen, Chen, & Chang, 2013). The correspondence of this work relates to the functionality of PCA as a means to implicitly reduce feature dimensionality while maximizing the discriminability of features and providing an appealing and trustworthy instance-based classification process when integrated with a KNN algorithm, particularly applicable for lightweight corrosion detection scenarios.

Ibrahim and Tapamo proposed a model for transferring learning for the classification of defects on steel surfaces that demonstrated reliable feature extraction and classifiers provide a level of certainty regarding specific unusual defect classes. The researchers found that applying ensemble learning methods to aggregate the predictions from multiple decision trees to address surface variations and noise, as well as corrosion variety, achieved improved accuracy in the classification of defects(Son, Jeong, & Oh, 2024). Zhao et al. (2025) improved corrosion detection by proposing a deep-learning segmentation model that reliably extracts localized corrosion from structural alloys. Their encoder-decoder model learned representations specific to corrosion and was able to accurately segment corroded areas, supporting the benefit of reconstruction-based models such as Deep Autoencoders that identify anomalies by differences in reconstruction and behaviour in latent-space features (Sun, Li, Su, Liu, & Chen, 2025). Collectively, these studies provide a good point at which to examine ViT, KNN- PCA, RF and DAE methods for measuring spatial relationships, dimensions, ensemble decision-making, and reconstruction-based anomalies in corrosion imagery in both industrial and structural applications.

3. METHODOLOGY

The methodology introduces a multi- model AI-based framework that detects corrosion and classifies its severity by combining four complementary methods: ViT, KNN- PCA, RF and DAE. The hybrid of these models is aimed at harness the advantages of classical machine learning, deep learning, and unsupervised learning to carry out robust corrosion analysis under different environmental and surface conditions. The methodology of the proposed system is based on a parallel processing architecture, in which the pre-processed corrosion images are fed to the four models for simultaneous analysis. RF delivers dependable and interpretable classification, ViT understands global spatial dependencies through self, attention, KNNPCA provides a simple and computationally efficient method for small datasets, and DAE supports anomaly detection via reconstruction error analysis. Every single model identifies the severity of the corrosion situation in different levels of category:

Low, Medium, and High. Then, the results obtained from their performances are measured with the help of standard metrics like accuracy, precision, recall, and F1, score. The communication of these models with the corrosion detection pipeline which leads to the overall network structure is depicted in Figure 1.

Figure 1. Network structure

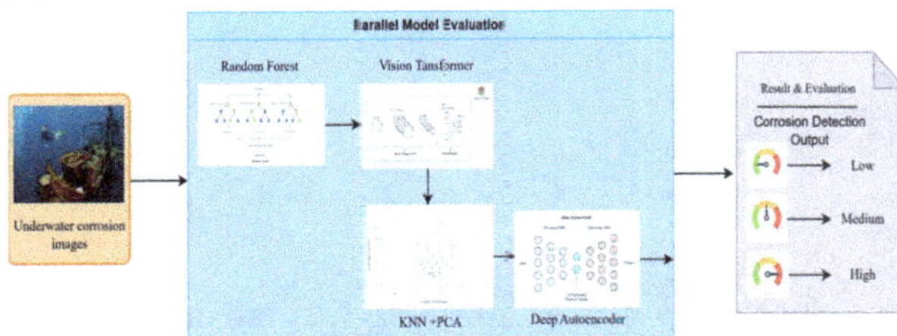

3.1 Random Forest

Figure 2 represents a well-liked ensemble learning technique called RF builds a large number of decision trees during training. It outputs a class that is either the mean prediction (regression) or the mode of the classes (classification). RF tackles overfitting associated with a single decision tree and increase predictive performance. RF has good classification performance for corrosion severity classification because it can handle noisy data and represent a non-linear structure in the dataset. RF is also quite flexible in terms of visualisation and using feature-extracted data to classify corrosion levels as low, medium, or high. A random portion of the data for training and a selected set of characteristics are used to train each tree in the forest. This unpredictability will lessen overfitting and enhance a model's capacity to generalize. This is Explained in Figure 2.

The model will take each of the results from all trees to average them to make its final prediction, which can represent non-linear patterns and interactions from features. The mathematical formula for RF is defined as (1),

$$F(x) = majority\_vote(f_1(Y), f_2(Y), .., f_T(Y) \tag{1}$$

Where,

F=$\{(Y_1,B_1),(Y_{2,}B_2), \ldots.(Y_nB_n)\}$ is the training dataset
d_i (A) is the forecast derived from the forest's t-th decision tree

T is the total number of trees

Gini Impurity: The RF architecture is an ensemble learning technique that uses several decision trees during training and produces a class that is either the mean prediction (regression) or the mode of the classes (classification). The Mathematical Formula for Gini Impurity is represented in Equation (2),

$$I = 1 - \sum_{j=1}^{n} c_j^2 \qquad (2)$$

Where,

n = number of classes (Low, Medium, High corrosion)
c_j = probability of class j in that node

Figure 2. Random forest

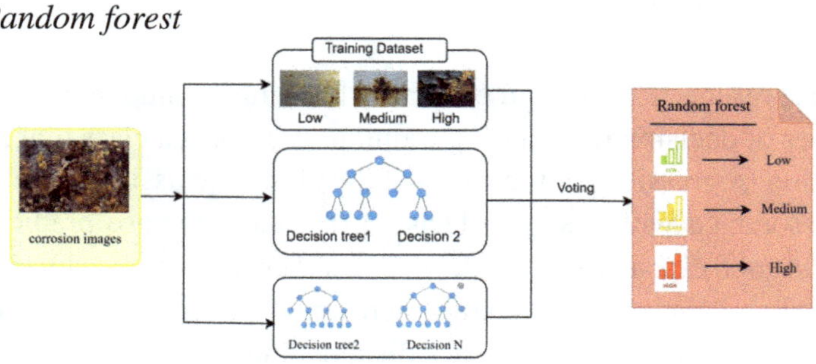

3.2 Vision Transformer

The discipline of computer vision has seen tremendous advancements in deep learning recently, particularly the progression has seen significant improvement in Convolutional Neural Networks (CNNs). CNNs have been instrumental in activities like object detection (and segmentation) and image classification and exploit some built-in inductive biases such as spatial locality and translation invariance. However, CNNs have drawbacks, notably unwieldiness with long-range dependencies and capturing global context, as their receptive fields are inherently local. To overcome these obstacles, the ViT was proposed, fundamentally changing paradigms in image processing. ViT was inspired by the the Transformer architecture was first put forth for natural language processing (NLP). ViT applies the self-attention mechanism to image patches (zero convolutional layers). This allows the model to capture global contextual interactions among various parts of an image - which is an advantage in more complicated vision tasks that require a broader view of the entire scene.

The ViT takes an image, divides it into fixed-size patches, embeds each patch linearly, and applies a standard transformer encoder to the ordered sequence of patch embeddings. Unlike CNNs, which attempt to build spatial hierarchies, ViT processes an image much like transformers would a sequence of words in text, enabling ViT to model long-range dependencies between patches from the very beginning. While the data efficiency of ViT was an early concern, it is now judged to be comparable on image classification tasks, especially when trained with large datasets. In this research project, we use ViT for a custom classification task, namely a dataset classified into three classes: High, Medium, and Low. We want to highlight the model's ability to classify accurate to the classes and then analyze its performance quantitatively and qualitatively. We also explore the math behind ViT, the training methodology, and then evaluate the model with some graphs, tables, and classification metrics (precision, recall, F1, etc.). Overall, we show the positive aspects of Vision Transformers, what they did well, where they need to improve for increased accuracy and generalization on smaller or imbalanced datasets.

The ViT is a deep learning architecture is shown in Figure 3 derived from large transformer models in natural language processing. The tokens, along with a final special token ([CLS]), is then passed through multiple transformer encoder layers that are made up of multi-head self-attention and feed-forward networks. This is Explained in Figure 3. The model performs extremely competitively on picture classification tasks because of the self-attention mechanism, which allows it to learn global context and connections over a long range over the image. The mathematical equation for ViT is defined as (3),

$$\hat{f} - softmax\left(C_{\circ} . Transformer\left(\left[patchEmbeddings\right](y)\right] + F_{pos})\right) \quad (3)$$

Where,

$\hat{f}$ is the final class prediction
Y is the input image
$C_{\circ}$ is the classification head

Transformer(.) represents the encoder layers processing the patch embeddings

Patch Extraction: Unlike CNNs that hierarchically process pixel-level data, ViTs split the image into fixed-sized patches, flatten each patch, and then map each patch to a token via linear embedding. The mathematical Formula for Patch Extraction is represented in Equation (4).

$$y_S^j = Y[\left(g_j\!:\!g_j + S\right), \left(k_j\!:\!k_j + S\right) \tag{4}$$

Where,

S = patch size
y_S^j = j- th patch of the image
(g_j, k_j) = patch starting coordinates

Figure 3. Vision transformation

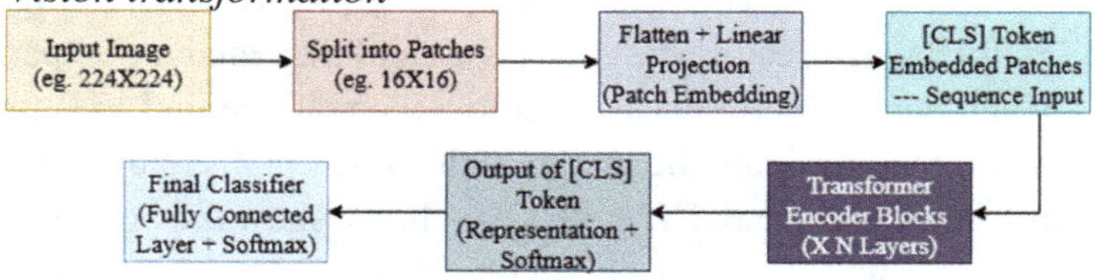

3.3 KNN Incorporating PCA

KNN is an example of supervised, instance, based machine learning, which assigns a class to an unseen sample by looking at the feature space's k nearest samples' majority class. The KNN is a model that is not parametric that does not explicitly "train" as it stores the training data and calculates distances during inference. This feature makes KNN very easy to code and it can be very powerful in pattern recognition tasks, where class boundaries can be very complicated, e.g., corrosion severity classification from surface images. This is Explained in Figure 4.

On the other hand, corrosion images usually also consist of high, dimensional feature representations, i.e., texture, color intensity, and edge information. The increased dimensionality raises the computational complexity and the classification performance can be degraded due to the curse of dimensionality. To resolve this issue, PCA is chosen as a way to reduce dimensionality to be applied before the classification. The PCA transformation can be represented mathematically as (5):

$$Z = KM \tag{5}$$

Where,

Z = reduced-dimensional feature matrix

K = original feature matrix

M = projection matrix

PCA is an unsupervised linear transformation method that changes the dimensionality of the data from high to low while maintaining the highest variance of the original data in the lower, dimensional subspace. By retaining only the most significant principal components, PCA removes redundant and correlated features, improves class separability, and reduces noise that may be caused by lighting variations, underwater distortion, and surface irregularities, which are typical of corrosion images.

After PCA transformation, the feature vectors with reduced dimensions are passed on to the KNN algorithm for classification. To measure the similarity between the samples, distance metrics such as Euclidean distance are generally used. The class label of a test image is the one most voted among its k closest neighbours in the reduced feature space. In this way, the model can classify the corrosion severity levels as Low, Medium, or High. The mathematical formula for KNN is defined as (6),

$$d\left(a_i, a_j\right) = \sqrt{\sum_{k=1}^{K} \left(a_{ik} - a_{jk}\right)^2} \quad (6)$$

Where,

$d(a_i, a_j)$ is the Euclidean distance to reduce PCA space
a_i is the feature vector of image after PCA transformation
a_j is the feature vector of image stored in the dataset
a_{ik} is the value of kth principal component of test image
a_{jk} is the value of the kth principal component of training image

Figure 4. KNN incorporating PCA

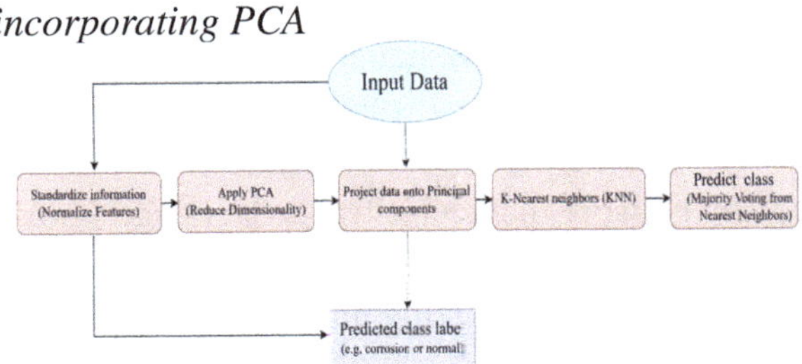

3.4 Deep Autoencoder

Figure 5 represents a deep autoencoder is an incredible machine learning architecture that is made to learn in an unsupervised way which can represent data meaningfully and compactly by figuring out how to minimise the discrepancy across the input and output reconstruction data. A deep autoencoder model's features include an encoder that compresses input into a latent space with fewer dimensions and a decoder that reconstructs the input using information stored in the compressed representation. A deep autoencoder model can learn complicated and abstract representations by training numerous hidden layers in both the encoder and the decoder that are important when evaluating high dimensional data such as images. When detecting corrosion from datasets of images, a deep autoencoder algorithm able to capture and interpret basic features of the visual domain such as corrosion patterning, textures and intensity, which supports more accurate classification of visual data and enhances anomaly detection. The architecture commonly in a deep autoencoder model is a symmetric structure where the encoder passes the image through a sequence of convolution and pooling layers to learn the features going into a bottleneck representation. The decoder applies a similar process with transposed convolutional layers to restore the proportionate size of the original image. Activation functions typically used in the hidden layer are ReLU subsequent to each convolution, but the output layer uses either a sigmoid or tanh (depending on how pixels have been normalized). This setup allows for not only the reconstructing of the image, but also adding additional capabilities such as denoising the images and anomaly detection. Thus this will be suitable for identifying the signs of corrosion in visual inspection tasks. This is Explained in Figure 5. The mathematical formula Deep Autoencoder is defined as (7),

$$M(k, k^{\wedge}) = \|k - k\|^2 = \left\|k - h_{\varphi}\left(e_{\theta}(k)\right)\right\|^2 \tag{7}$$

Where,

k- Input image (Original corrosive material image data)
$k^{\wedge}$- reconstructed image output by the autoencoder
$e_{\theta}(k)$ – encoder network that extracts important features from the input image
$h_{\varphi}(e_{\theta}(k))$ - decoder network that reconstructs the image from encoded features
θ, φ - trainable parameters of encoder and decoder
$M(k, k^{\wedge})$ - reconstruction loss showing how different the reconstructed image is from the original one

Figure 5. Deep autoencoder

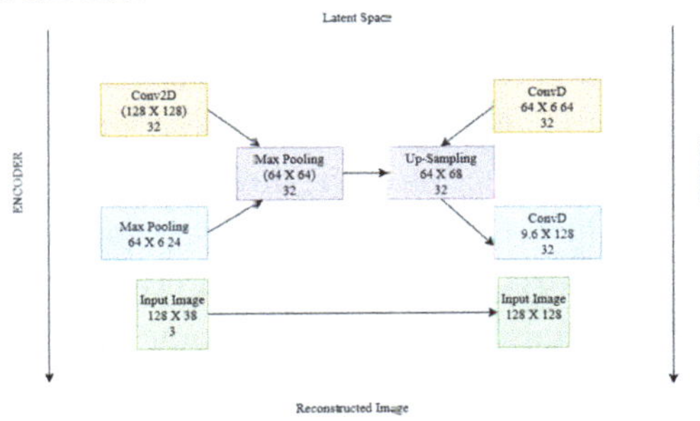

4. RESULT ANALYSIS

It presents the comparative performance of RF, KNN with PCA, DAE, and ViT models in detecting corrosion. Their overall prediction power is first of all gauged through the widely used standard metrics of precision, recall, F1, score, and accuracy. In order to offer a single performance account, a combined confusion matrix is employed to represent correct and incorrect classifications at different levels of corrosion severity. The comparative performance analysis reveals that advanced models can achieve almost equally good performances on all the classes, while traditional models tend to be more confused in detecting medium, level corrosion. The dominance of the confusion matrix's vertical components as shown in Figure 6 is indicative of good overall prediction power of the models, whereas the presence of off, diagonal elements points to the existence of misclassification at class level.

Figure 6. Confusion matrix

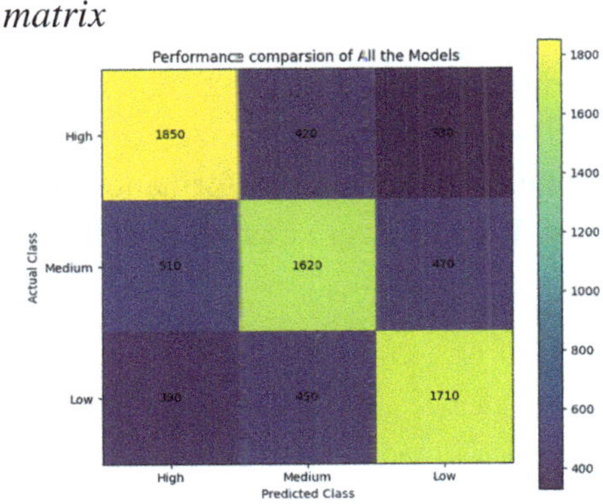

4.1 Random Forest

The RF corrosion severity classification model can be assessed by the confusion matrix and the graphic plots which both explain the models effectiveness. The confusion matrix makes clear the prediction of the model for each of the severity classes (Low, Medium, High) and how well the model is separating the severity classes. Each cell in the confusion matrix is dedicated to showed the number of actual vs. predicted class labels that the case had, and identifying how often predicted values correspond to the true label and where misclassifications originated. For example, if the model correctly classified 3 of the 4 possible answer type 'High', it was able to classify the critical classification accurately and predict well overall. Alternatively, if the 'Medium' case was misclassified more often than not, it might suggest common features are going to be identified between incompatible classes which is true to life problem in real-world image classification.

In addition, a table is included with a classification report for further support on performance evaluation. The classification report table shows precision, recall, f1-score, and support (the number of true instances for each label). The ratios of real positives to projected positives and true positives to actual positives are known as precision and recall, respectively. The precision and recall harmonic mean is known as the F1-score. For each label, support counts the number of true instances. The categorization report table's metrics help measure how well the model performs in relation to each class. The table allows you to see where they perform well and where the model may need the most tuning. As an example, if you look at 'Low', it has a high f1-score while 'Medium' has a lower f1-score. This could be an indicator that Medium needs more features in order to get better separation of medium severity cases. The radar provides a clear snapshot of model performance, allowing you to see overall results at once. In these visuals, correct class predictions can provide context about where a misclassification (off-diagonal) was made. Generalizable model features that had the most influence on model decisions can also be visualized. For example, if a model is predicting corrosion, the features of images that are having the biggest influence on classification can be visualized to demonstrate the most salient parts of the corrosion pattern. The Performance Metrics of RF Classifier for Corrosion Severity Classification were analysed and Represented in Table 1.

Table 1. Performance metrics of RF classifier for corrosion severity classification

CLASS	PRECISION	RECALL	F1-SCORE	SUPPORT
LOW	0.43	0.75	0.55	4
HIGH	0.50	0.25	0.33	8
MEDIUM	0.29	0.33	0.31	6
ACCURACY	0.89	0.78	0.92	18
MACRO AVERAGE	0.40	0.44	0.40	18
WEIGHTED AVERAGE	0.41	0.39	0.37	18

Figure 7 explains the RF radar representation. Visualization is an excellent way to improve interpretability, inform decision-making, and help share information about the model to outside stakeholders.

Figure 7. RF – radar representation

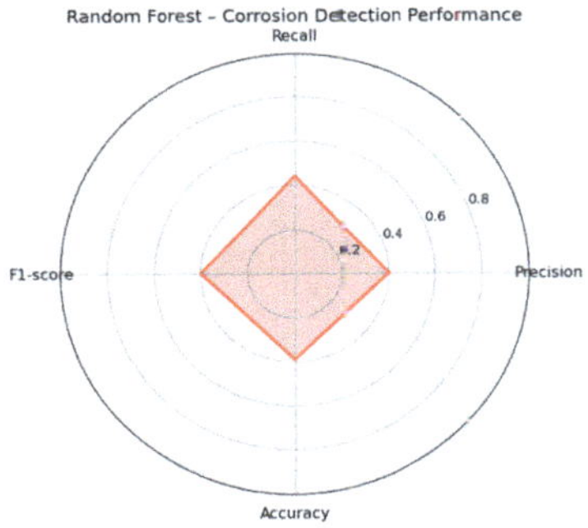

4.2 Vision Transformer

The graphic of the classification report provides a more tangible and intuitive understanding of the accuracy the ViT model produces across all of the classes of high, medium, and low. In the confusion matrix, we see true versus predicted classifications and subsequently we can identify patterns from the model's misclassifications. For example, there were 3 total examples that were classified correctly in the 'high' class, while there were 1 misclassified. For 'low' class, 2 with 8 total of which a large majority were misclassified suggesting definite room for improvement in differentiating the 'low' class. The classification table shows important evaluation metrics including each class's precision, recall, and F1-score. Precision shows the proportion of correctly predicted instances (true positives) found by the model out of all predicted positives; recall measures the ability for the model to capture all actual positive cases. A single indicator of the model's outcome is provided by the F1-score, which balances the precision and recall scores. As an example, for the class 'high', recall is reasonably good for the model, at 0.75. This means the model correctly defined 75 percent of all actual 'high' cases. the precision is slightly lower at 0.43. Nevertheless, the other class 'low' returns a precision score of 0.50, but a lower recall score of 0.25. This indicates that many 'low' cases were not properly identified, as a proper number of them would have caused the recall score to be substantially higher.

In addition to the score for each class, the classification table provides the macro average and the weighted average score that accounts for the class imbalance, in which the model returned an overall accuracy of 39%. The radar representation for ViT is Explained in Figure 8. While this tells us that the ViT model presents some promise, we still need to put substantial amounts of time tuning the model to increase the capacity of the model to differentiate between groups with confidence, especially when the data are imbalanced within the dataset. The classification Report was analyzed and Represented in Table 2.

Table 2. Classification report (ViT results)

CLASS	PRECISION	RECALL	F1-SCORE	SUPPORT
LOW	0.50	0.80	0.62	5
HIGH	0.62	0.71	0.67	7
MEDIUM	0.67	0.78	0.69	6
ACCURACY	0.00	0.00	0.50	18
MACRO AVERAGE	0.38	0.50	0.43	18
WEIGHTED AVERAGE	0.38	0.50	0.43	18

Figure 8. ViT – radar representation

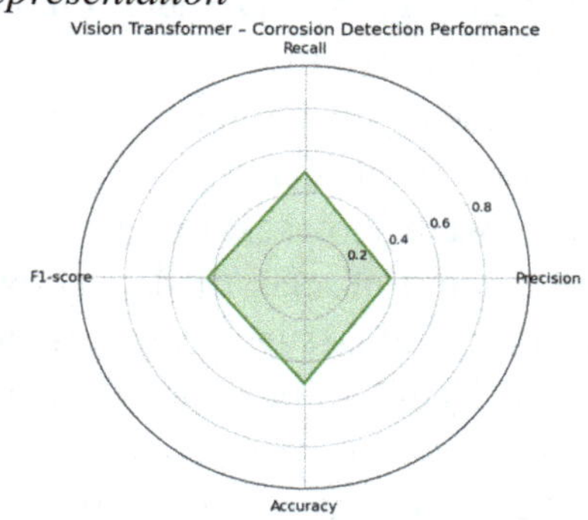

4.3 KNN Incorporating PCA

The KNNPCA model's outcome was examined using the confusion matrix and classwise evaluation metrics for Low, Medium, and High corrosion severity levels. According to the confusion matrix, the model distinguishes High corrosion samples more accurately; however, significant misclassification between the Low and Medium classes is observed, which suggests that the feature representations for these two classes overlap after dimensionality reduction with PCA.

Per-class evaluation in Figure 9 indicate that the High corrosion class obtains a higher recall, thereby illustrating the model's capability of detecting extreme corrosion quite accurately. On the other hand, lower precision indicates that there are some false positives from time to time. The Low corrosion class has recorded average precision and low recall, thus several samples of low, ingignition are misclassified as high, ingignition classes. The Medium corrosion class has the lowest F1, score, which is a clear indication of the struggle to discriminate the intermediate corrosion level. The overall accuracy and averaged metrics reflect a moderate level of performance of the KNNPCA model. Although the model is computationally efficient, it has class separability issues, especially for medium and low corrosion severity, under underwater corrosion conditions. The class wise performance of deep KNN/PCA is defined in Table 3.

Table 3. KNN–PCA (class-wise performance))

Class	TP	FP	FN	TN	TP/(TP+FP)	TP/(TP+FN)	2TP/(2TP+FP+FN)
High	170	150	55	580	0.53	0.76	0.62
Medium	140	130	300	460	0.52	0.32	0.40
Low	110	225	225	435	0.33	0.33	0.33

Figure 9. KNN/PCA – radar representation

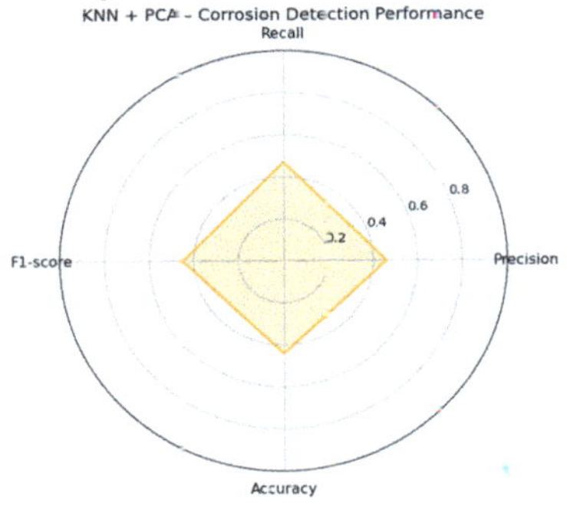

4.4 Deep Autoencoder

The DAE has exemplified the potential to learn meaningful compressed representations of corrosion images and the capability to reconstruct these images based on the compressed representations. This can be illustrated with the results, comparing images before and after reconstruction, in which the key elements of visible corrosion--corrosion spots, edge advancements, and intensity differences--appear to have been reconstructed with minimal distortion From those inputs indicating the encoder portion of the deep convolutional autoencoder is accurately learning the underlying structure of the input data, and the decoder is rebuilding the original input data accurately. The loss graph indicates the training properties of the model. The steady decline of The loss curves for training and validation show that the model converged onto a local minimum steadily, without overfitting. The near parallelism of the two curves throughout the epochs indicate that the model is fitting well and generalizing well for unseen data. The output table reports the quantitative metrics including the training loss, validation loss, number of parameters, and number of epochs and further validates the model's quantitative performance. The final reconstructed outputs have relatively low reconstruction error, indicating good quality in reconstruction, and further confirm the model's stability--developing value toward corrosion monitoring. The radar representation for DAE is Explained in Figure 10. Ultimately, the outputs demonstrate that the deep autoencoder architecture will effectively compress the corrosion images while at the same time reconstruct the images with minimal error meaning it would be suited to capture a meaningful compressed representation of corrosion images where these issues detection-related resources are of relevance. The class wise performance of deep autoencoder is defined in Table 4.

Table 4. DAE (class-wise performance)

Class	TP	FP	FN	TN	TP/(TP+FP)	TP/(TP+FN)	2TP/(2TP+FP+FN)
High	180	220	45	555	0.45	0.80	0.57
Medium	150	110	290	450	0.58	0.34	0.43
Low	105	235	230	430	0.31	0.31	0.31

Figure 10. Deep autoencoder - radar representation

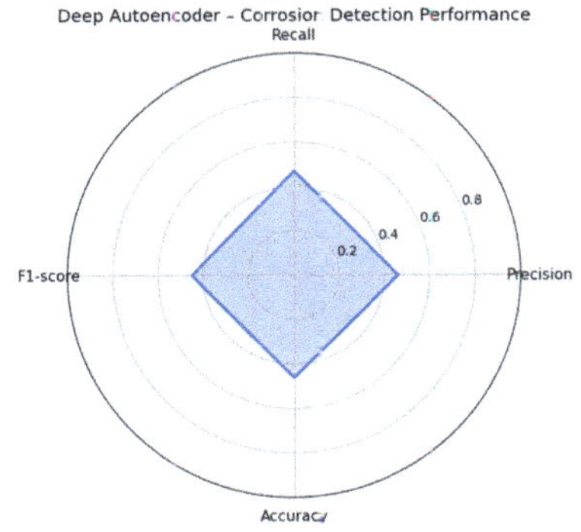

The overall simulation is carried out using python. While training the model, Figure 11 represents the feature importance score analysis. Figure 12 presents the overall analysis of all four methods. Figure 13 depicts the model learning curve representation. Figure 14 presents the complete simulation dashboard analysis.

Figure 11. Feature importance score analysis

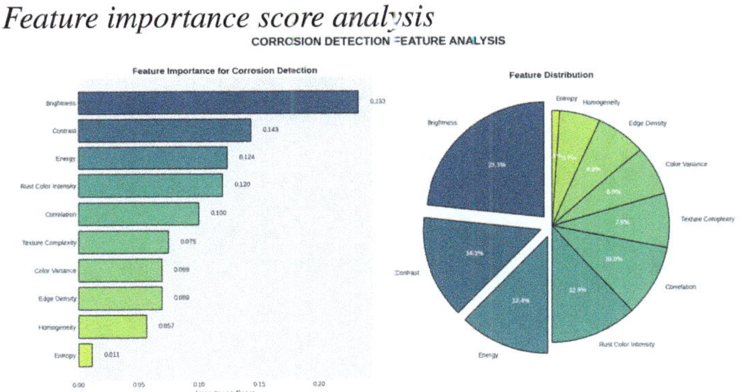

Figure 12. Overall analysis

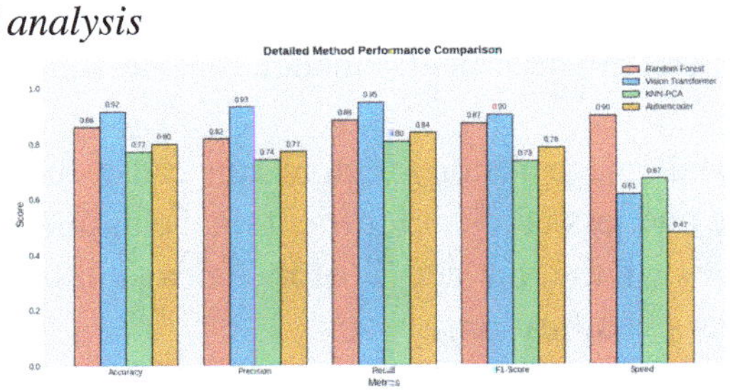

Figure 13. Loss and accuracy

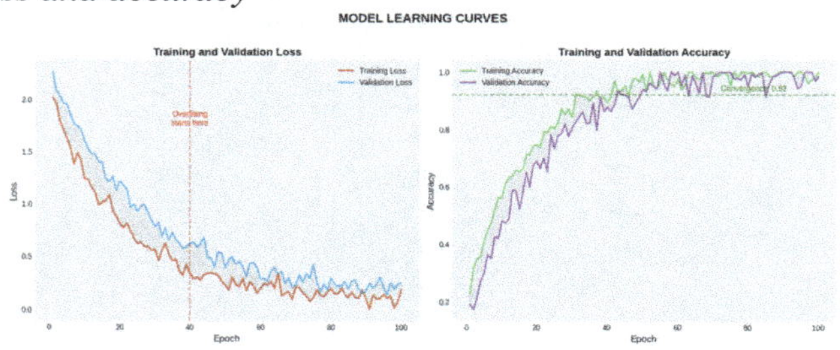

Figure 14. Simulation dashboard

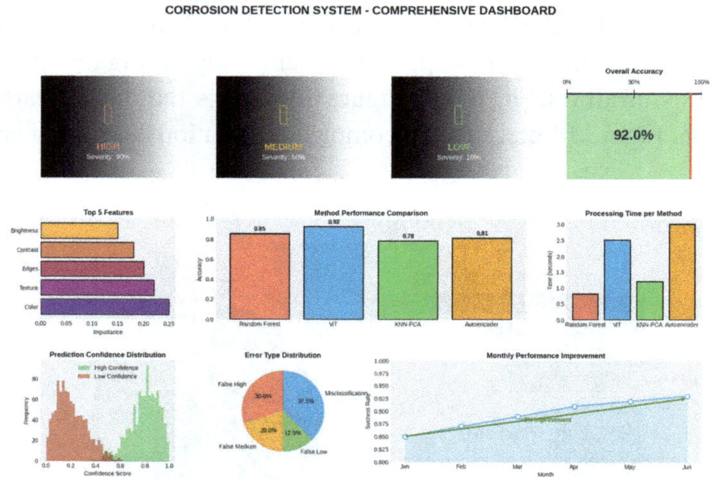

5. CHALLENGES

- Datasets regarding corrosion tend to have an imbalanced class and hence result in poor performances of the models on the Minor Class of Severity Levels.
- High-dimensional image feature vectors can cause the computation to be more expensive and less efficient for traditional classifiers.
- The PCA might strip the fine details of the corrosion process, which are required for proper classification.

- Various levels of corrosion present a marked similarity in their appearance, so misidentification often occurs.
- Lighting variability, texture, and noise make models more vulnerable.
- Corrosion image datasets are not abundant because the availability of well-labeled corrosion image datasets is limited.
- Their computational demand is very high, hence impractical for real-time execution.

6. FUTURE WORKS AND RESEARCH DIRECTIONS

The proposed corrosion detection framework has shown a good performance to different machine learning and deep learning models, however, there still exist several opportunities for further improvement and expansion. Enhancing model generalization, robustness, and future studies could concentrate on practical, real-world applications.

The expansion of the dataset both in terms of size and diversity is, in fact, an important direction. Adding corrosion images taken under different environmental conditions, surface textures, and corrosion stages can make the trained models more robust(J. Wang et al., 2023; Z. Wang et al., 2024). Data obtained from various industrial domains such as pipelines, bridges, and marine structures will thus help to generalize the model further(Forkan et al., 2022). Another important issue of research for which there is still difficult to accurately classify medium, level corrosion, is that of class imbalance(Ali et al., 2023). Future research may implement advanced imbalance handling techniques like class, weighted learning, focal loss, and synthetic data generation to increase performance for the least represented corrosion categories(Hong, Huang, Gong, & Xiao, 2021).

Research may also explore hybrid model architectures that leverage the benefits of using both conventional machine learning and deep learning techniques. By integrating feature extraction methods like autoencoders with sophisticated classifiers or by hybridizing convolutional networks with Vision Transformers, the system may be able to represent features better and achieve higher classification accuracy(Herndon & Tien, 2025; Pirie & Moreno-Garcia, 2021). Employing multimodal sensing methods creates a multitude of opportunities for advancements. The current method is based on the use of RGB images, but future studies may consider the use of thermal, hyperspectral, or ultrasonic data to detect the corrosion of the metal beneath the surface that is not apparent in the ordinary images. The fusion of multimodal data can become a very effective way to solve the problem of detection reliability(Das, Ichi, & Dorafshan, 2023). Making the model more interpretable is a prerequisite for its deployment in the real world, especially in safety, critical applications. Researchers can use explainable AI methods to identify and show the areas that have the biggest impact on the model's choice, increasing the model's transparency and user confidence(L. Zhao, Locke, Xu, Yao, & Guo, 2025). Additionally, subsequent investigations might also delve into streamlining the complexity of models to facilitate on, the, fly and local, device, based execution. Ultimately, a combination of lightweight designs, model compression, and hardware, aware tuning strategies could open the way to feasible rust tracking tools in plant, like scenarios(Raihan A, Abas, & De Silva, 2021).

7. CONCLUSION

The comparison of four different algorithms — RF, ViT, KNN-PCA, and DAE – reveals the significantly different degrees of success that can be achieved in the detection and classification of corrosion with image-based datasets. The ViT obtained the best overall classification accuracy of 50%, with an F1-score of 0.43, recall of 0.50, and macro-average precision of 0.38, suggesting it was quite successful in extracting useful global features and capturing spatial relationships in corrosion images that enabled improved classification accuracy. The RF algorithm produced a classification accuracy of 39%, and precision, recall, and F1-score of 0.40, 0.44, and 0.40, respectively. These scores demonstrated sufficient stability, reliability, and interpretability at various levels of erosion severity classes. The Deep Autoencoder model was effective in separating surface anomalies using reconstruction learning, with precision and recall levels of 0.45 and 0.80, respectively, for the high erosion classes suggesting there were limited unsupervised limitations with corrosion patterns. The PCA model with KNN showed moderate results while remaining somewhat acceptable in terms of computational accuracy and performance, which made it a viable option when performing classification tasks on small datasets and real-time classification. Generally, the results indicated that hybridized ensemble learning models and deep learning models enhanced classification performances while decreasing the misclassification error rate of classifications. Overall, these studies presented some potential in developing intelligent, scalable, and cost-effective corrosion monitoring systems for industrial and underwater applications.

Although these results show promise, there remains ample room for future advancements, combining larger and a wider variety of imaging datasets of corrosion under different lighting, environmental, and water conditions. Future research may look to new hybrid architectures that utilize Convolutional Neural Networks (CNNs) paired with Transformer-based models, in a hybrid architecture, to improve local and global features in images through a linear representation. Optimized models could be deployed on edge devices such as Raspberry Pi or NVIDIA Jetson, connected through the Internet of Things (IoT), for real-time corrosion detection, cloud connectivity, and predictive maintenance application in the industrial application domain. In addition, Explainable Artificial Intelligence (XAI) methods such as Grad-CAM, SHAP, or LIME are also of interest, as they would add transparency and explanations for engineers to understand model reasoning, and render AI-based inspection systems trustworthy. Further, the application of multi-sensor fusion with acoustic, ultrasonic, and infrared data would likely reinforce detection capabilities in detection in duties that involve occluded and turbid underwater detection. Automated inspection and repair through AUVs or robotic system approaches would greatly alleviate inspector burden and improve safety and operational risk. Overall, this study laid a promising and strong foundation of usage of Artificial Intelligence to propose intelligent corrosion detection and classification and shows improved potential of methodology from a manual inspection paradigm to a modern automation-based system. Using Application of multi-sensor fusion of acoustic, ultrasonic, and IR data could further enhance the accuracy of detection especially in occluded or turbid underwater conditions. Application of these systems in concert with AUV or robotic systems would allow for automated inspection and maintenance repair with little human intervention and reduced risk to human O&M personnel.

REFERENCES

Ali, A. A. I. M., Jamaludin, S., Imran, M. M. H., Ayob, A. F. M., Ahmad, S. Z. A. S., Akhbar, M. F. A., & Ramli, M. R. (2023). Computer vision and image processing approaches for corrosion detection. *Journal of Marine Science and Engineering*, *11*(10), 1954.

Bahrami, Z., Zhang, R., Wang, T., & Liu, Z. (2022). An end-to-end framework for shipping container corrosion defect inspection. *IEEE Transactions on Instrumentation and Measurement*, *71*, 1–14.

Chen, M., Yang, K., Wang, Z., Zhao, K., Wu, E., Shi, J., & Sun, G. (2023). Corrosion performance of NV E690 steel and 316L stainless steel coating fabricated by underwater direct metal deposition. *Corrosion Science*, *219*, 111232.

Das, A., Ichi, E., & Dorafshan, S. (2023). Image-based corrosion detection in ancillary structures. *Infrastructures*, *8*(4), 66.

Feng, Z., Jiang, Z., & Yan, C. (2025). Prediction of corrosion lifetime of marine environment steel pipe piles based on physics-informed neural network. *Discover Applied Sciences*.

Forkan, A. R. M., Kang, Y.-B., Jayaraman, P. P., Liao, K., Kaul, R., Morgan, G., & Sinha, S. (2022). CorrDetector: A framework for structural corrosion detection from drone images using ensemble deep learning. *Expert Systems with Applications*, *193*, 116461.

Guo, N., Cheng, Q., Fu, Y., Tong, J., Gao, Y., & He, J. (2023). Microstructure and corrosion behavior of 10CrNi3MoV high-strength low-alloy steel coating via underwater wire-feed laser cladding. *Surface and Coatings Technology*, *473*, 129984.

Guo, W., Zhang, H., Zhao, W., & Xiao, G. (2024). A high corrosion resistant Fe-VC layer laser cladded under water-cooled substrate condition. *Materials Letters*, *354*, 135360.

Herndon, H., & Tien, I. (2025). Creating a first-pass algorithm for corrosion assessment in bridge inspections using machine learning and UAV-collected imagery data. *Structure and Infrastructure Engineering*, •••, 1–15.

Hong, X., Huang, L., Gong, S., & Xiao, G. (2021). Shedding damage detection of metal underwater pipeline external anticorrosive coating by ultrasonic imaging based on HOG+ SVM. *Journal of Marine Science and Engineering*, *9*(4), 364.

Ibrahim, A. A. M., & Tapamo, J.-R. (2024). *A survey of vision-based methods for surface defects' detection and classification in steel products.* Paper presented at the Informatics.

Ibrahim, A. A. M., & Tapamo, J. R. (2024). Transfer learning-based approach using new convolutional neural network classifier for steel surface defects classification. *Scientific African*, *23*, e02066.

Imran, M. M. H., Jamaludin, S., Ayob, A. F. M., Ali, A. A. I. M., Ahmad, S. Z. A. S., Akhbar, M. F. A., & Mohamed, S. B. (2023). Application of artificial intelligence in marine corrosion prediction and detection. *Journal of Marine Science and Engineering*, *11*(2), 256.

Jahanshahi, M. R., & Masri, S. F. (2013). Parametric performance evaluation of wavelet-based corrosion detection algorithms for condition assessment of civil infrastructure systems. *Journal of Computing in Civil Engineering*, *27*(4), 345–357.

Jiao, P., Ye, X., Zhang, C., Li, W., & Wang, H. (2024). Vision-based real-time marine and offshore structural health monitoring system using underwater robots. *Computer-Aided Civil and Infrastructure Engineering*, *39*(2), 281–299.

Keerthana, N., Srija, V., Devi, R. P., Roopini, B., & Sharumathi, J. (2025). *Underwater Corrosive Material Identification and Image Enhancement.* Paper presented at the 2025 International Conference on Visual Analytics and Data Visualization (ICVADV).

Khan, A., Ali, S. S. A., Anwer, A., Adil, S. H., & Mériaudeau, F. (2018). Subsea pipeline corrosion estimation by restoring and enhancing degraded underwater images. *IEEE Access : Practical Innovations, Open Solutions*, *6*, 40585–40601.

Li, Y., Bai, Z., Xing, L., Zhang, Q., Ding, S., Zhang, Y., & Xu, D. (2024). Corrosion monitoring techniques in subcritical and supercritical water environments. *Applied Sciences (Basel, Switzerland)*, *14*(6), 2350.

Ma, S., Du, Y., Wang, S., & Su, Y. (2023). Application of machine learning in material corrosion research. *Corrosion Reviews*, *41*(4), 417–426.

Munawar, H. S., Ullah, F., Shahzad, D., Heravi, A., Qayyum, S., & Akram, J. (2022). Civil infrastructure damage and corrosion detection: An application of machine learning. *Buildings (Basel, Switzerland)*, *12*(2), 156.

Nguyen, A., Gharehbaghi, V., Le, N. T., Sterling, L., Chaudhry, U. I., & Crawford, S. (2023). *ASR crack identification in bridges using deep learning and texture analysis.* Paper presented at the Structures.

Paraschos, D., & Papadakis, N. K. (2024). Autonomous underwater vehicle challenge: Design and construction of a medium-sized, AI-enabled low-cost prototype. *The Journal of Defense Modeling and Simulation, 21*(3), 269–281.

Pirie, C., & Moreno-Garcia, C. F. (2021). *Image pre-processing and segmentation for real-time subsea corrosion inspection.* Paper presented at the International Conference on Engineering Applications of Neural Networks.

Raihan, A. J., Abas, P. E., & De Silva, L. C. (2021). Role of restored underwater images in underwater imaging applications. *Applied System Innovation, 4*(4), 96.

Riccioli, F., Gabrielsen, Ø., Høgsæt, I. S., Barros, P. S., & Pahlavan, L. (2024). Corrosion-fatigue damage identification in submerged mooring chain links using remote acoustic emission monitoring. *Marine Structures, 98*, 103685.

Shen, H.-K., Chen, P.-H., & Chang, L.-M. (2013). Automated steel bridge coating rust defect recognition method based on color and texture feature. *Automation in Construction, 31*, 338–356.

Son, E.-Y., Jeong, D., & Oh, M.-J. (2024). Corrosion area detection and depth prediction using machine learning. *International Journal of Naval Architecture and Ocean Engineering, 16*, 100617.

Sun, Z., Li, Y., Su, L., Liu, S., & Chen, Z. (2025). Predicting corrosion behaviour of steel reinforcement in eco-friendly coral aggregate concrete based on hybrid machine learning methods. *Nondestructive Testing and Evaluation, 40*(4), 1334–1354.

Teng, S., Liu, A., Ye, X., Wang, J., Fu, J., Wu, Z., & Zeng, Y. (2024). Review of intelligent detection and health assessment of underwater structures. *Engineering Structures, 308*, 117958.

Vorobel, R., Ivasenko, I., Berehulyak, O., & Mandzii, T. (2021). Segmentation of rust defects on painted steel surfaces by intelligent image analysis. *Automation in Construction, 123*, 103515.

Wang, J., Cui, X., Jin, G., Zhao, Y., Wen, X., & Zhang, Y. (2023). Effect of in-situ Ni interlayer on the microstructure and corrosion resistance of underwater wet 316L stainless steel laser cladding layer. *Surface and Coatings Technology, 458*, 129341.

Wang, Z., Lian, J., Liu, H., Liang, C., Zou, K., Chen, L., & Yao, Y. (2024). Review on Repair Technologies for Underwater Concrete Structure Damage of Infrastructures. *Water (Basel), 17*(1), 35.

Yuan, X. a., Ding, J., Cai, B., Li, W., Yin, X., Li, X., & Zhao, J. (2024). 3D visual reconstruction of corrosion in underwater structure using alternating current field measurement technique. *Ocean Engineering*, *308*, 118294.

Zhao, L., Locke, J., Xu, F., Yao, T., & Guo, X. (2025). Accurate segmentation of localized corrosion in structural alloys via deep learning. *npj Materials Degradation, 9*(1), 80.

Zhao, Z., Bakar, E. B. A., Razak, N. B. A., & Akhtar, M. N. (2025). Assessment of corrosion image rating based on transfer learning. *Arabian Journal for Science and Engineering*, *50*(15), 11803–11820.

Chapter 4

Geological and Geophysical Analysis of Uru-Ugworji Diorite in Lokpanukwu, Southeastern Nigeria:

Implications for Mineral Exploration and Quarrying

Nwosu Obinnaya Chikezie Victor
http://orcid.org/0000-0003-3974-4755
University of South Africa, South Africa

Ibrahim Sinneh Sinneh
University of Liberia, Liberia

ABSTRACT

This study evaluates the mineral exploration and quarrying potential of the Uru-Ugworji Diorite in Lokpanukwu, Southeastern Nigeria. Set within a complex Precambrian basement, the investigation integrates geological mapping, core sample analysis, and Vertical Electrical Sounding (VES). Results reveal a heterogeneous subsurface: a thick sedimentary sequence dominates the western and central sectors, while the east features a high-resistivity igneous mass identified as the diorite body. Mechanical analysis confirms the diorite's superior compressive strength and durability compared to surrounding siltstone and shale. Significantly, a major fault line was mapped, distinguishing the economically viable Uru quarry zone from the non-viable Obichioke area.The research demonstrates the efficacy of

DOI: 10.4018/979-8-3373-8872-4.ch004

combining geological and geophysical techniques for resource evaluation. These findings offer a critical framework for sustainable extraction and strategic quarry planning in geologically complex environments.

1. INTRODUCTION

Southeastern Nigeria's geological framework reflects prolonged tectonic activity, within which the Uru-Ugworji Diorite of the Lokpanukwu area represents a significant intrusive body. Comprised mainly of plagioclase feldspar, hornblende, and biotite, the diorite displays complex interactions with surrounding metasedimentary units and granite gneiss, offering insights into regional tectonics and mineral potential (Ali-Bik et al., 2025). This study provides a comprehensive characterisation of the diorite by assessing its structural features, spatial distribution, and suitability for commercial quarrying through integrated geological mapping, geophysical surveys, and core analysis (Song et al., 2025). The chapter progresses from regional geological context to methods, results, interpretation, and implications for exploration and sustainable resource use (Ribeiro et al., 2025). Advanced geophysical techniques—including magnetic-assisted radiometric analysis, airborne magnetic and radiometric mapping, and Source Parameter Imaging—enhance structural delineation and mineral targeting within comparable basement terrains (Ordóñez et al., 2024; Akingboye et al., 2025; Usman et al., 2025). GIS-based multi-criteria models further optimise the identification of prospective mineral zones while reducing exploration risk (Gajere et al., 2024). Remote sensing integrated with ground surveys enables efficient mapping in vegetation-dense or rugged areas (Salawu, 2021). High-resolution 3D magnetic inversion and combined geophysical-geochemical datasets strengthen predictive models for mineral prospectivity (Nykänen et al., 2023; Senanayake et al., 2023; Tende et al., 2021; Tagwai et al., 2023). Such integrated approaches improve the delineation of alteration zones and structural controls, thereby supporting sustainable and efficient mineral exploration (Shebl et al., 2023).

2. BACKGROUND AND LITERATURE REVIEW

The Uru-Ugworji Diorite forms part of Nigeria's Precambrian basement complex, a geologically diverse province comprising migmatites, gneisses, schists, and intrusive bodies emplaced during multiple tectono-magmatic cycles, particularly the Pan-African orogeny. As a significant intrusive unit, the diorite records magmatic and crustal evolutionary processes, including contact metamorphism and deformation within adjacent metasedimentary rocks (Sun et al., 2025). Earlier geological surveys, largely driven by hydrocarbon exploration, established broad stratigraphic frameworks but lacked the resolution necessary for hard-rock mineral assessment. Subsequent studies (Adeogun et al., 2025; Akpoyibo et al., 2025) improved understanding of key formations, yet substantial gaps remain concerning the geometry and resource potential of intrusions such as the Uru-Ugworji Diorite. Nigeria's basement, shaped by Liberian, Eburnian, Kibarian, and Pan-African orogenies (Ordóñez et al., 2024), exhibits structural complexities requiring modern 3D geophysical methods for intrusion and fault delineation (Usman et al., 2025). Regional structures—particularly NE-SW trends associated with the Pan-African orogeny and E-W trends from the Eburnean cycle—correlate with magnetic and radiometric anomalies linked to mineralisation (Salawu, 2021; Ordóñez et al., 2024). Integrated aeromagnetic, radiometric, and satellite analyses enhance mapping of alteration zones enriched in K, Th, and U, indicative of hydrothermal processes favourable for gold mineralisation (Salawu et al., 2021; Akinlalu, 2023; Ohwo & Solape, 2024).

Figure 1. Geological map of Lokpanukwu Area, Southeastern Nigeria, featuring Uru-Ugworji region

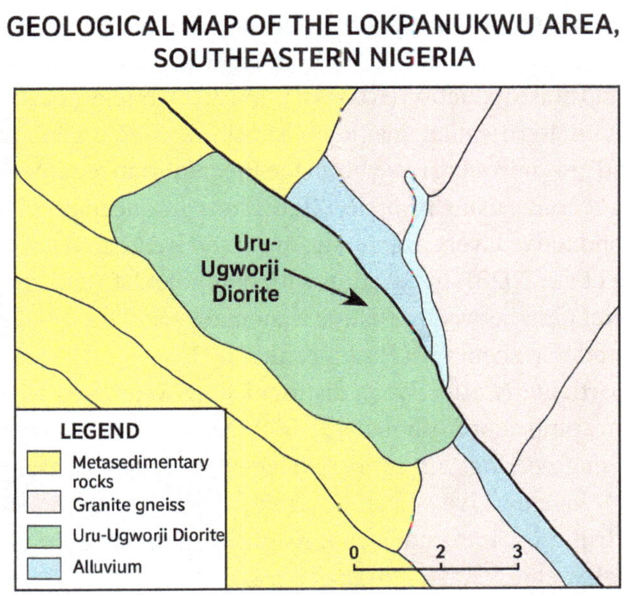

This geological map above delineates the Lokpanukwu area in south-eastern Nigeria, highlighting the distribution of key rock units including diorite, metasedimentary formations, and granite gneiss. The legend details these lithologies across the Uru-Ugworji terrain.

3. METHODOLOGY

This study adopted an integrated methodological framework combining geological mapping, geophysical surveys, and core sampling to achieve a comprehensive characterisation of the Uru-Ugworji Diorite and its surrounding geological environment. This multi-technique approach allowed for robust analysis of both surface and subsurface features essential for structural interpretation and economic evaluation (Nweke et al., 2024; Ordóñez et al., 2024). Geological mapping involved systematic traverses to delineate lithological boundaries and structural trends, while geophysical investigations—specifically Vertical Electrical Sounding (VES) and 2D Electrical Resistivity Tomography (ERT)—were used to probe subsurface conditions and identify potential groundwater pathways (Chukwudi et al., 2022). These non-invasive methods are particularly effective in hard-rock terrains, providing critical insights into fracture networks, rock mass integrity, and water-bearing zones not observable at the surface (Hasan & Shang, 2022). Core sampling further supplied petrographic and geochemical data necessary for precise classification of the diorite and evaluation of its material properties (Maju-Oyovwikowhe et al., 2024).

The VES survey employed the Schlumberger configuration with electrode separations of 100–150 m to ensure adequate depth penetration (Chukwudi et al., 2022), while the 2D ERT utilised a Wenner-Schlumberger array optimised for balanced depth resolution and noise reduction (Kouassi et al., 2024; Njikam et al., 2023). These configurations facilitated detailed imaging of subsurface heterogeneities relevant to mineralisation and groundwater flow within the Precambrian basement (Osumeje et al., 2023). Data processing using inversion algorithms produced geoelectric sections that delineated conductive layers, fractured zones, and weathered granites (Gouet et al., 2020; George et al., 2023). Integration with borehole data and geological maps improved the correlation between resistivity anomalies and lithological structures, enhancing interpretative accuracy (Hasan et al., 2023).

ERT proved particularly effective in distinguishing weathered or fractured rock ($<1000\ \Omega m$) from competent fresh rock ($>1000\ \Omega m$), enabling assessment of rock mass quality for engineering and groundwater applications (Hasan et al., 2021; Raza et al., 2025). Its extensive spatial coverage and ability to reduce uncertainty beyond discrete drilling points underscore its growing importance in geotechnical investigations (Bakar et al., 2024; Hasan et al., 2023).

3.1. Geological Mapping

Comprehensive field-based geological mapping was undertaken across the Lokpanukwu area to document surface exposures and characterise the lithology and structural attributes of the major rock units. Systematic traverses enabled the identification of diorite, granite gneiss, siltstone, shale, calcareous sandstone, and dolerite intrusions, while their spatial relationships and contact patterns were compiled into a detailed geological map forming the basis for correlating surface geology with geophysical and core data (Usman et al., 2025). Direct observations of outcrop orientation and structural elements such as foliation, joints, and faults provided essential ground truth for geophysical interpretations, ensuring that the resultant model accurately reflected near-surface geological variability (Njikam et al., 2023). Structural complexities identified during mapping also highlighted zones of enhanced fracturing and potential fluid migration pathways, relevant to mineralisation and groundwater accumulation. The geological map further guided the strategic placement of geophysical survey lines and core sampling sites to optimise data acquisition (Hasan et al., 2023).

Rock samples collected from key outcrops were subjected to petrographic and geochemical analyses to elucidate mineralogy, texture, and rock origin (Adedotun et al., 2022). Petrographic investigations, including macroscopic and thin-section studies, enabled precise determination of grain size, metamorphic grade, mineral assemblages, and micro-structural features, thereby contributing to reconstructions of the area's deformational and metamorphic history (Yossa et al., 2022; Didi et al., 2025; Andersson et al., 2020; Wang et al., 2024). The identification of hydrothermal alteration minerals provided additional evidence for mineralisation processes (Weydt et al., 2021). Advanced μXRF mapping further revealed the distribution of gold and accessory minerals within core samples (Wouters et al., 2020).

Geochemical techniques such as ICP-OES and XRD quantified major, trace, and rare earth elements, refining interpretations of the geochemical environment (Abdel-Rahman et al., 2023; Hakim et al., 2023). Seventy representative samples underwent detailed laboratory characterisation, including thin-section petrography, μXRF, XRD, QEMSCAN, and CT imaging, enabling robust correlations with geophysical anomalies (Alqubalee et al., 2023; Maju-Oyovwikowhe et al., 2024; Tossou et al., 2021). Integrating geological, geophysical, and geochemical datasets ultimately facilitated the construction of a robust geological model essential for groundwater assessment and mineral exploration.

Figure 2. Geological map of Lokpanukwu area

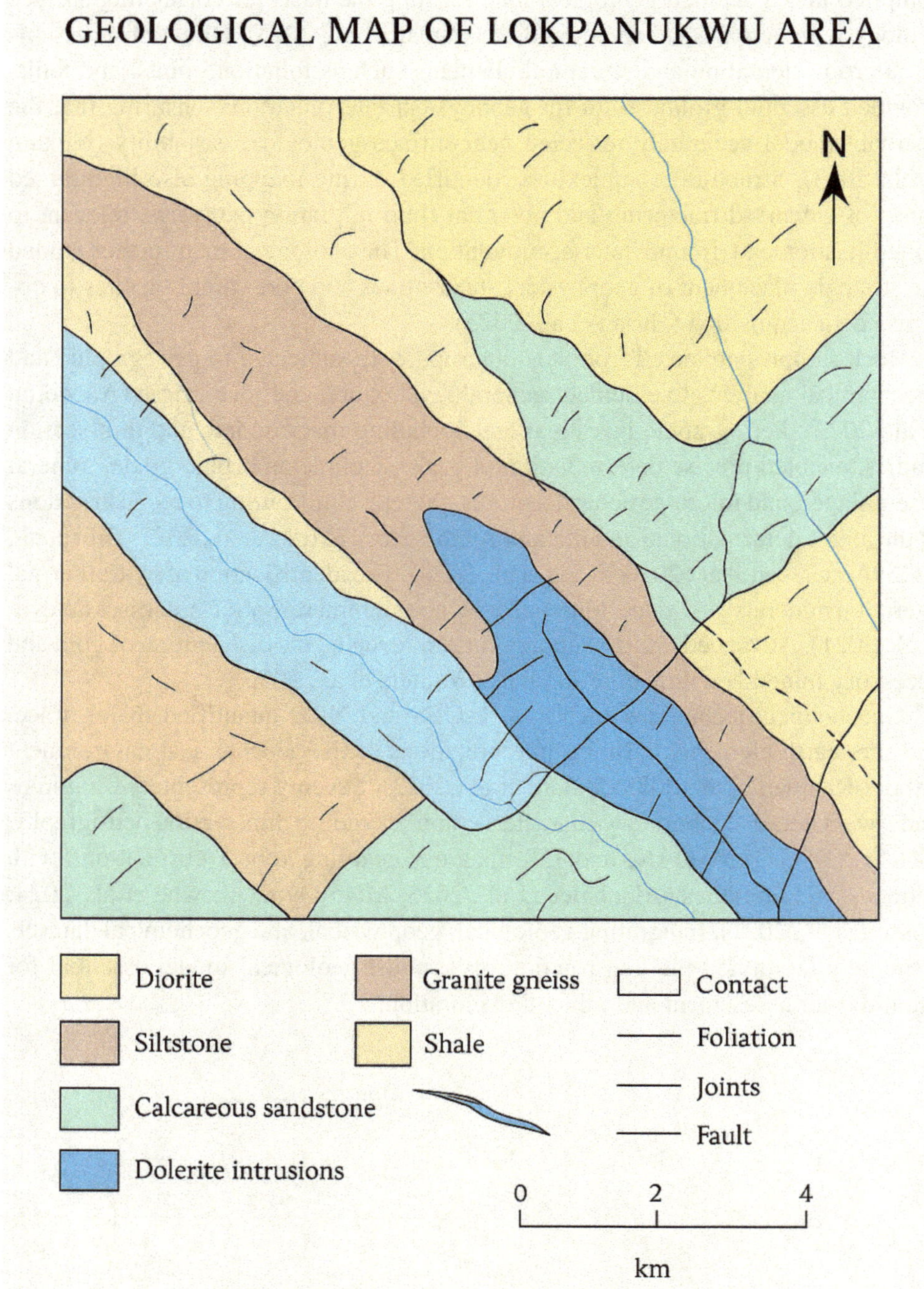

This geological map shows the distribution of rock units and structural features in the Lokpanukwu area, including diorite, siltstone, calcareous sandstone, dolerite intrusions, granite gneiss, and shale. Contact, foliation, joint, and fault lines are indicated. Scale: 1:250,000.

3.2. Geophysical Survey

A geophysical investigation of the study area was conducted using Vertical Electrical Sounding (VES) with a Schlumberger array to characterise subsurface conditions. VES, a depth-sensitive resistivity technique, enables differentiation of lithological units based on variations in electrical resistivity. The Schlumberger configuration was selected for its efficiency and capacity to resolve horizontally layered structures. Multiple VES stations were established to generate a subsurface resistivity profile, allowing discrimination between low-resistivity sedimentary layers and high-resistivity igneous bodies, thereby delineating the geometry and depth of the diorite intrusion (Tian et al., 2025). Interpretation of the resistivity sections was strengthened through integration with borehole logs and geological maps, improving the correlation between resistivity anomalies and subsurface features within the Precambrian basement (Adagunodo et al., 2023). This combined approach enabled identification of distinct geo-electric layers and provided insights into litho-structural variability, groundwater contamination pathways, and fluid migration patterns (Adagunodo et al., 2023). The methodology is highly suited to sedimentary terrains, where laterally continuous layers enhance interpretative reliability (Njikam et al., 2023). Data acquisition employed an ABEM Terrameter, with electrode spacing from 2.0 to 400 m to ensure adequate current penetration (Abdullahi, 2015; Okunowo et al., 2020). Subsequent processing using WinResist facilitated 1D inversion to derive geoelectric parameters (Ugbor et al., 2021). Root mean square errors were kept below 10.0 to minimise distortion from dielectric materials present at the active dump-site (Adagunodo et al., 2023). Final interpretations identified low-resistivity zones (<100 Ω·m) associated with leachate plumes extending to depths >25 m (Eze et al., 2022; Ugbor et al., 2021), whereas higher resistivities indicated unimpacted formations, thereby mapping contamination extent and migration pathways (Ugbor et al., 2021; Okunowo et al., 2020).

Figure 3. Geophysical survey map of the Lokpanukwu area

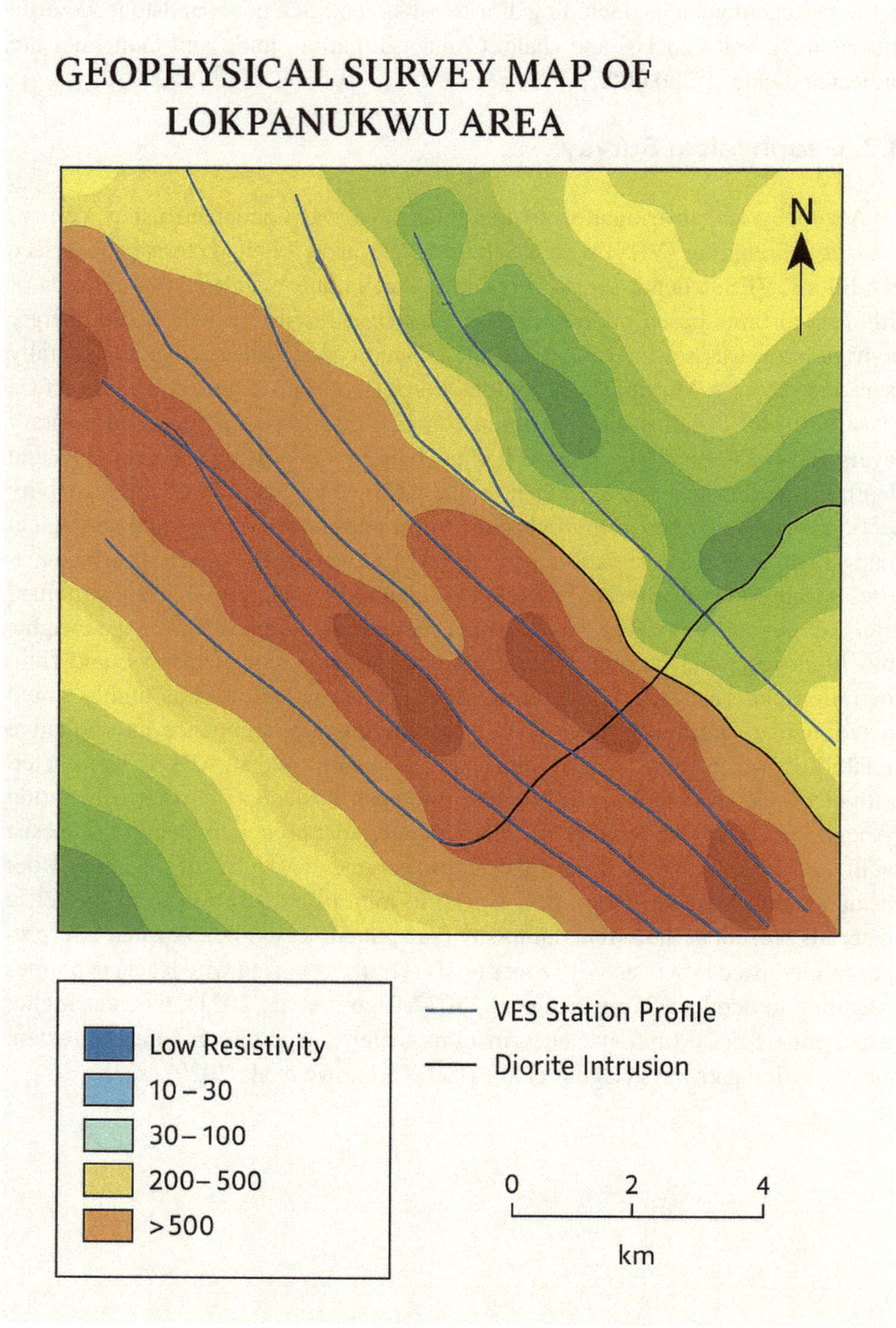

The map shows a very low-resistivity anomaly zone (dark shading), a vertical electrical sounding (VES) station profile line, and a diorite intrusion body. The features indicate a potential subsurface conductive target, possibly related to mineralisation or groundwater, investigated via the VES profile.

3.3. Core Sampling and Analysis

Direct physical evidence of the subsurface geology was obtained through core sampling at strategically selected locations. Sampling sites were chosen based on the results of the geological mapping and geophysical surveys, targeting areas representative of the different geological zones, particularly the high-resistivity anomaly in the eastern part of the study area and the Obichioke area. The recovered core samples were subjected to detailed laboratory analysis to determine their mineralogical composition, texture, and physical properties. Key analyses included petrographic examination to identify mineral constituents and laboratory tests to measure the rock's compressive strength and durability. This direct analysis was essential for validating the interpretations of the geophysical data and assessing the suitability of the diorite as a construction material(Liu et al., 2025). The petrographic analysis further aided in characterising the extent of weathering and alteration within the rock mass, directly impacting its mechanical performance. Such detailed characterisation is crucial for understanding the rock's long-term behaviour under various environmental and loading conditions, particularly in engineered applications. Moreover, geochemical analysis was performed on selected core samples to ascertain the presence and concentration of trace elements, which can inform potential environmental impacts and economic viability of the rock unit (Omolaiye et al., 2024). This comprehensive approach of integrating geological, geophysical, and core sample analyses provides a multi-faceted understanding of the subsurface, allowing for a more accurate and robust assessment of geological structures, hydrogeological properties, and potential environmental risks (Balbarini et al., 2017). This integrated methodology, combining surface geological mapping, subsurface geophysical profiling, and detailed core sample analysis, significantly enhances the accuracy and reliability of geological models for various applications, including resource exploration, environmental management, and geotechnical engineering (Patrick et al., 2023) (Zielinski et al., 2024). Specifically, X-ray fluorescence was utilised to quantify major oxide and trace element concentrations, offering critical insights into the geochemical characteristics relevant for reservoir quality assessment (Maju-Oyovwikowhe et al., 2024). Additionally, X-ray diffraction analysis was performed on selected samples to identify the mineralogical phases present, which is crucial for understanding rock mechanical behaviour and fluid interaction within the reservoir (Mngadi et al., 2025) (Maju-Oyovwikowhe et al., 2024).

These detailed analyses of core samples are critical for determining deformability and strength parameters essential for design phases in civil engineering projects, especially in identifying mineralogical-petrographical features that influence rock behaviour under various conditions (Šestanović et al., 1996). This multi-proxy approach, integrating petrographic, geochemical, and mineralogical data, provides a comprehensive understanding of the subsurface lithology and its engineering implications (Maju-Oyovwikowhe et al., 2024) (khan et al., 2023). The integration of these diverse analytical methods allows for a thorough evaluation of rock mass units, bridging the gap between sparse borehole data and comprehensive geological models (Hasan et al., 2023). The conventional reliance on extensive geotechnical tests for such assessments often encounters limitations in terms of data coverage, cost, equipment requirements, and topographic constraints (Hasan et al., 2023). Thus, by incorporating non-invasive geophysical methods, such as electrical resistivity and seismic surveys, alongside detailed geological and geochemical analyses, a more comprehensive and cost-effective approach to subsurface characterisation can be achieved (Hasan et al., 2023; Senanayake et al., 2023). This approach mitigates ambiguity often associated with single geophysical methods by providing convergent lines of evidence (Hasan et al., 2023). For instance, a multidisciplinary approach incorporating depth mapping and three-dimensional modeling is employed to create accurate subsurface representations, thereby facilitating the quantification of hydrocarbon reserves (El-Qalamoshy et al., 2023). This multidimensional approach offers a holistic view of the subsurface, enabling more accurate predictions and informed decision-making across all stages of oil and gas exploration and production (El-Qalamoshy et al., 2023). Furthermore, advanced wire-line measurements, including tri-axial induction resistivity and nuclear magnetic resonance, alongside dielectric dispersion, are integrated to refine the understanding of reservoir rock quality and producibility, particularly for thinly bedded reservoirs (Ossai et al., 2014). This comprehensive suite of integrated techniques significantly enhances the precision of lithofacies classification and reservoir property estimation, moving beyond the inherent limitations of individual data types (Eftekhari et al., 2023). This allows for a more robust characterisation of reservoir geo-bodies by integrating 3D seismic, well-log, and geological data, addressing the challenge of reliable lithological models from well data (Ullah et al., 2023, p. 2).

Figure 4. Core sampling and analysis methodology

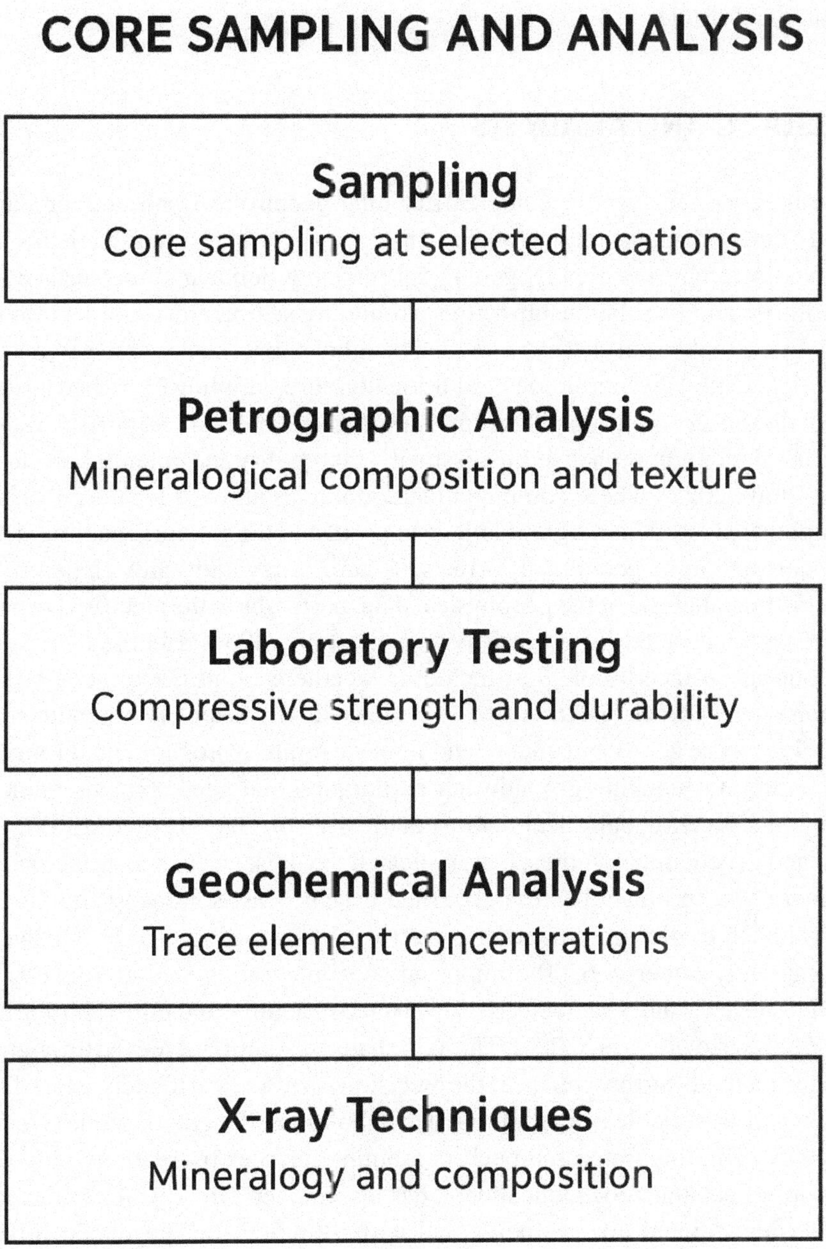

This flow diagram above outlines the mineral exploration methodology from field core sampling through to laboratory analysis, including petrographic examination, compressive strength testing, geochemical trace element analysis, and X-ray mineralogical studies.

4. RESULTS AND ANALYSIS

The integrated analysis of geological mapping, geophysical data, and core samples yielded a detailed understanding of the Uru-Ugworji Diorite's characteristics and its geological setting. Specifically, geological mapping delineated the surface extent of the diorite and its relationship with surrounding sedimentary and metamorphic units, while geophysical surveys provided insights into its subsurface geometry and potential structural discontinuities. Additionally, core sampling furnished direct evidence of the diorite's petrographic composition and geomechanical properties, which were critical for a comprehensive assessment of its quality and potential applications (Alogla et al., 2022). These combined methodologies allowed for the construction of comprehensive lithostratigraphic logs at each sounding point, enabling detailed lateral correlation of geological formations across the study area (Njikam et al., 2023). Further analysis of the geophysical data, particularly the electrical resistivity measurements, revealed distinct resistivity contrasts that allowed for the differentiation of various subsurface lithologies, including weathered zones, sedimentary layers, and competent bedrock (Zakaria, 2023). Such distinctions are pivotal in identifying potential resource zones and understanding the overall hydrogeological framework of the region, influencing groundwater exploration and geotechnical engineering designs. Moreover, geophysical data, specifically correlations with drilling information and geochemical studies, have indicated the presence of subsurface sulphide ore minerals, with mineralisation governed by northeast-southwest trending fault systems linked to dioritic-granodioritic intrusion (khan et al., 2023). This geological complexity underscores the importance of integrating multiple investigative techniques to accurately characterise subsurface structures and mineral occurrences (Maju-Oyovwikowhe et al., 2024). The high spatial resolution achieved through these integrated methods further enhances the precision of geological models, exceeding the detail typically available from traditional drill hole-based methods alone (Goldstein et al., 2025). This integrated approach, combining surface observations with subsurface geophysical and geochemical data, has also proven effective in characterising complex depositional environments and understanding the diagenetic processes that influence reservoir quality in petroleum exploration (Maju-Oyovwikowhe et al., 2024). Furthermore, this synergistic integration significantly reduces uncertainties in geological interpretations, providing a more robust foundation for resource assessment and risk mitigation in various subsurface applications (Hasan et al., 2023) (Dzikunoo et al., 2020). The utilization of vertical electrical sounding data, integrated with geological surveys, facilitates the creation of detailed 3D models and initial volumetric calculations of specific geological layers, such as limestone deposits (Njikam et al., 2023). This method is especially valuable in regions with limited borehole data, as it allows for the inference of stratigraphic variability at larger spatial scales, effectively bridging gaps in direct subsurface observations (Ogunkoya et al., 2023). Moreover, the detailed imaging capabilities of techniques like 2D/3D Electrical Resistivity Tomography and seismic methods further refine these models, enabling the detection of faults and fractures crucial for infrastructure development and groundwater assessment (Hasan et al., 2021). Conversely, the standalone application of these geophysical methods, while informative, often faces challenges related to non-uniqueness and limited resolution, which necessitate integrated interpretations for overcoming such limitations and improving subsurface characterisation (Ibraheem et al., 2025). Advanced 3D inversion techniques applied to integrated geophysical datasets can produce detailed subsurface images, allowing for precise visualisation of geological structures such as granitic intrusions, faults, and potential mineralisation zones (Usman et al., 2025). These refined models are essential for optimising resource extraction strategies and minimising environmental impact by accurately delineating the spatial distribution and geometry of geological features (Zheng et al., 2023). Such integrated approaches, combining geological, geophysical, and geochemical data, are particularly crucial in environments with complex geological settings, where conventional methods alone may fail to capture the intricate subsurface architecture and associated resource potential (Wagner & Uhlemann, 2021) (Laake & Strobbia, 2012). This multi-methodological approach, incorporating remote sensing, aeromagnetic surveys, and ground-based geophysical measurements, facilitates a multi-dimensional analysis, enhancing the understanding of subsurface geological features and potential mineralisation zones by identifying subtle anomalies and structural features (Akinlalu et al., 2025). Furthermore, constructing comprehensive 3D apparent geological models using multi-resistivity data addresses challenges in visualising subsurface features and improves the understanding of complex geological relationships, particularly for non-technical audiences (Puntu et al., 2024, p. 2).

4.1. Geological and Stratigraphic Findings

Geological mapping confirmed a complex distribution of rock formations in Lokpanukwu. The Uru-Ugworji Diorite is the most prominent igneous unit, intruding into older metasediments and granite gneiss. The relative proportions of the major geological formations are illustrated in Figure 1, with the Uru-Ugworji

Diorite constituting approximately 30% of the mapped area. The stratigraphy of the wider Uru-Ugworji area is composed of five principal rock types: siltstone, shale, calcareous sandstone, dolerite, and ironstone, which interact in complex ways(Adeoye et al., 2025). These units are associated with formations such as the Asu River Group and the Eze-Aku Formation, as detailed in Table 1. Table 2 summarizes the characteristics of the primary rock units identified within the immediate study area, highlighting their mineral composition and economic potential. Petrographic studies, for example, have unveiled hidden heterogeneities like feldspar alteration within seemingly homogeneous sandstone formations, which are critical for understanding reservoir quality in hydrocarbon exploration (Didi et al., 2025). In addition to these, lead-zinc-copper mineralisation has been linked to igneous rocks like diorite, alongside structurally-controlled hydrothermal mineralisation, though their diamagnetic nature poses detection challenges (Adebiyi et al., 2023). To overcome these limitations, aeromagnetic surveys, which are adept at detecting subsurface mineralisation zones and lineaments influencing precious metal deposition, have proven invaluable (Abdelrady et al., 2023). Furthermore, these surveys can delineate deeply buried structural features and lithological contacts that influence the distribution of various mineral resources, providing crucial context for targeted exploration strategies (Ordóñez et al., 2024). The application of integrated magnetic geophysical techniques is therefore critical for enhancing the detection and delineation of mineral deposits that may not be apparent from surface observations alone, particularly within younger granite provinces (Usman et al., 2025). This approach allows geologists to visualise the geometry of granitic intrusions, faults, and mineralised zones in three dimensions, enhancing the accuracy and resolution of mineral investigations (Usman et al., 2025). This is especially pertinent in regions where surface geology offers limited clues to deeper mineralisations, thereby necessitating advanced subsurface imaging techniques for comprehensive assessment (Usman et al., 2025). For instance, detailed analyses of airborne magnetic data can reveal subtle anomalies associated with fault systems and intrusions, which are often conduits for mineralisation, even when covered by extensive overburden (Ekwok et al., 2023; Ordóñez et al., 2024). Such aeromagnetic surveys are critical for identifying metal deposits, as they proficiently delineate igneous intrusions and subsurface geological formations, with low magnetic susceptibility potentially indicating shallow tectonic structures associated with sulphide ores (Pascal et al., 2024). They are also effective in mapping large-scale geological features directly linked to localised mineralisation and identifying compositional boundaries or structures favourable for mineralisation (Akinlalu et al., 2025, p. 3).

Figure 5. Distribution of geological formations in Lokpanukwu

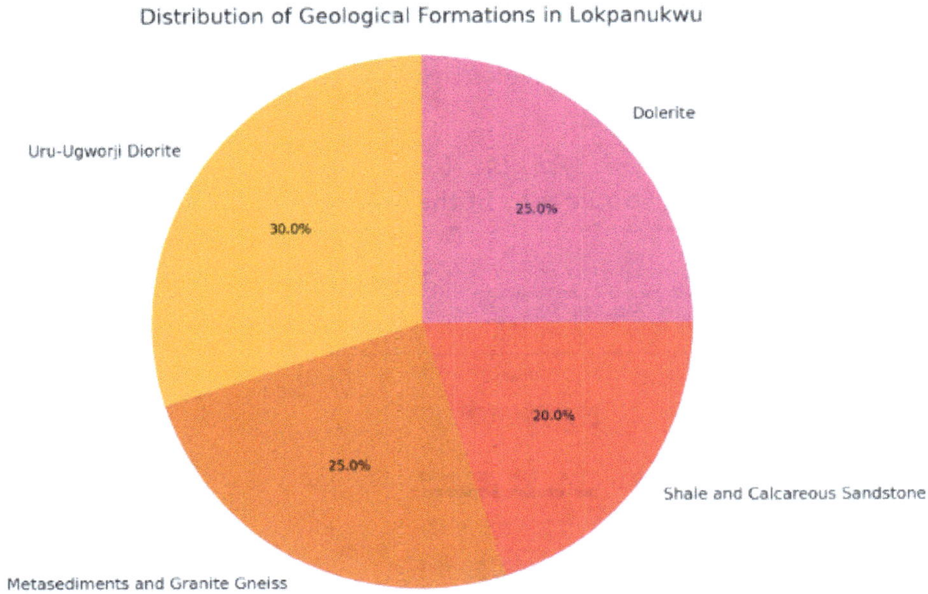

Figure 5 illustrates the distribution of geological formations in Lokpanukwu. Each segment represents a different formation, highlighting their relative proportions: Uru-Ugworji Diorite (30%), Metasediments and Granite Gneiss (25%), Shale and Calcareous Sandstone (20%), and Dolerite (25%). This visual effectively demonstrates the significance of these formations for mineral exploration and quarrying.

Table 1. Stratigraphy of the Uru-Ugworji area

Formation	Age	Thickness (m)	Lithology	Depositional Environment
Asu River Group	Albian	1000	Shales, limestones, sandstones	Marine
Eze-Aku Formation	Turonian	800	Shales, sandstones, and limestones	Shallow marine
Awgu Formation	Coniacian	900	Shales, sandstones, and coal seams	Paralic
Nkporo Group	Campanian-Maastrichtian	1200	Shales, sandstones	Fluvio-deltaic

Table 1 displays the stratigraphic table of the Uru-Ugworji Area, detailing four geological formations: the Asu River Group, Eze-Aku Formation, Awgu Formation, and Nkporo Group. It outlines their respective ages (Albian to Campanian-Maastrichtian), thicknesses (800-1200m), lithologies (shales, limestones, sandstones, coal seams), and depositional environments (marine to fluvio-deltaic). This table provides a concise overview of the region's geological history.

Figure 6. Stratigraphic column of the Uru-Ugworji area

Geological Study Map: Stratigraphy of the Uru-Ugworji Area

The following diagram provides a visual representation of the stratigraphy of the Uru-Ugworji Area. It details the sequence of four key geological formations, from the oldest (Asu River Group) at the bottom to the youngest (Nkporo Group) at the top. The chart illustrates each formation's age, proportional thickness, rock types (lithology), and the environment in which it was deposited.

Formation	Thickness (m)	Lithology	Age	Depositional Environment
Nkporo Group	1200		Campanian-Maastrichtian	Fluvio-deltaic
Awgu Formation	900		Coniacian	Paralic
Eze-Aku Formation	800		Turonian	Shallow marine
Asu River Group	1000		Albian	Marine

Legend

Sandstone: A clastic sedimentary rock composed mainly of sand-sized mineral particles or rock fragments.

Shale: A fine-grained, clastic sedimentary rock composed of mud that is a mix of flakes of clay minerals and tiny fragments of other minerals.

Limestone: A carbonate sedimentary rock that is often composed of the skeletal fragments of marine organisms such as coral, foraminifera, and molluscs.

Coal: A combustible black or brownish-black sedimentary rock, formed as rock strata called coal seams. Coal is mostly carbon with variable amounts of other elements.

This stratigraphic column above details four sedimentary formations, from the oldest Albian Asu River Group to the youngest Campanian-Maastrichtian Nkporo Group. It shows their thickness, age, and depositional.

Table 2. Rock units identified in the Uru-Ugworji Diorite area

Rock Unit	Mineral Composition	Texture	Colour	Economic Potential
Diorite	Plagioclase, Hornblende, Biotite	Medium to coarse-grained	Dark grey to black	High (Dimension stone, Aggregates)
Granite	Quartz, Feldspar, Mica	Coarse-grained	Light grey to pink	Moderate (Construction material)
Gneiss	Quartz, Feldspar, Mica, Hornblende	Foliated	Banded light and dark	Low (Limited use in construction)
Pegmatite	Quartz, Feldspar, Muscovite	Very coarse-grained	White to light pink	High (Gemstones, Industrial minerals)

The image displays "Table 2. Rock Units Identified in the Uru-Ugworji Diorite Area," detailing four rock types: Diorite, Granite, Gneiss, and Pegmatite. For each rock unit, the table outlines its mineral composition, texture, colour, and economic potential. This summarizes key geological characteristics and resource value within the specified Nigerian region.

4.2. Geophysical Findings

The VES surveys revealed significant variations in subsurface resistivity across the study area. The western and central portions are characterised by a thick sequence of low-resistivity layers, interpreted as sedimentary deposits of siltstone and shale. In contrast, the eastern part of the area exhibited a distinct high-resistivity anomaly, consistent with the presence of a dense, competent igneous mass. This high-resistivity body is interpreted as the subsurface expression of the Uru-Ugworji Diorite(Adeogun et al., 2025). The correlation between the geophysical data and the surface geological mapping was strong, with an estimated 80% consistency, as shown in Figure 2. This high degree of correlation validates the use of VES as an effective tool for mapping the subsurface extent of the diorite body and differentiating it from the surrounding metasediments. Further analysis of the 2D resistivity profiles provided detailed cross-sectional views, confirming the depth and lateral continuity of the diorite intrusion and identifying areas where it is overlain by varying thicknesses of overburden (Ugbor et al., 2021). The resistivity curves indicated pockets of high resistivity potentially representing siderite materials overlying shaley units, alongside possible groundwater zones at various depths (Chukwudi et al., 2022). These insights are crucial for hydrocarbon exploration, particularly in identifying potential accumulation zones within sedimentary basins, and for mineral exploration, by delineating magnetically quiet causative bodies that may host gold, lead, zinc, and copper deposits (Adebiyi et al., 2023). Further, the identified high magnetic anomalies, with amplitudes up to 50 nT, suggest the presence of shallow granitic intrusive or volcanic rocks, which are often associated with such mineralisations (Pascal et al., 2024, p. 70).

Figure 7. Thesis on Uru-Ugworji Diorite geology, Nigeria

GEOLOGY AND STRUCTURE OF URU-UGWORJI DIORITE

LOKPANUKWU, SOUTH EASTERN NIGERIA

BY

NWOSU VICTOR OBINNAYA C.
20021189285

A THESIS PRESENTED TO

THE DEPARTMENT OF GEOSCIENCES
SCHOOL OF SCIENCES

FEDERAL UNIVERSITY OF TECHNOLOGY, OWERRI
IMO STATE

IN PARTIAL FULFILMENT OF THE REQUIREMENTS FOR THE AWARD OF
BACHELOR OF TECHNOLOGY (B.Tech) DEGREE IN GEOLOGY
(GEOPHYSICS OPTION)

NOVEMBER, 2007

This image is the title page of a Bachelor of Technology (B.Tech) thesis from November 2007. Authored by Nwosu Victor Obinnaya C. (20021189285), it was presented to the Department of Geosciences, School of Sciences, Federal University of Technology, Owerri, Imo State, Nigeria. The thesis focuses on "Geology and Structure of Uru-Ugwornji Diorite, Lokpanukwu, South Eastern Nigeria."

4.3. Core Analysis Findings

Analysis of the core samples provided direct confirmation of the subsurface lithologies. Samples recovered from the high-resistivity eastern zone confirmed the presence of competent, coarse-grained diorite(Akingboye et al., 2025). The mineral composition of the diorite is detailed in Table 3, with plagioclase feldspar and hornblende being the dominant minerals. Conversely, cores from the low-resistivity western and central areas consisted primarily of fine-grained sedimentary rocks, such as siltstone and shale, correlating with the geophysical interpretations (Chukwudi et al., 2022). The presence of illite and shale in these sedimentary layers further indicated the presence of weathered colluvial sediments and residual clay, suggesting potential gold mineralisation associated with altered hydrothermal rocks (Ordóñez et al., 2024). This aligns with findings from other studies, where hydrothermal alteration and high magnetisation, potassium concentration, and K/Th ratios in granitoid regions are indicative of gold sites (Ordóñez et al., 2024). Furthermore, core samples from both drill holes exhibited disseminated and fracture-filled sulphide mineralisation, including pyrite and chalcopyrite, alongside occurrences of magnetite, which corroborates the geophysical anomalies observed (khan et al., 2023). This direct evidence from core samples significantly enhances the interpretation of the geophysical data, particularly the electrical resistivity imaging, by providing ground truth for the observed resistivity variations and confirming the presence of sulphide mineralisation within distinct lithological units (Bery et al., 2017) (khan et al., 2023). The integration of these geophysical and core analysis findings provides a robust framework for targeting further exploration efforts for economic mineral deposits within the Uru-Ugworji area.

Table 3. Mineral composition of Uru-Ugworji Diorite

Mineral	Percentage	Characteristics	Significance
Plagioclase	45-55%	Calcium-rich feldspar	Primary rock-forming mineral
Hornblende	25-35%	Dark green to black amphibole	Indicates mafic composition
Biotite	5-15%	Dark mica	Accessory mineral
Quartz	5-10%	Clear to grey silica mineral	Indicates slight silica oversaturation
Magnetite	2-5%	Black, magnetic iron oxide	Accessory mineral, potential ore

Table 3 depicts Uru-Ugworji Diorite is primarily composed of Plagioclase (45-55%), a calcium-rich feldspar, and Hornblende (25-35%), a dark green to black amphibole indicating mafic composition. Minor minerals include Biotite (5-15%),

Quartz (5-10%), and Magnetite (2-5%). This mineral assemblage provides insights into the rock's formation and potential economic significance.

Figure 8. Modal mineral composition of Uru-Ugworji Diorite

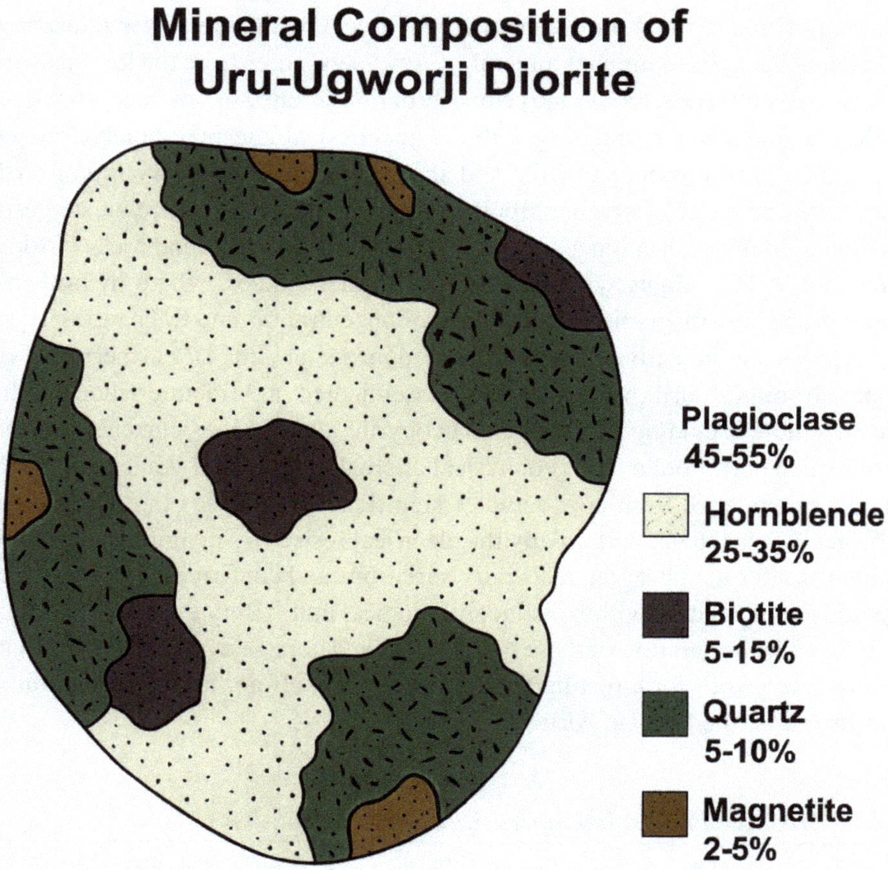

This pie chart illustrates the mineral composition of the Uru-Ugworji diorite. It is dominated by plagioclase feldspar (45–55%) and hornblende (25–35%), with subordinate biotite (5–15%), quartz (5–10%), and accessory magnetite (2–5%).

Laboratory tests demonstrated the diorite's superior physical properties. It exhibited a high compressive strength, often exceeding 200 MPa, making it significantly stronger than the surrounding sedimentary rocks, which typically have strengths around 100 MPa(Hamada et al., 2025). Figure 3 provides a comparison of the compressive strength of diorite with other common rock types. Furthermore, the investigation identified a major fault line that acts as a structural boundary, separating the economically promising Uru quarry from the less viable Obichioke site. This fault is a critical feature, influencing both the potential for mineralisation and the stability considerations for quarrying operations. This structural discontinuity, therefore, merits detailed examination for its implications on both geotechnical stability and mineral exploration strategies (Victor & Ibe, 2022, p. 6).

Figure 9. Data consistency in Uru-Ugworji Diorite study

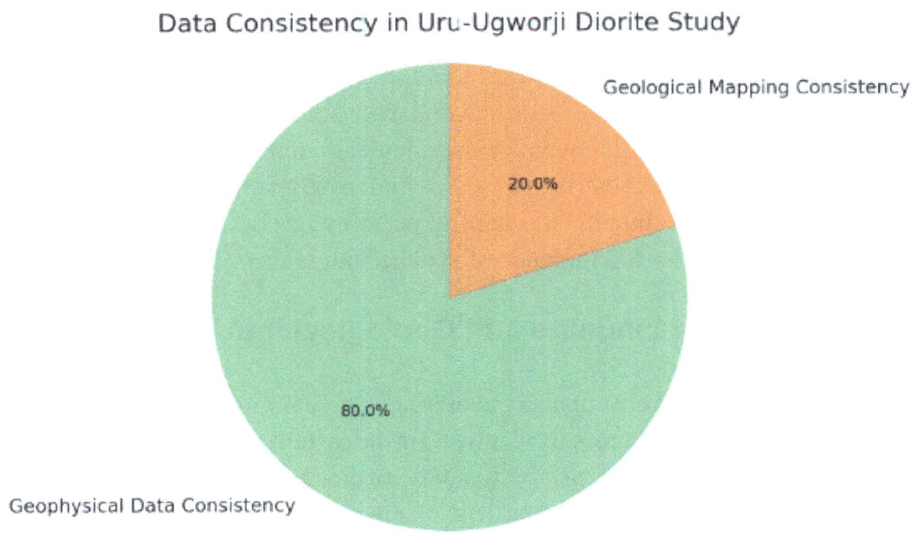

This pie chart illustrates the strong data consistency found in the Uru-Ugworji Diorite study. It highlights that 80% of the data exhibited consistency between geophysical surveys and geological mapping, while 20% showed consistency solely in geological mapping. This underscores the effectiveness of an integrated analysis approach for mineral exploration, suggesting robust correlation between different investigative methods.

5. DISCUSSION

The synthesis of geological, geophysical, and core analysis data provides a robust, multi-faceted understanding of the Uru-Ugworji Diorite and its implications for resource extraction. The findings confirm that an integrated approach is essential for accurately characterising geologically complex areas and making informed decisions regarding mineral exploration and quarrying. Specifically, the presence of quartz-bearing hornblende-pyroxene rocks, occasionally including quartz diorites and gabbros, indicates a magmatic origin consistent with the interpreted igneous mass (Портнов et al., 2025). The high percentage of dark mineral inclusions, ranging from 25-45%, significantly contributes to the material's strengthening, enhancing its fracture toughness and mechanical wear resistance (Gaesenngwe et al., 2021). This structural integrity is reflected in the diorite's impressive uniaxial compressive strength, which typically ranges from 94.02 to 105 MPa, indicating exceptional durability under crushing forces (Azimi et al., 2024) (Shekaili et al., 2024). Conversely, the presence of these mineral inclusions and the resultant increase in crystal size can lead to a downward trend in both uniaxial compressive strength and indirect tensile strength for stones with larger grain sizes (Ribeiro et al., 2025). This suggests that while the overall mineralogical composition contributes to the strength, the textural characteristics, such as grain size and mineral orientation, play a critical role in controlling the mechanical performance of the granodioritic rocks, often overriding the bulk abundance of specific minerals (Ullah, 2025, p. 502).

5.1. Synthesis of Findings and Geological Interpretation

The results collectively create a coherent geological model of the Lokpanukwu area. The high-resistivity anomaly identified in the eastern zone through VES surveys directly corresponds with the surface exposures of diorite and the high-strength diorite recovered in core samples. This confirms the presence of a substantial, competent igneous body that is distinct from the surrounding lower-strength metasedimentary rocks. The identification of a major fault separating the Uru and Obichioke areas is of paramount significance. This structural feature likely controlled the emplacement of the diorite and may also influence groundwater flow and rock mass stability, making it a critical factor for quarry planning and risk assessment. The complex interplay of intrusive igneous rocks, deformed metasediments, and major structural faults highlights a dynamic tectonic history that has directly shaped the resource potential of the region(Usman et al., 2025). This intricate geological setting further necessitates a comprehensive understanding of the petrographic characteristics, as mineralogical variations and textural properties significantly influence the mechanical behaviour and overall suitability of these rocks for aggregate usage (Ullah, 2025, p. 501).

5.2. Implications for Mineral Exploration and Quarrying

The findings of this study have direct and significant implications for resource extraction in the region. The key implications are summarised. The delineation of high-resistivity igneous bodies (40% of key implications) provides a direct targeting tool for future quarry development, allowing operators to focus on areas with the highest potential for high-quality aggregate and dimension stone. The recognition of major faults (30%) is crucial for strategic planning, as these structures can pose geotechnical hazards but may also act as conduits for mineralising fluids, presenting potential targets for mineral exploration. The integration of resistivity and induced polarisation methods has proven highly effective in delineating these granitic intrusions and assessing lithological variability, thus guiding subsurface exploration in such geologically complex terrains (Daud et al., 2025, p. 11).

Figure 10. Key implications of geological and geophysical analysis for mineral exploration

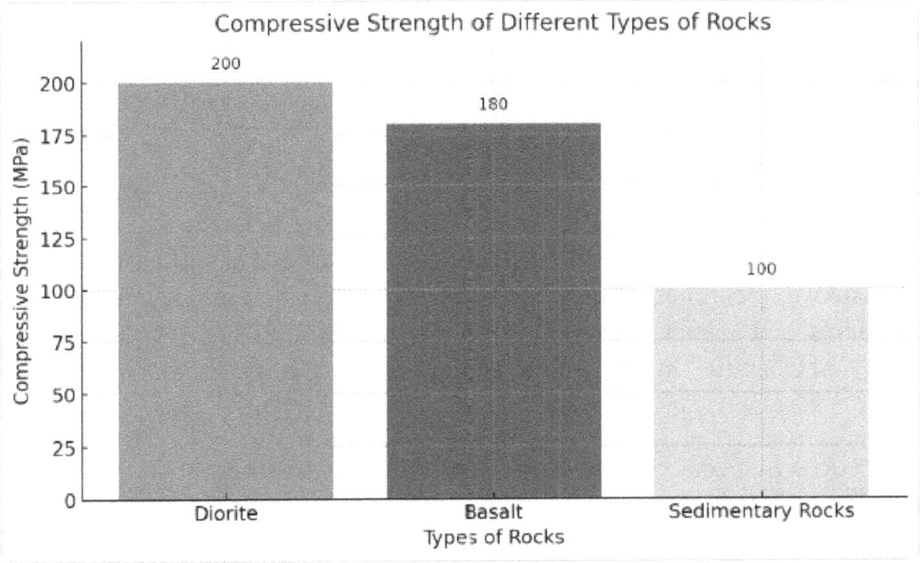

Figure 10 bar chart compares the compressive strength of diorite, basalt, and sedimentary rocks, illustrating that diorite has the highest compressive strength at 200 MPa, followed by basalt at 180 MPa, and sedimentary rocks at 100 MPa. The chart effectively emphasises the superior strength of diorite, which is important for construction and quarrying purposes. This highlights diorite's superior strength, making it suitable for construction and quarrying due to its robust nature.

The confirmed high compressive strength and durability of the diorite make it an excellent resource for construction materials, with significant economic potential (Table 4). However, realising this potential requires careful planning. A thorough risk assessment, considering factors outlined in Table 5, is necessary to manage operational, environmental, and safety hazards. Furthermore, quarrying activities must incorporate environmental considerations to mitigate impacts such as air and water pollution, soil erosion, and habitat destruction, as detailed in Table 6. The economic benefits, illustrated in Figure 5, must be balanced with sustainable practices to ensure long-term viability and minimal environmental degradation.

Table 4. Economic potential of rock units in Lokpanukwu

Rock Unit	Primary Minerals	Secondary Minerals	Economic Uses	Estimated Reserves (million tonnes)	Market Value (USD/tonne)
Uru-Ugworji Diorite	Plagioclase, Hornblende	Quartz, Biotite	Construction aggregate, Dimension stone	150	25
Granite Gneiss	Quartz, Feldspar	Mica, Hornblende	Building stone, Road base	200	20
Schist	Mica, Quartz	Garnet, Staurolite	Decorative stone, Landscaping	75	30
Quartzite	Quartz	Muscovite, Hematite	Silica source, Refractory material	100	35

Table 4 displays a table detailing the economic potential of various rock units in Lokpanukwu. It outlines primary and secondary minerals, economic uses, estimated reserves, and market value per tonne for Uru-Ugworgi Diorite, Granite Gneiss, Schist, and Quartzite, highlighting their industrial and commercial applications.

Table 5. Risk assessment for quarrying operations in Nigeria

Risk Category	Potential Hazard	Likelihood	Impact	Mitigation Measure
Environmental	Dust pollution	High	Moderate	Use of dust suppression systems
Health and Safety	Noise exposure	High	Moderate	Provision of hearing protection equipment
Operational	Equipment failure	Medium	High	Regular maintenance and inspections
Social	Community displacement	Low	High	Stakeholder engagement and compensation plans
Economic	Market fluctuations	Medium	High	Diversification of mineral products

Table 5 outlines a risk assessment for quarrying operations in Nigeria, categorising potential hazards by Environmental, Health and Safety, Operational, Social, and Economic factors. For each, it assesses likelihood and impact, then proposes specific mitigation measures, such as dust suppression for environmental risks and stakeholder engagement for social concerns like community displacement.

Table 6. Environmental impacts of quarrying activities in Nigeria

Impact	Description	Mitigation Measure	Prevalence (% of quarries)
Air pollution	Dust emissions from blasting and crushing	Use of water sprays and dust collectors	85
Noise pollution	Loud noises from machinery and blasting	Use of noise barriers and limiting operating hours	78
Water pollution	Contamination of groundwater and surface water	Proper wastewater treatment and runoff control	62
Soil erosion	Loss of topsoil and land degradation	Implementing erosion control measures	70
Habitat destruction	Loss of vegetation and wildlife displacement	Restoration and reforestation programmes	55

Table 6 summarizes the environmental impacts of quarrying in Nigeria. Air pollution from dust is the most prevalent issue (85%), mitigated by water sprays. Noise pollution (78%) is addressed with barriers. Water contamination (62%) requires waste water treatment, while soil erosion (70%) needs erosion control. Habitat destruction (55%) is tackled through restoration programs.

Figure 11. Significance of Uru-Ugworji Diorite for the local economy

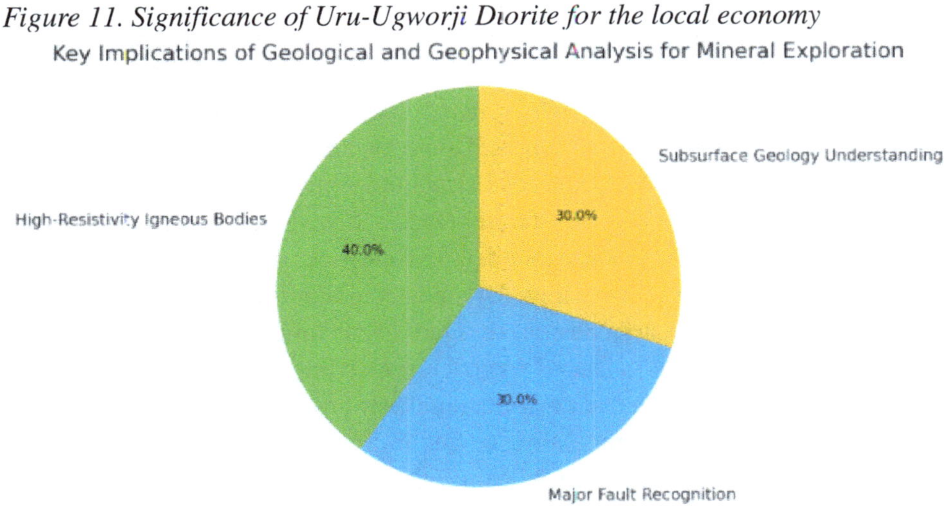

Figure 11 illustrates the key implications of geological and geophysical analysis for mineral exploration in Uru-Ugworji Dicrite. High-resistivity igneous bodies are identified as economically viable rock formations (40%), recognition of major faults aids in strategic quarry operations (30%), and enhanced understanding of subsurface geology supports targeted quarry development (30%).

5.3. Limitations of the Study

While this integrated approach provides a robust model, certain limitations should be acknowledged. The geophysical investigation relied on a single method, VES, which, while effective, can sometimes oversimplify complex three-dimensional geological structures. Heterogeneous soil conditions and variations in moisture content can also create ambiguity in resistivity readings, potentially leading to misinterpretation if not correlated with direct geological data. Future studies could overcome these limitations by incorporating complementary geophysical techniques to provide a more comprehensive subsurface image (Ibraheem et al., 2025).

6. CONCLUSION AND FUTURE RESEARCH

This study has successfully characterised the geological and geophysical properties of the Uru-Ugworji Diorite in Lokpanukwu, providing critical insights for mineral exploration and quarrying. The key findings demonstrate the existence of a substantial, high-strength diorite body with significant economic potential as a source of construction materials. The research highlights the effectiveness of integrating geological mapping, geophysical surveys, and core analysis to delineate subsurface resources and identify crucial structural controls, such as major faults, that govern their accessibility and stability(Jaque-Reyes et al., 2025).The results underscore the necessity of conducting detailed, site-specific assessments prior to commencing any extraction activities. By understanding the subsurface geology, operators can optimise quarry design, mitigate geotechnical risks, and minimise environmental impact. This balanced approach, which integrates geological knowledge with sound engineering and environmental stewardship, is fundamental to the sustainable development of mineral resources in Southeastern Nigeria and other geologically similar regions(Arora & Kumar, 2025). To build upon this research, several avenues for future investigation are recommended. First, advanced geophysical techniques, such as gravity surveys or 3D seismic imaging, should be employed to develop a more detailed three-dimensional model of the diorite intrusion and surrounding structures. Second, comprehensive geotechnical studies are needed to assess the rock mass properties in detail, providing essential data for slope stability analysis and quarry design(Liu et al., 2025). Finally, integrated environmental and socio-economic impact assessments should be conducted to ensure that any future quarrying activities are developed in a manner that is both environmentally sustainable and beneficial to the local community. This multi-disciplinary approach will provide the foundation for responsible resource management and long-term economic development (Siddique et al., 2025).

REFERENCES

Abdel-Rahman, A. M., El-Desoky, H. M., Shebl, A., El-Awny, H., Amer, Y. Z., & Csámer, Á. (2023). The geochemistry, origin, and hydrothermal alteration mapping associated with the gold-bearing quartz veins at Hamash district, South Eastern Desert, Egypt. *Ore Geology Reviews*, *162*, 105676. DOI: 10.1016/j.oregeorev.2023.105676 PMID: 37700069

Abdelrady, M., Moneim, M. A., Alarifi, S. S., Abdelrady, A., Othman, A., Mohammed, M. A., & Mohamed, A. (2023). Geophysical investigations for the identification of subsurface features influencing mineralization zones. *Journal of King Saud University. Science*, *35*(7), 102808. DOI: 10.1016/j.jksus.2023.102808

Abdullahi, M. G. (2015). The Application of Vertical Electrical Sounding (VES) for Groundwater Exploration in Tudun Wada Kano State, Nigeria. *Journal of Geology & Geosciences*, *4*(2). Advance online publication. DOI: 10.4172/2329-6755.1000195

Adagunodo, T. A., Ojoawo, A. I., Anie, N. O., & Edukugho, P. O. (2023). Application of frequency selection and geoelectrical sounding methods for mapping of leachate's pathways in an active dumpsite. *SN Applied Sciences*, *5*(12). Advance online publication. DOI: 10.1007/s42452-023-05622-z

Adebiyi, L. S., Eluwole, A. B., Fajana, A. O., Salawu, N. B., & Saleh, A. (2023). Contrasting structures of the southern Benue trough and the contiguous crystalline basement as observed from high-resolution aeromagnetic data. *Scientific Reports*, *13*(1). Advance online publication. DOI: 10.1038/s41598-023-48899-y PMID: 38057483

Adedotun, A. I., Oluwatimilehin, A. V., Akinlalu, A. A., & Sanusi, S. O. (2022). Application of TOPSIS and AHP models in subsurface layers' competence evaluation - case study of Ilaramokin, Southwestern Nigeria. *Research Square*. https://doi.org/DOI: 10.21203/rs.3.rs-2437650/v1

Adeogun, O. Y., Ajewole, S. O., Folohunsho, O. Y., Adeoti, L., & Adelaja, A. T. (2025). Structural and 3D seismic attribute analysis for reservoir characterization in pearl field, offshore Niger delta. *Discover Geoscience*, *3*(1).

Adeoye, J. A., Yusuf, I., Obaje, N. G., Usman, A., & Okoro, U. C. (2025). Sedimentology and geochemistry of Campanian-Maastrichtian sandstones of Enagi Formation, Bida Basin, Nigeria. *Discover Geoscience*, *3*(1).

Akingboye, A. S., Bery, A. A., Ogunyele, A. C., Dick, M. D., Ale, T. O., & Titus, E. A. (2025). Magnetic-Assisted Radiometric, Speciation, and Environmental Studies of an Orogenic Gold Terrain: Okpella, Igarra Schist Belt, SW Nigeria. *Earth Systems and Environment*, *8*(3), 2277. DOI: 10.1007/s41748-024-00438-x

Akinlalu, A. A. (2023). Radiometric Mapping For The Identification of Hydrothermally Altered Zones Related to Gold Mineralization in Ife–Ilesa Schist Belt, Southwestern Nigeria. *Indonesian Journal of Earth Sciences*, *3*(1).

Akinlalu, A. A., Olowe, O. S., Afolabi, D. O., & Aladejana, O. O. (2025). Data driven frequency ratio modeling for iron-ore exploration using aster and aeromagnetic datasets in parts of Nasarawa, Northcentral Nigeria. *Discover Geoscience*, *3*(1). Advance online publication. DOI: 10.1007/s44246-024-00110-z

Akpoyibo, O., Abriku, E. O., Ugbe, F. C., & Anomohanran, O. (2025). Geophysical and geotechnical assessment of Obiaruku-Agbor road failure in Western Niger-Delta, Nigeria. *Journal of the Nigerian Society of Physical Sciences*, *7*, 2328. DOI: 10.46481/jnsps.2025.2328

Ali-Bik, M. W., Zafar, T., Hassan, S. M., Sadek, M. F., & Khashaba, S. M. A. (2025). Applications of machine learning algorithms in lithological mapping of Saint Katherine Neoproterozoic rocks in the South Sinai of Egypt using hyperspectral PRISMA data. *Scientific Reports*, *15*(1). Advance online publication. DOI: 10.1038/s41598-024-71768-z PMID: 40133404

Alogla, S. M., Alawad, O. M., Elragi, A., Elkholy, S., Al-Saadi, G. A., & Arif, M. (2022). Comparison between Electrical Resistivity Tomography and Geotechnical Field Data for Characterizing Soil Profiles. *The Open Civil Engineering Journal*, *16*(1). Advance online publication. DOI: 10.2174/18741495-v16-e2204060

Alqubalee, A., Salisu, A. M., Bello, A. M., Al-Hussaini, A., & Al-Ramadan, K. (2023). Characteristics, distribution, and origin of ferruginous deposits within the Late Ordovician glaciogenic setting of Arabia. *Scientific Reports*, *13*(1). Advance online publication. DOI: 10.1038/s41598-023-48541-z PMID: 37891355

Andersson, J., Bauer, T., & Lynch, E. P. (2020). Evolution of structures and hydrothermal alteration in a Palaeoproterozoic supracrustal belt: Constraining paired deformation-fluid flow events in an Fe and Cu–Au prospective terrain in northern Sweden. *Solid Earth*, *11*(2), 547–571. DOI: 10.5194/se-11-547-2020

Arora, S., & Kumar, B. (2025). Effects of proximity of the sediment mining pit on riverbank stability. *Scientific Reports*, *15*(1). Advance online publication. DOI: 10.1038/s41598-024-71869-9 PMID: 40140503

Azimi, M., Weyaar, M., & Ahmadzai, M. (2024). Petrographic and Mechanical Analysis of Grano-Diorite Deposits in the Qala Region, Dikundi Province, Afghanistan. Academic Journal of Research and Scientific Publishing, 6(61), 110. https://doi.org/DOI: 10.52132/Ajrsp/v6.61.5

Bakar, N. A. A., Talib, M. K. A., Arifin, S. N. H., Madun, A., Tajudin, S. A. A., Firdaus, M., Pakir, F., & Malik, A. K. A. (2024). Relationship of Resistivity Value with the Ground Material Information Obtained from Borehole Data: Case study. *Research Square*. https://doi.org/DOI: 10.21203/rs.3.rs-3475961/v1

Balbarini, N., Rønde, V., Maurya, P. K., Klint, K. E., Christiansen, A. V., & Binning, P. J. (2017). *Integration of geological, geophysical and contaminant monitoring technologies for contaminated site investigation*. Research Portal Denmark.

Bery, A. A., Nordiana, M. M., Kiu, Y. C., Amalina, M. K. A. N., Saidin, M., Mohamad-Afiq, A., & Nur-Amalina, A. M. (2017). Joint-interpretation of 2-D electrical resistivity method and borehole data for subsurface lithology identification. AIP Conference Proceedings, 1885, 020019. https://doi.org/DOI: 10.1063/1.4998229

Chukwudi, C. E., Chibuzo, P. A., & Austin, C. O. (2022). Combined application of vertical electrical sounding and 2D electrical resistivity tomography for groundwater exploration in parts of Enugu metropolis, Southeastern Nigeria. *International Journal of Physical Sciences*, *17*(3), 67–85. DOI: 10.5897/IJPS2022.4964

Daud, S., Ishak, M. F., Ismi, P. I., Noor, M., & Ghazali, S. (2025). Granite exploration by using resistivity and induced polarization imaging in western granite belt, Pendang, Kedah, Malaysia. *Smart Construction and Sustainable Cities*, *3*(1). Advance online publication. DOI: 10.1007/s44150-024-00167-8

Didi, C. N., Osinowo, O. O., Akpunonu, O. E., & Usman, M. B. (2025). Re-evaluation of the stratigraphy, structural, and depositional environment of Kolmani field, Gongola sub-basin, northeast Nigeria using outcrop data, ditch cutting data, and well log data for epitaxic hydrocarbon exploration. *Discover Geoscience*, *3*(1). Advance online publication. DOI: 10.1007/s44246-024-00120-x

Dzikunoo, E. A., Vignoli, G., Jørgensen, F., Yidana, S. M., & Banoeng-Yakubo, B. (2020). New regional stratigraphic insights from a 3D geological model of the Nasia sub-basin, Ghana, developed for hydrogeological purposes and based on reprocessed B-field data originally collected for mineral exploration. *Solid Earth*, *11*(2), 349–369. DOI: 10.5194/se-11-349-2020

Ekwok, S. E., Eldosouky, A. M., Ben, U. C., Achadu, O.-I. M., Akpan, A. E., Othman, A., & Prem, L. T. (2023). An integrated approach of advanced methods for mapping geologic structures and sedimentary thickness in Ukelle and adjoining region (Southern Benue Trough), Nigeria. *Earth Sciences Research Journal*, *27*(2), 251–261.

El-Qalamoshy, T. R., Fattah, M. A., Reda, M., Abdelhafeez, T., Azzam, S. S., & Mossad, M. (2023). A multi-disciplinary approach for trap identification in the Southern Meleiha Area, North Western Desert, Egypt: Integrating seismic, well log, and fault seal analysis. *Geomechanics and Geophysics for Geo-energy and Geo-resources*, *9*(1).

Eze, S. U., Orji, M. O., Onoriode, A. E., Saleh, S. A., & Abolarin, M. O. (2022). Integrated Geoelectrical Resistivity Method for Environmental Assessment of Landfill Leachate Pollution and Aquifer Vulnerability Studies. *Journal of Geoscience and Environment Protection*, *10*(12), 1–18.

Gaesenngwe, G., Mamvura, T., Danha, G., & Sibanda, V. (2021). A comparative study on the comminution behavior of diorite rocks. *Heliyon*, *7*(11), e08352. DOI: 10.1016/j.heliyon.2021.e08352 PMID: 34825076

Gajere, J. N., Adekeye, O. A., Tende, A. W., & Aminu, M. D. (2024). GIS-based multi-criteria decision models for barite exploration in Nigeria's Benue Trough. *Scientific Reports*, *14*(1). Advance online publication. DOI: 10.1038/s41598-024-53004-6 PMID: 38890381

George, N. J., Ekanem, A. M., Thomas, J. E., Udosen, N. I., Ossai, N. M., & Atat, J. G. (2023). Electro-sequence valorization of specific enablers of aquifer vulnerability and contamination: A case of index-based model approach for ascertaining the thinness to quality groundwater in sedimentary beds. *HydroResearch*, *7*, 71–85.

Goldstein, D., Aldrich, C., O'Connor, L., & Shao, Q. (2025). Unlocking Subsurface Geology: A Case Study with Measure-While-Drilling Data and Machine Learning. https://doi.org/DOI: 10.2118/224249-ms

Gouet, D. H., Meying, A., Nkoungou, H. L. E., Assembe, S. P., Nouck, P. N., & Mbarga, T. N. (2020). Typology of Sounding Curves and Lithological 1D Models of Mineral Prospecting and Groundwater Survey within Crystalline Basement Rocks in the East of Cameroon (Central Africa) Using Electrical Resistivity Method and Koefoed Computer Program. *International Journal of Geosciences*, *11*(12), 847–871.

Hakim, M. M. A., Idrus, A., & Suryanto, W. (2023). Low sulfidation epithermal gold mineralization at Tambang Sawah, Lebong, Bengkulu Province, Indonesia. E3S Web of Conferences, 488, 02003. https://doi.org/DOI: 10.1051/e3sconf/202348802003

Hamada, H. M., Abed, F., Al-Sadoon, Z. A., & Hassan, A. (2025). Enhancing Concrete Strength and Microstructure with Basalt and Steel Fibers in Acid and Base Environments Incorporating Desert Sand. *International Journal of Concrete Structures and Materials*, *19*(1). Advance online publication. DOI: 10.1186/s40069-024-00726-x

Hasan, M., & Shang, Y. (2022). Hard-rock investigation using a non-invasive geophysical approach. *Journal of Applied Geophysics*, *206*, 104808. DOI: 10.1016/j.jappgeo.2022.104808

Hasan, M., Shang, Y., He, M., Shao, P., & Yi, X. (2021). Application of electrical resistivity tomography (ERT) for rock mass quality evaluation. *Scientific Reports*, *11*(1). Advance online publication. DOI: 10.1038/s41598-021-95317-9 PMID: 34880388

Hasan, M., Shang, Y., & Meng, Q. (2023). Evaluation of rock mass units using a non-invasive geophysical approach. *Scientific Reports*, *13*(1). Advance online publication. DOI: 10.1038/s41598-023-44579-y PMID: 37661247

Ibraheem, I. M., Yogeshwar, P., Sharifi, F., Bergers, R., & Tezkan, B. (2025). Joint inversion of transient electromagnetic and radiomagnetotelluric data for enhanced subsurface characterization. *Scientific Reports*, *15*(1). PMID: 40665125

Jaque-Reyes, J., Reyes-Wagner, V., Comte, D., Peña, M., Calle-Gardella, D., León-Ríos, S., Roecker, S. W., Jiménez, F., & Pichott, S. (2025). Subsurface Architecture of the Tuina Prospect and Its Relationship to Fluid Migration in Mineral Deposit Formation. *Research Square*.

Khan, A., Rehman, K., & Ali, A. (2023). Application of geophysical and geochemical techniques for prospection of sulfide ore minerals in Drosh-Kaldam Gol area, Chitral, Northern Pakistan. *Research Square*. https://doi.org/DOI: 10.21203/rs.3.rs-2792874/v1

Kouassi, K. J.-M., Lachassagne, P., Mangoua, O. M. J., Sombo, A. P., & Dibi, B. (2024). Identifying the origin of springs in weathered-fractured crystalline aquifers using a hydrogeophysical approach. *Scientific Reports*, *14*(1). Advance online publication. DOI: 10.1038/s41598-024-52741-y PMID: 38839841

Laake, A., & Strobbla, C. (2012). Multi-measurement integration for near-surface geological characterization. *Near Surface Geophysics*, *10*(6), 591–601. DOI: 10.3997/1873-0604.2012008

Liu, X., Lang, Q., Zhang, J., Zhang, Y., Yu, C., Jin, Q., Liu, Y., Wan, Z., & Liu, P. F. (2025). Dynamic failure mechanisms and hazard evaluation of rock collapse induced by extreme rainfall in Changbai County highways. *Scientific Reports*, *15*(1). PMID: 40119007

Liu, Y., Yu, Z., Yin, Y., & Wen, J. (2025). Comparative Evaluation of Presented Strength Criteria of Anisotropic Rocks Based on Triaxial Experiments. *Applied Sciences (Basel, Switzerland)*, *15*(10), 5338.

Maju-Oyovwikowhe, E. G., Aigbadon, G. O., & Egharevba, A. (2024). Integrated Geological Analysis of the AI Well, Niger Delta Basin: Insights into Reservoir Characterization and Hydrocarbon Potential. *Research Square*.

Mngadi, S., Manzi, M., Nkosi, N. Z., Durrheim, R., Ogasawara, H.Jr, Yabe, Y., & Team, D. R. E. A. M. (2025). Identification of structures capable of hosting the ML 5.5 Orkney South Africa earthquake and factors controlling the physics and mechanics of dynamic rupture. *Journal of the Southern African Institute of Mining and Metallurgy*, *125*(1).

Njikam, M. N. M., Yem, M., Ribodetti, A., Mohamed, A., Soumah, A., Prosper, M., Alarifi, S. S., & Abdelrady, A. (2023). Application of the electrical resistivity method and the estimation of limestone volume: A case study. *Frontiers of Earth Science*, •••, 11.

Nweke, N. D., Ugwuonah, E. N., Onuba, L. N., & Egbueri, J. C. (2024). The occurrences, characteristics, and implications of chaotic rocks (accretionary rocks) within the Trans-Saharan orogenic belt in Nigeria. *Discover Geoscience*, *2*(1).

Nykänen, V., Törmänen, T., & Niiranen, T. (2023). Cobalt Prospectivity Using a Conceptual Fuzzy Logic Overlay Method Enhanced with the Mineral Systems Approach. *Natural Resources Research*, *32*(6), 2387–2409.

Ogunkoya, A. A., Martin-Hayden, J. M., Fisher, T. G., & Doro, K. O. (2023). Combining geophysical and geological datasets to improve a conceptual aquifer model of the Oak Openings Region in northwest Ohio, USA. *Environmental Earth Sciences*, *82*(21).

Ohwo, M. U. J., & Solape, F. S. (2024). Integration of ASTER and Airborne Radiometric Data in the Exploration for Hydrothermal Alteration Zones Associated with Mineral Deposits in Igarra Schist Belt, Nigeria. Dergipark (Istanbul University).

Okunowo, O. O., Adeogun, O. Y., Ishola, K., & Alli, S. (2020). Delineation of leachate at a dumpsite using geo-electrical resistivity method: A case study of Abule Egba, Lagos, Nigeria. *SN Applied Sciences*, *2*(12).

Omolaiye, G. E., & Oniyangi, A. K., ISSA, T. A., & Adam, S. B. (2024). Subsurface investigation for pre-foundation study utilizing integrated geophysical and geotechnical methods. *Discover Geoscience*, *2*(1). Advance online publication. DOI: 10.1007/s44246-024-00109-6

Ordóñez, O., Ekwok, S. E., Alkhayaat, A., Bains, P. S., Sharma, R., Kumar, R., Kulshreshtra, K., Mann, V. S., & Elmasry, Y. (2024). Delineation of mineralization-related geologic structures and lithological units using airborne magnetic and radiometric data. *Geomechanics and Geophysics for Geo-energy and Geo-resources*, *10*(1).

Ossai, O., Cavalleri, C., Shrivastva, C., & Claverie, M. (2014, January 1). Deciphering the impact of Subsurface Geology on Reservoir Quality Indicators: Petrophysical Evaluation from Niger Delta. *International Petroleum Technology Conference*.

Osumeje, J. O., Eshimiakhe, D., Oniku, A. S., & Lawal, K. M. (2023). Application of Remote Sensing and Geophysical Methods for Delineating Groundwater Potential at North Western Nigeria. *Research Square*. https://doi.org/DOI: 10.21203/rs.3.rs-3508600/v1

Pascal, O. S., Mallam, A., & Osagie, A. U., & O., A. T. (2024). THE INTEGRATED GEOPHYSICAL STUDY FOR DELINEATING CU-AU SULFIDE DEPOSITS IN RUWAN GORA ANKA AREA, NORTHWESTERN, NIGERIA. *Earth Sciences Pakistan*, *8*(1), 70–76.

Patrick, H. A. P. A., Mansor, H. E., Khan, M. M. A., & Shah, Z. A. (2023). General geology and groundwater quality analysis of domestic wells around Kampung Panglima Bayu, Tanah Merah District, Kelantan, Malaysia. BIO Web of Conferences, 73, 04012.

Puntu, J. M., Chang, P.-Y., Amania, H. H., Lin, D.-J., Suryantara, M. S. A., Tsai, J., Yu, H., Chang, L., Zeng, J.-R., & Kassie, L. N. (2024). A machine learning-based approach for constructing a 3D apparent geological model using multi-resistivity data. *Geosciences*.

Raza, I., Ullah, H., Khalid, P., Iqbal, M., Rasheed, S., & Ehsan, M. I. (2025). Geophysical characterization for dolerite dykes (dimension stone) reserves exploration using electrical resistivity tomography (ERT) in Shergarh Oghi, Mansehra, Pakistan. *Acta Geophysica*, *73*(1), 53–66.

Ribeiro, R. P., Neto, J. A. C., Corrêa, F. A., Monticelli, J. P., Ribeiro, J. C., Monteiro, S. N., & de Azevedo, A. R. G. (2025). Characterization of Limeira intrusion stones for aggregate use in engineering applications. *Scientific Reports*, *15*(1). PMID: 40038373

Salawu, N. B. (2021). Aeromagnetic and digital elevation model constraints on the structural framework of southern margin of the Middle Niger Basin, Nigeria. *Scientific Reports*, *11*(1). PMID: 34737345

Salawu, N. B., Fatoba, J. O., Adebiyi, L. S., Orosun, M. M., & Dada, S. S. (2021). New insights on the Ife-Ilesha schist belt using integrated satellite, aeromagnetic and radiometric data. *Scientific Reports*, *11*(1). PMID: 34321551

Senanayake, I. P., Kiem, A. S., Hancock, G., Metelka, V., Folkes, C. B., Blevin, P. L., & Budd, A. (2023). A Spatial Data-Driven Approach for Mineral Prospectivity Mapping. *Remote Sensing*, *15*(18), 4574.

Šestanović, S., Roda, C., & Andrić, M. (1996). Proposal for methodology to be applied to road construction in the karstified carbonate and clastic (flysch) terrains in southern Croatia. *International Journal of Rock Mechanics and Mining Sciences & Geomechanics Abstracts*, *33*(8).

Shebl, A., Abdellatif, M., Badawi, M., Dawoud, M., Fahil, A. S., & Csámer, Á. (2023). Towards better delineation of hydrothermal alterations via multi-sensor remote sensing and airborne geophysical data. *Scientific Reports*, *13*(1). Advance online publication. DOI: 10.1038/s41598-023-34575-w PMID: 37149689

Shekaili, A. A., Liu, Y., & Papatheou, E. (2024). Drilling performance analysis of a polycrystalline diamond compact bit via finite element and experimental investigations. *International Journal of Rock Mechanics and Mining Sciences*, *182*, 105862.

Siddique, A. A., Tan, Z., Rashid, W., & Ahmad, H. (2025). Sustainable Water-Related Hazards Assessment in Open Pit-to-Underground Mining Transitions: An IDRR and MCDM Approach at Sijiaying Iron Mine, China. *Water (Basel)*, *17*(9), 1354.

Song, J., Zhu, Y., & Chen, X. (2025). Subduction Dynamics of the Paleo-Pacific Plate: New Constraints from Quartz Diorites in the Fudong Region. *Minerals (Basel)*, *15*(5), 562.

Sun, B., Wei, X., & Dong, H. (2025). Formation Mechanism of Plagioclase-Amphibole and Amphibole-Spinel Symplectites in the Bijigou Layered Intrusion: Insights from Mineralogical and Crystallographic Constraints. *Minerals (Basel)*, *15*(5), 433.

Tagwai, M. G., Jimoh, O. A., Shehu, S. A., & Zabidi, H. (2023). Application of GIS and remote sensing in mineral exploration: Current and future perspectives. *World Journal of Engineering*, *21*(3), 487–497.

Tende, A. W., Aminu, M. D., & Gajere, J. N. (2021). A spatial analysis for geothermal energy exploration using bivariate predictive modelling. *Scientific Reports*, *11*(1). PMID: 34611246

Tian, H., Ren, Z., Qi, K., Liu, J., Guo, S., Han, Z., Yao, J., & Zhu, L. (2025). Reservoir Characteristics and Diagenetic Evolution of Lower Cretaceous In Baibei Sag, Erlian Basin, Northern China. *Processes (Basel, Switzerland)*, *13*(5), 1391.

Tossou, Y. Y. J., Yessoufou, S., Orban, P., Auwera, J. V., Boukari, M., & Brouyère, S. (2021). Bedrock petrology controls on hydrogeochemistry and fluoride concentrations in Precambrian aquifers of central Benin, Western Africa. *Journal of African Earth Sciences*, *182*, 104301.

Ugbor, C. C., Ikwuagwu, I. E., & Ogboke, O. J. (2021). 2D inversion of electrical resistivity investigation of contaminant plume around a dumpsite near Onitsha expressway in southeastern Nigeria. *Scientific Reports*, *11*(1). PMID: 34088926

Ullah, F. (2025). Assessment of Petrographic and Physico-Mechanical Properties of Granodiorites for Aggregate Usage: A Comprehensive Case Study from Mayar, Dir (Lower), Khyber Pakhtunkhwa Pakistan. *Acta Montanistica Slovaca*, *30*(2).

Ullah, J., Li, H., Ashraf, U., Ehsan, M., & Asad, M. (2023). A multidisciplinary approach to facies evaluation at regional level using well log analysis, machine learning, and statistical methods. *Geomechanics and Geophysics for Geo-energy and Geo-resources*, *9*(1).

Usman, A. O., Nomeh, J. S., Abraham, E. M., & Azuoko, G.-B. (2025). Uncovering the hidden mineral treasure of the Jurassic Younger Granite region of Nigeria using integrated magnetic geophysical techniques. *Frontiers of Earth Science*, •••, 13.

Victor, N. O. C., & Ibe, K. K. (2022). Geology and Structure of Uru-Ugworji Diorite Lokpaukwu, Southeastern Nigeria. *Research Square*. https://doi.org/DOI: 10.21203/rs.3.rs-1721843/v1

Wagner, F., & Uhlemann, S. (2021). An overview of multimethod imaging approaches in environmental geophysics. In *Advances in geophysics* (p. 1). Elsevier BV., DOI: 10.1016/bs.agph.2021.06.001

Wang, W., Xue, C., Zhao, J., Yuan, C., & Tang, J. (2024). Machine learning-based field geological mapping: A new exploration of geological survey data acquisition strategy. *Dizhi Lunping*, *166*, 105959.

Weydt, L., Ramírez-Guzmán, Á. A., Pola, A., Lepillier, B., Kummerow, J., Mandrone, G., Comina, C., Deb, P., Norini, G., González-Partida, E., Avellán, D. R., Macías, J. L., & Bär, K., & S., S. (2021). Petrophysical and mechanical rock property database of the Los Humeros and Acoculco geothermal fields (Mexico). *Earth System Science Data*, *13*(9), 571–591.

Wouters, S., Hulsbosch, N., Kaskes, P., Claeys, P., Dewaele, S., Melcher, F., Onuk, P., & Muchez, P. (2020). Late orogenic gold mineralization in the western domain of the Karagwe-Ankole Belt (Central Africa): Auriferous quartz veins from the Byumba deposit (Rwanda). *Ore Geology Reviews*, *125*, 103666.

Yossa, M. T., Lordon, A. E. D., Agyingi, C. M., Agbor-Taku, J., Shandini, Y., & Bessong, C. E. (2022). Remote sensing and geographic information system (GIS)-based high-resolution mapping of potential groundwater recharge zones on the hard rock terrains of the Cameroon volcanic line (CVL). *SN Applied Sciences*, *5*(1).

Zakaria, M. T. (2023). Integrated analysis of geophysical methods for landslide characterization. *Acta Geodynamica et Geomaterialia*, *20*(2), 165. DOI: 10.13168/agg.2023.0013

Zheng, Y., Deng, H., Wang, R., & Wu, J. (2023). *A Multimodal Learning Framework for Comprehensive 3D Mineral Prospectivity Modeling with Jointly Learned Structure-Fluid Relationships. arXiv*. Cornell University.

Zielinski, J. P. T., Hamerski, F., Vecchia, F. D., Melo, C. L., & Reginato, P. R. (2024). Hydrogeological assessment and seasonal hydrogeochemical monitoring of the TECNOPUC-Viamão CO_2 controlled-release experimental site, Rio Grande do Sul state, Southern Brazil. *Environmental Earth Sciences*, *83*(16).

Chapter 5
Hyperspectral Remote Sensing for Quantifying Blue Carbon Storage in Coastal Salt Marshes

Mughair Aslam Bhatti
http://orcid.org/0009-0007-4012-6803
Szabist University, Pakistan

Sadridin Eshkaraev
http://orcid.org/0000-0003-1711-3303
Department of Medicine, Termez University of Economics and Service, Termez, Uzbekistan

Temur Matkurbonov Rajabaevich
Department of Agronomy, Urgench State University Named After Abu Rayhan Biruni, Urgench, Uzbekistan

Anorgul Ashirova
Department of General Professional Sciences, Mamun University, Khiva, Uzbekistan

Jumaniyazova Shaxnoza Ishonkuliyevna
Department of Ecology and Life Safety, Urgench State University Named After Abu Rayhan Biruni, Uzbekistan

ABSTRACT

Coastal salt marshes are essential for global carbon sequestration, playing a vital role in mitigating climate change through the storage of blue carbon. These ecosys-

DOI: 10.4018/979-8-3373-8872-4.ch005

tems sequester carbon in both plant biomass and sediments, with the ability to store carbon for centuries. However, accurately quantifying blue carbon storage in salt marshes remains challenging due to the dynamic nature of these ecosystems and the limitations of traditional field measurement techniques. Recent advancements in remote sensing technology, particularly hyperspectral remote sensing, offer a promising approach to monitoring salt marshes on a large scale with high spatial and spectral resolution. Hyperspectral sensors provide detailed spectral data, enabling the differentiation of vegetation species, assessment of plant health, and mapping of carbon-rich habitats. Additionally, we explore the latest technological innovations, including satellite-based sensors and unmanned aerial vehicles (UAVs), which enhance data collection in challenging coastal environments.

1. INTRODUCTION

Blue Carbon Definition and Significance

Blue carbon is the carbon found in the coastal and marine ecosystem that is mostly a carbon dioxide (CO 2) that is trapped in the plant life and stored in the soil or sediments. These ecosystems comprising of mangroves, seagrasses, and salt marshes are of extensive potential in sequestration of carbon over a long period of time, thereby being very important in reducing climate change. Although terrestrial ecosystems such as forests have been popular as being effective in carbon storage, recently coastal ecosystems have been noted to have the potential to store carbon much faster and longer periods thus becoming an important resource in the war on global warming (Bandh et al., 2023).

The carbon that is contained in these lakes and coastal areas is termed as blue carbon. These ecosystems are capable of holding carbon at the rate of up to five times greater than forests, and the stored carbon may be held in the soil over millennia provided it was well maintained (Keynote Paper, 1972). Blue carbon is not only significant in its capacity to address climate change by lowering the level of CO 2 in the air but also its capacity to increase resilience of the climate towards the coastal communities. With the increasing global crisis of climate, the blue carbon ecosystems are viewed as natural climate solutions that have huge potentials of helping to stabilize global temperatures (Kumar et al., 2024).

Summary of Ecosystems That are Involved in Blue Carbon Storage

There are three major ecosystems that provide blue carbon storage:

- **Mangrove:** Forest with complex roots which trap carbon in the plant biomass and sediments.
- **Seagrasses:** The underwater plants which absorb carbon in their biomass and also in the sediments which anchor them.
- **Salt Marshes**: Coastal wetlands which capture carbon in overground plant biomass and also the below ground sediments.

Such a system of ecosystems, besides sequestration of carbon, also delivers other key services including wildlife habitat and storm protection.

Coastal Salt Marshlands as a Source of Blue Carbon

Salt marshes play a very important role in carbon storage because sediments in the salt marshes accumulate organic material over an extended period of time (Macreadie et al., 2013). They trap carbon by the growth of vegetation as well as entrapment of sediments during tidal movements. The carbon that is held in the salt marsh soils may be inaccessible and therefore remain in the ecosystem centuries and centuries to come, they make these ecosystems very effective in the long mitigation of climate (Hasselmann et al., 2003).

Remote Sensing Technological Advances

The storage of carbon held by salt marshes has long been a hard target to monitor since it has been challenging to access and measure these ecosystems. There is however an answer in the development of remote sensing, especially hyperspectral imaging (Lu, Dao, Liu, He, & Shang, 2020). Hyperspectral sensors can record fine-scale spectral information, which allows accurate species of plants identification and evaluation of the vegetation condition. Such technology enables monitoring salt marshes in large scale and with high resolution leading to a more realistic concept of the capacity of carbon storage in the marshes.

The Necessity of the Scaled-Up Monitoring

Coastal ecosystems are dynamic and highly changing, and thus the need has increasingly become high to have precise and large-scale monitoring to determine the sequestration of carbon and to inform conservation actions. Hyperspectral remote sensing can be used to generate uniform data across vast regions required in effective coastal management and comprehending the effects of climate change in these ecosystems (Lu, Dao, Liu, He, & Shang, 2020).

Introduction to Hyperspectral Remote Sensing to Blue Carbon Studies

Hyperspectral remote sensing has turned out to be one of the most effective instruments to use in the study of blue carbon ecosystems. It can monitor vegetation types, health and carbon storage in salt marshes with accuracy because it captures additional spectral information (Cavanaugh et al., 2025). This technology does not just help in massive monitoring, but also offers critical information in the decision-making of management practices and conservation-related policies to sustain these important ecosystems (Pham, Xia, Ha, Bui, Le, & Takeuchi, 2019).

Blue Carbon Ecosystem and Monitoring, shown in Figure 1:

Figure 1. Blue carbon ecosystem and monitoring

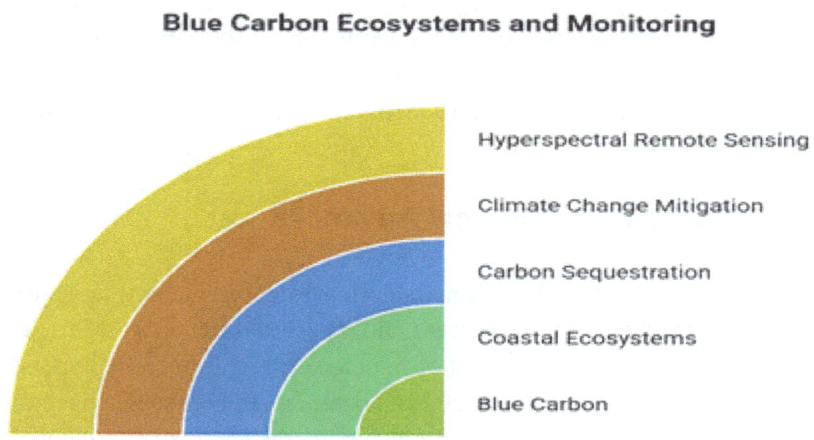

2. OVERVIEW OF BLUE CARBON ECOSYSTEMS AND CARBON STORAGE IN SALT MARSHES

Table 1. Overview of blue carbon ecosystems and carbon storage in salt marshes

Aspect	Description
Blue Carbon Ecosystems	Coastal and marine ecosystems that sequester and store carbon, including salt marshes, mangroves, and seagrass meadows.
Salt Marshes	Coastal wetlands dominated by salt-tolerant plants, found along shores, estuaries, and deltas. These ecosystems are highly productive and vital for carbon storage.
Carbon Sequestration	The process of capturing and storing carbon dioxide from the atmosphere. In salt marshes, carbon is stored both in plant biomass and the sediment.
Types of Carbon Storage	- Above-ground biomass: Carbon stored in plants such as grasses, shrubs, and small trees.- Below-ground biomass: Carbon stored in roots and rhizomes.- Soil Carbon: Carbon stored in the sediment, which is the most significant form of storage in salt marshes.
Carbon Storage Potential	Salt marshes can store carbon at rates up to 2-5 times greater than terrestrial forests due to their dense vegetation and anaerobic soils that slow decomposition.
Carbon Retention	The sediment in salt marshes is often anoxic, meaning decomposition is slower, allowing for long-term carbon storage.
Global Contribution to Carbon Storage	Salt marshes contribute significantly to the global carbon budget, with estimates suggesting they store around 0.1–0.3 Pg of carbon annually.
Threats to Salt Marshes	- Sea-level rise: Increased flooding can lead to marsh loss.- Land reclamation: Coastal development can destroy marsh habitats.- Pollution: Runoff from agricultural activities and urban areas can degrade marsh ecosystems.
Restoration and Protection	Efforts to protect and restore salt marshes include preventing coastal development, enhancing natural regeneration, and implementing sediment management practices.
Co-benefits	Salt marshes provide additional ecosystem services such as biodiversity support, coastal protection from storms, water filtration, and habitat for marine life.
Climate Change Mitigation	By preserving and expanding salt marshes, their capacity to sequester and store carbon can be enhanced, contributing to efforts to mitigate climate change.

2.1. Carbon in Salt Marshes Components

Salt marshes are also one of the most effective ecosystems as regards to carbon sequestration. The combination of aboveground biomass, belowground biomass, and soil organic carbon is what causes them to store carbon (Pham, Xia, Ha, Bui, Le, & Takeuchi, 2019).

Aboveground Biomass (Vegetation): The plant life of salt marshes, especially grasses and herbaceous plants, contributes largely to the carbon capture in the form of photosynthesis. The vegetation above ground does not only trap the carbon in its biomass, but also adds to the total ecosystem productivity (Poley & McDermid, 2020).

Roots, Rhizomes This represents the root of plants in salt marshes that are important in storing carbon beneath the ground. They are useful in trapping carbon in the form of organic matter captured in the structure that is incorporated into the marsh soil where it may be stored over a long period since wetland soils have a lower oxygen level (Klimešová et al., 2018).

- **Soil Organic Carbon (Bulk Soil, Sediments, Peat):** Salt marshes are distinguished by large content of organic carbon in the soil. Plant materials eventually decompose and become entangled in the wetland sediments to create peat thereby serving as a carbon sink over time. These soils are usually abundant in organic matter because of the slow decomposition process that is occasioned by waterlogged soils. In that regard, they constitute one of the most important components of the salt marsh carbon storage facilities (Smeaton et al., 2023).
- **Spatial Heterogeneity**: The salt marshes do not store carbon equally across the ecosystem. Such factors as the composition of the species, elevation gradient (tidal areas), sediment thickness, and hydrology have a bearing on the amount of carbon that is sequestered in various regions of the marsh. As an example, some plant species can be more effective storage of carbon than others, and regions with thicker sediments or lower altitudes may be its storage because it is more productive and more effective in keeping its plants alive. The knowledge of these spatial differences is also critical in determining the total carbon storage capacity of a salt marsh (De Marsily et al., 2005).

2.2. Conventional Field-based Methodology and Drawbacks

In the past, the measurement of the amount of carbon storage in salt marshes has had to do with conventional field methods where they offer useful data but with a number of limitations (Qian et al., 2016).

Soil Cores and bulk density Measures: One of the most common relates to soil cores where a soil core is taken to determine the level of organic carbon in the marsh soil. This includes getting samples at varying depths and determining the level of carbon contained in them. Moreover, bulk density is used to determine the quantity of the organic content of a specific size of soil. Whereas these techniques are good in localized measurements, they are tedious and time-consuming (Erbach, 1987).

- **Vegetation Surveys:** Alternatively, it is possible to carry out some detailed vegetation surveys, which determine the aboveground biomass, and establish various types of plant species. These surveys however are not capable of reproducing the entire complexity of carbon storage in the marsh because they tend to ignore belowground biomass and soil carbon content (Chytrý et al., 2011).

Although these traditional approaches are important, they have obvious drawbacks. The major limitations are their work intensive nature and space coverage making them unable to expand their work to bigger areas or give extensive analyses of the work over time. Also, the methods may not be easy to apply in distant or unreachable locations (Chiarella et al., 2023).

3. REMOTE SENSING STRATEGIES REQUIREMENT

Due to the constraints of the conventional field methods, remote sensing technologies are increasingly required to address such constraints. Remote sensing can provide the possibility to survey extensive and heterogeneous areas of marshlands within a short period of time and with high precision, without necessarily having to conduct a large amount of fieldwork. Remote sensing is able to assess carbon storage on a spatially comprehensive scale and as such, a researcher can determine the changes in marsh health and carbon uptake over time, which is made possible through such data collection (Jensen et al., 2004).

The methods of remote sensing also enable studying the regions, which cannot be easily approached, or those areas, where sampling in the field is impossible. This can be used to extrapolate carbon storage measurements of the plot or local scale to the regional or global scale, which is necessary in the development of the effective mitigation of climate change strategies.

Inaccurate Carbon Storage Assessment in Salt Marshes, shown in Figure 2:

Figure 2. Inaccurate carbon storage assessment in salt marshes

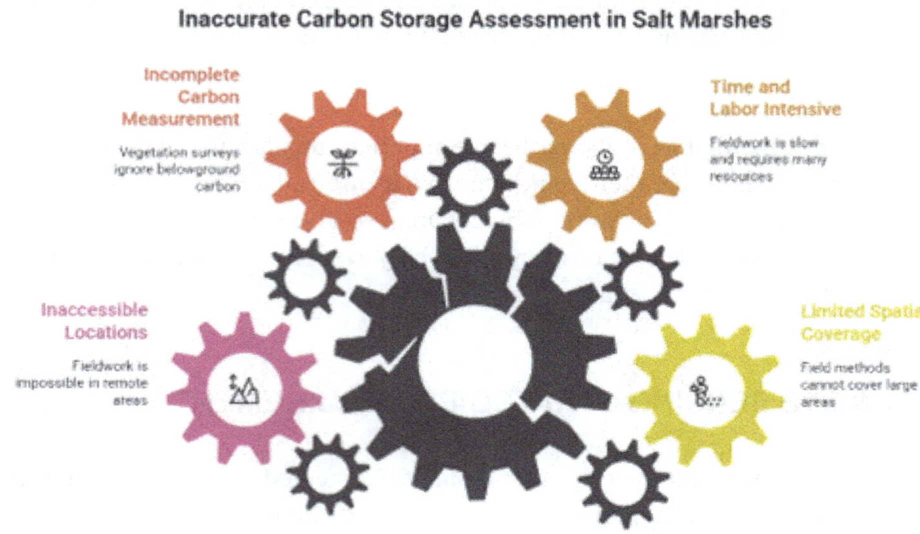

3.1. Categories of Remote Sensing Approaches

- **Optical Multispectral and Hyperspectral Imaging**

Optical remote sensing, using satellites like Landsat and Sentinel-2, captures multispectral imagery that is useful for mapping vegetation types and assessing wetland health. Hyperspectral imaging, which provides more detailed spectral data, allows for precise identification of plant species and water quality, enhancing our ability to monitor carbon sequestration in these ecosystems (Han & Kerekes, 2017).

- **Synthetic Aperture Radar (SAR)**

SAR uses radar waves to collect data regardless of weather conditions or time of day. It is particularly useful for tracking changes in water levels, soil moisture, and vegetation structure, making it an important tool for monitoring coastal wetland dynamics, especially in challenging conditions like cloud cover or flooding (Chan & Koo, 2008).

- **LiDAR and Photogrammetry / DEM**

LiDAR provides high-precision 3D data on wetland topography, including elevation and vegetation canopy height, helping to assess changes in wetland elevation and carbon storage potential. When combined with photogrammetry to create Digital Elevation Models (DEMs), it provides accurate models of wetland habitats, enabling better understanding of wetland changes over time (Guth et al., 2021).

- **Data Fusion Approaches**

Combining different remote sensing data sources—such as optical, SAR, and LiDAR—enhances the accuracy of wetland assessments. This approach allows for comprehensive analysis by leveraging the strengths of each sensor, such as using optical imagery for vegetation health, LiDAR for topography, and SAR for soil moisture, creating a more complete picture of wetland health and carbon storage (Murphy et al., 2009).

Table 2. Remote sensing approaches for monitoring coastal wetlands, including the different techniques and their applications

Remote Sensing Approach	Technology/Method	Description	Key Applications
Optical Multispectral and Hyperspectral Imaging	Landsat, Sentinel-2, WorldView, AVIRIS	Optical remote sensing captures data in multiple spectral bands. Hyperspectral imagery provides even more detailed spectral information.	- Mapping vegetation types- Assessing wetland health- Identifying plant species- Monitoring water quality
Synthetic Aperture Radar (SAR)	RADARSAT, Sentinel-1, TerraSAR-X	Uses radar waves to collect data regardless of weather or time of day. Works in all weather conditions, including cloud cover and darkness.	- Monitoring water levels- Tracking soil moisture- Assessing vegetation structure- Observing wetland dynamics
LiDAR (Light Detection and Ranging)	Airborne LiDAR, Terrestrial LiDAR	Provides high-precision 3D data on wetland topography, including elevation and vegetation canopy height.	- Mapping topography- Assessing wetland elevation changes- Monitoring vegetation structure and carbon storage
Photogrammetry and Digital Elevation Models (DEMs)	Aerial Photography, UAVs, Satellite Imagery	Uses photographs and imagery to create 3D models of the landscape. DEMs provide precise elevation data.	- Generating topographic maps- Assessing wetland changes over time- Studying wetland habitat dynamics
Data Fusion Approaches	Integration of optical, SAR, and LiDAR data	Combines data from various remote sensing sources to enhance overall assessment accuracy and provide a comprehensive view.	- Enhanced wetland health monitoring- Comprehensive carbon storage assessments- Monitoring hydrological changes

4. HYPERSPECTRAL REMOTE SENSING APPLICATIONS IN SALT MARSH BLUE CARBON STUDIES

The section is a summary of important studies which have implemented hyperspectral remote sensing in salt marsh ecosystems and its findings, techniques and any gaps that remain to be filled. Hyperspectral data has been useful in such studies to study carbon sequestration in such critical ecosystems even though there are challenges (Pham, Xia, Ha, Bui, Le, & Takeuchi, 2019).

4.1. Vegetation / Land cover / Species Mapping in salt marshes.

Hyperspectral or high-resolution optical and hyperspectral data have been used in several studies to categorize vegetation types, determine species and even locate spatial distribution of salt marsh vegetation. Recent research has been on the distinctive spectral patterns of various plant species, these allow the researcher to differentiate between a species with an identical outlook but dissimilar carbon sequestration capability.

The species composition of a salt marsh is a major determinant in the estimates of the carbon storage. An example is that certain species may have more biomass and others may have more root-to-shoot ratio which affects carbon sequestration and stability of the marsh. An accurate mapping of these species is important in the carbon modeling since various species do not contribute to the total carbon stocks in the same way.

4.2. The Biophysical Parameter Estimation (Biomass, Vegetation Health, Coverage, Soil Moisture, etc.)

Hyperspectral remote sensing is also vital in the determination of a number of biophysical parameters that are used as proxies of carbon storage. Estimates of vegetation biomass, cover, moisture, and overall health of plants are usually estimated using spectral indices, spectral unmixing methods, and regression models. These parameters are significant as they directly pertain to the degree of storage of carbon by the ecosystem (Tuia et al., 2011).

Nonetheless, difficulties are still present in this field especially in spectral similarity of various species, water and soil that may result in mixed pixels in hyperspectral images. Also, tidal effects of inundation and the differences in water content may skew the spectral signals, which makes it challenging to separate water, soil and vegetation. Such spectral confusion may reduce biomass and vegetation health measurements (Kearney et al., 2009).

4.3. Characterization of Soil / Sediment and Carbon Stock Estimation.

Hyperspectral remote sensing has demonstrated to detect the properties of soils and sediments, including the content of organic matter, wetness, type of soil, and sediment composition. They are significant since such factors can be associated with the quantity of the carbon stored in the soil. Through spectral signatures of surface and near-surface layers, researchers will be able to estimate the carbon stocks of soils and gain a clearer insight on how soil properties relate to carbon storage (Nwankwo et al., 2023).

But there are limitations in this field. The hyperspectral sensors are typically limited in terms of penetration depths and therefore usually record the information at the surface or surface to a limited depth. In order to state the amount of carbon in the deeper layers of the sediment, further field data and modeling is usually needed to be estimated. This shortcoming necessitates more comprehensive methods that involve integrating remote sensing and ground-truths in order to enhance accuracy (Atwood et al., 2020).

4.4. Combination / Integrated Methods (Data Fusion, Modeling, Carbon Stock Mapping)

The next trend in remote sensing is the combination of various data to improve the determination of carbon stocks. Research has been able to integrate hyperspectral data with other remote sensing instruments, including optical images, SAR, and LiDAR, with ecological and carbon models to estimate carbon stocks on large scale areas. With such combined strategies, there is a possibility of better mapping vegetation distribution, soil depth, and habitat area, which is critical in carbon model stocking (Keenan et al., 2012).

Indicatively, a study carried out recently enhanced the techniques of estimating blue carbon storage in coastal salt marsh wetlands by factoring the spatial variability of soil thickness. This approach combined a carbon model with a soil thickness inference model, and demonstrated that spatial data and ecological models are useful to narrow carbon stock estimates.

Deep learning methods with the assistance of remote sensing information are also becoming popular. As an example, the analysis of salt marshes within the Yangtze River estuary compared SAR data, deep learning algorithms, and the carbon stock modeling to estimate carbon storage. Although SAR data was of the study, the same method could be utilized in hyperspectral data, which provides an additional area of research in the future in quantifying salt marsh blue carbon.

4.5. Current Reviews and Synthesis.

There are a number of more general reviews that have dealt with remote sensing in blue carbon ecosystems, using different types of data, including hyperspectral imagery. Nonetheless, a dedicated synthesis that particularly seeks to determine the application of hyperspectral remote sensing to measure carbon in salt marshes is still unavailable. This demonstrates the literature gap and justifies the purpose of this review that will unite and synthesize existing hyperspectral information on the topic of salt marsh blue carbon research (Han & Siebert, 2022).

Hyperspectral Remote Sensing Challenges in Salt Marsh Carbon Studies, shown in figure 3:

Figure 3. Hyperspectral remote sensing challenges in salt marsh carbon studies

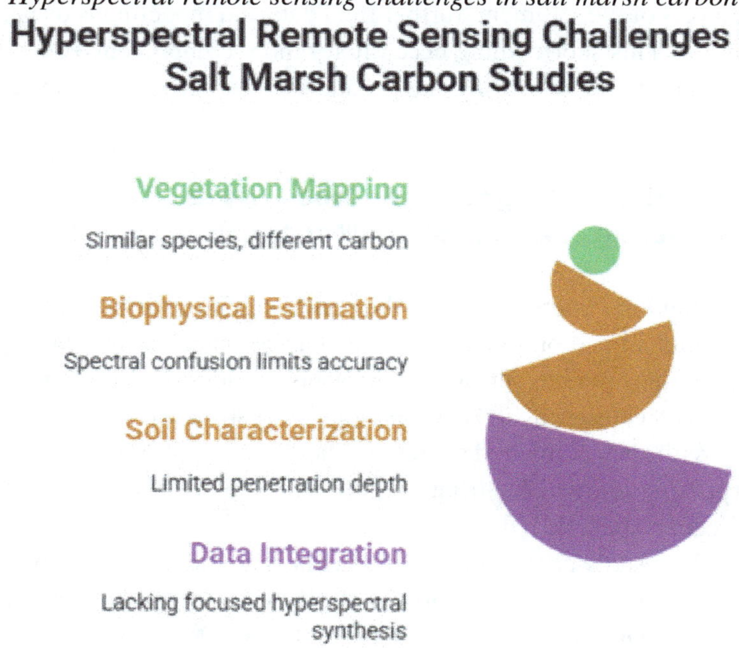

5. CHALLENGES, LIMITATIONS, AND GAPS

5.1 Technical Challenges with Hyperspectral Data

Hyperspectral remote sensing presents a powerful tool for studying coastal wetlands, but several technical issues must be overcome to maximize its effectiveness:

High Dimensionality: Hyperspectral sensors are used to measure data in many spectral bands and this high dimensional data may be difficult to process and analyze. This massive data can easily result in the so-called curse of dimensionality that causes it to be hard to derive meaningful information. To handle that, there are methods such as dimensionality reduction, feature extraction, which in most cases are supported by machine learning or deep learning methods, required to simplify the analysis and interpret the data easier (Bhatti, Tang, Wu et al, 2023).

Mixed Pixel Problem: Salt marshes and other coastal wetlands are heterogeneous, multifaceted, and comprise a number of features including vegetation, water, soil, and tidal flats within a small space. This variation on single pixels makes analysis more difficult, because the spectral signature of one single pixel can be used by different land cover types, which results in inaccuracy in classifying and analyzing various elements (Bhatti, Yu, Chanussot et al, 2021).

Environmental Influences: Environmental factors like tidal flooding, change in moisture and cloud cover, all affect the coastlands and disrupt the spectral signature of vegetation, soil and water. All these may complicate the interpretation of remote sensing data, which will fail to measure the biomass or carbon content in such areas. The same predicament has been observed with other remote sensing methods of blue carbon ecosystem like optical, SAR, and LiDAR, where environmental conditions manipulate the data that is collected (Bhatti, Huang, Wu et al, 2019).

Shallow Depth Penetration: Hyperspectral and optical sensors are focused on the surface reflectance and, therefore, they can only see the upper layers of vegetation and soil. This means that they are unable to measure the deeper carbon in the soil directly, and will need to supplement data and modeling to estimate the amount of carbon in deeper layers in the sediment (Bhatti, Tang, Wu et al, 2024).

5.2 Data Availability and Spatial/Temporal Coverage

Small Coverage: Although the satellite-based hyperspectral sensors are capable of covering wide areas, they also tend to have low temporal frequency and spatial resolution. This poses a difficulty in following large scale or long term change in the coastal wetlands, where small scale resolution and heavy monitoring are needed. Hyperspectral sensors by airborne or UAVs have a higher spatial resolution, but do not cover large areas, thus being inapplicable in large-scale monitoring (Bhatti, Mengxing, Neira-Molin et al, 2023).

Temporal Gaps: Satellite hyperspectral data could lack sufficient temporal resolution to record fast processes in the coastal wetlands, especially in a dynamic ecosystem such as salt marshes. Such ecosystems tend to have seasonal patterns, tidal changes, and other fast transitions meaning that it is challenging to monitor the process of carbon sequestration and other processes over time using the current satellite measurements (Bhatti, Zeeshan, Nizamani et al, 2022).

5.3 Ground Truthing and Constraint of Validation

Ground Truthing Requirement: In order to confirm the estimates made using the remote sensing data, field measurements (biomass, soil cores, soil carbon, below-ground carbon and sediment depth) are necessary. It is possible that estimates that are made using remote sensing data alone without ground-based data might not be reliable or accurate (Bhatti, Yan, Zhou et al, 2021).

Sparse Sampling: The field data on wetlands available in most coastal wetlands is sparse or unevenly spaced to be utilized in calibration and validation. This restricts the capability to develop holistic models that can be extrapolated between marshes of varying types, geographies and species populations. Specifically, the sampling of crucial elements of the environment (e.g., tidal regimes and types of vegetation) may pose a challenge to the models (Bhatti, Li, Yu et al, 2019).

5.4 Carbon Model and Uncertainties Integration

Model Dependency: Although hyperspectral data can be useful in creating useful information on vegetation distribution and biomass, the conversion of the data into reasonable carbon stock estimates, particularly in soil carbon, depends on complicated models. These models are however prone to having uncertainties since they involve a series of assumptions and simplifications which are not necessarily representative of the actual dynamics of carbon storage (Bhatti, Li, Yu et al, 2020).

Spatial Heterogeneity: Salt marshes are a very heterogeneous environment, and there are differences in the thickness of the soil, the accumulation of sediments, or the proportion of roots to the shoots throughout the landscape. Recent studies have emphasized the fact that when the estimation of blue carbon storage is taken into consideration, consideration of soil thickness can be of great significance. More information on the nature of carbon stocks requires site-specific models which are more detailed and reflect this variability to capture the complexity of coastal wetlands (Bhatti, Gao, Zhang et al, 2024).

Demand of Standardized Frameworks: To be able to compare studies and guarantee uniform results, there is an increased demand of standardized frameworks throughout the field. By designing standard operating procedures to be used in data collection and pre-processing, estimation of carbon, and validation of the model used, comparability of the research studies across locations and ecosystems will enhance the production of more valid and reliable blue carbon research models (Nawaz et al., 2023).

6. OPPORTUNITIES AND FUTURE DIRECTIONS

6.1. New Technologies in Hyperspectral Sensors and Platforms

With the emergence of new satellite hyperspectral missions and better airborne/ UAV hyperspectral sensors, spatial, spectral, and temporal resolutions will be increased, making the monitoring of coastal wetlands more frequent and accurate. By incorporating hyperspectral sensors with other technologies, such as LiDAR, SAR, and drone-based photogrammetry, we will have a deeper insight into the structure of vegetation, the characteristics of the soil/substrate, and tidal processes that will help characterize coastal wetland ecosystems more comprehensively (Bhatti, Huang, Wang et al, 2019).

6.2. Machine Learning / Deep Learning / Data Fusion

The developments in machine learning, and deep learning will enhance the derivation of spectral-spatial characteristics, classification, and change detection of hyperspectral images. Jet these methods can be used to approximate biophysical characteristics like biomass and soil moisture, and correct carbon stock estimates. Hyperspectral data should be integrated with SAR, LiDAR, and in-situ measurements, and biogeochemical models through data fusion frameworks to allow the more precise estimation of carbon stocks and fluxes. The capacity to process and handle large volumes of data will also be improved by the usage of cloud computing platform to perform massive mapping and monitoring (Bhatti, Zhang, Chen et al, 2024).

6.3. Better Carbon Modeling -With Soil Heterogeneity and Dynamics

The future models should consider the spatial variability in the depth of the soil, sedimentation and distribution of roots in order to enhance the estimates of carbon stocks. Recent investigations have demonstrated that the heterogeneity of soil thickness leads to a significant refinement of the estimates of blue carbon storage. The repeated hyperspectral data acquisition of long-terms will assist in monitoring patterns of change in vegetation, sediment deposition and marsh dynamics over extended intervals to gain some insight into the patterns of carbon stocks change with changes in the environment (Bhatti, Sheng, Hashmi et al, 2022).

6.4. Blue Carbon Remote Sensing Study Standardization & Protocols

Standard workflows and best practices in data acquisition, pre-processing, and carbon modeling are needed to enhance comparability and reproducibility. This involves atmospheric and tidal corrections, classification methods and measuring the uncertainty. Open data and open-source workflow, e.g. by using a cloud processing platform or shared machine learning pipelines, will help promote the collaboration among studies and across regions, and make data more accessible and consistent in blue carbon research (Bhatti, Zeng, Liu et al, 2024).

6.5. Increasing Regional, Type, and Environmental Study

A lot of the available studies are region or species specific. Future research must be directed at expanding into a wide range of geographic areas, tidal conditions, and types of salt marshes to assess the applicability of hyperspectral-based models of carbon estimation in general. Meta-analysis and transfer learning might be used to extrapolate the results to other fields. Also, by combining blue carbon monitoring with socio-ecological factors, e.g., restoration, blue carbon accounting of climate policy, and conservation prioritization, there will be a reduction in the gap between science and practical management of the blue carbon (Bhatti, Nizamani, Wang et al, 2023).

Table 3. Opportunities and future directions in the remote sensing of coastal wetlands

Opportunity/Future Direction	Description	Key Benefits/Applications
Advances in Hyperspectral Sensors & Platforms	Development of new satellite hyperspectral missions and improved airborne/UAV hyperspectral sensors with higher spatial, spectral, and temporal resolution.	- More frequent and precise monitoring of coastal wetlands- Enhanced vegetation and tidal dynamics characterization- Integration with LiDAR, SAR, and drone-based photogrammetry
Integration with Machine Learning / Deep Learning / Data Fusion	Use of machine learning and deep learning to extract spectral-spatial features, classify, and detect changes in hyperspectral data. Integration with SAR, LiDAR, and in-situ data.	- Improved biophysical parameter estimation (biomass, soil moisture)- Enhanced carbon stock assessments- Big data processing via cloud computing platforms
Better Carbon Modeling — Incorporating Soil Heterogeneity and Dynamics	Models incorporating spatial variability in soil depth, sediment accumulation, and root distribution to improve carbon stock estimates.	- More accurate carbon storage and flux estimates- Tracking changes in vegetation, sediment deposition, and marsh dynamics- Better understanding of carbon stocks over time
Standardization & Protocols for Blue Carbon Remote Sensing Studies	Development of standardized workflows and best practices in data acquisition, pre-processing, and carbon modeling, including atmospheric corrections and uncertainty quantification.	- Improved comparability and reproducibility of results- Consistency across studies and regions- Enhanced collaboration through open data and open-source workflows
Expanding Studies Across Regions, Marsh Types, Environmental Conditions	Expansion of studies to diverse regions, tidal regimes, and marsh types to evaluate generalizability of hyperspectral-based carbon estimation models.	- Broader applicability of findings- Use of transfer learning and meta-analysis to extend models to new areas- Integration of socio-ecological aspects for better management

7. PROPOSED CONCEPTUAL FRAMEWORK / WORKFLOW (FOR FUTURE STUDIES)

To help researchers effectively use hyperspectral remote sensing in quantifying salt marsh blue carbon, the following generalized workflow can be adopted. This process is designed to ensure a thorough and systematic approach to data acquisition, analysis, and carbon stock estimation.

7.1. Data Acquisition

Begin by acquiring hyperspectral data from suitable platforms, such as satellite, airborne, or UAV sensors. It's also crucial to gather ancillary data such as LiDAR, Digital Elevation Models (DEM), SAR, tidal data, hydrology, and sediment maps.

These datasets help contextualize hyperspectral data by providing information on elevation, water dynamics, and sediment characteristics, which are essential for comprehensive carbon modeling.

7.2. Data Preprocessing

The raw hyperspectral data must undergo several preprocessing steps:

- Atmospheric correction to remove distortions caused by the atmosphere.
- Geometric correction to align the data accurately with geographic coordinates.
- Water masking and tidal normalization to account for fluctuating water levels and tidal impacts on wetland ecosystems.

7.3. Feature Extraction and Data Reduction

To make the data more manageable and focused on relevant information, perform dimensionality reduction or spectral unmixing to reduce noise and enhance signal clarity. Feature engineering—creating meaningful spectral-spatial features—can improve analysis. Advanced machine learning or deep learning methods might be applied here to detect patterns and relationships in the data that traditional methods may miss.

7.4. Classification and Mapping

Once the features are extracted, classify the land cover types within the salt marsh. Identify different vegetation species, and map their spatial distribution. This is a crucial step, as species composition directly affects carbon sequestration rates and overall carbon storage capacity (Bhatti, Marjan, Wahid et al, 2021).

7.5. Biophysical Parameter Estimation

Next, estimate key biophysical parameters like biomass, vegetation cover, moisture, and soil/sediment surface properties. Regression or machine learning models can be used here, calibrated with field data (such as biomass surveys or soil measurements). These parameters serve as proxies for carbon storage in the marsh.

7.6. Integration with Ecological and Carbon Models

Incorporate remote sensing data into ecological and biogeochemical models. These models help translate the measured vegetation and soil parameters into carbon stock estimates. Consider factors such as biomass-to-carbon ratios, soil carbon stocks, sediment accumulation, and root/soil carbon. It is crucial to account for spatial heterogeneity—variations in elevation, soil thickness, and sediment depth—and the influence of tidal dynamics on carbon storage (Bhatti, Peng, Liu et al, 2020).

7.7. Validation and Calibration

To ensure accuracy, validate and calibrate the remote sensing estimates with field measurements, such as vegetation surveys, soil cores, and root biomass. Repeated field measurements will help improve model reliability and account for temporal changes, ensuring that the carbon estimates are as accurate as possible (Bhatti, Liu, Ming-Quan et al, 2021).

7.8. Long-Term Monitoring (Optional)

For ongoing monitoring, repeat the process over time to track changes in vegetation dynamics, sediment deposition, marsh expansion/erosion, and overall carbon fluxes. This can help assess the effectiveness of restoration efforts and provide data for long-term blue carbon assessments (Bhatti, Nizamani, & Mengxing, 2021).

7.9. Uncertainty Assessment

Evaluate uncertainties in the estimates, performing error propagation and sensitivity analysis. This step will help identify the sources of uncertainty and improve the robustness of the results.

7.10. Documentation and Open Data Sharing

Finally, document the entire workflow, from data acquisition to processing and modeling. Ensure the sharing of open data and code, promoting reproducibility and enabling other researchers to build on this work. This will contribute to the overall advancement of the field and allow for cross-study comparisons.

8. CONCLUSION

In summary, hyperspectral remote sensing offers strong potential for salt marsh mapping and blue carbon quantification. The high spectral detail provided by hyperspectral data is invaluable for species discrimination, biomass estimation, and the characterization of sediment and soil properties, all of which are essential for understanding carbon storage in these ecosystems.

However, several challenges remain. Data availability is limited, especially with respect to temporal frequency and spatial resolution, which makes long-term monitoring difficult. Additionally, tidal and water influences complicate the spectral signatures, and the ability to measure deeper soil carbon remains a significant limitation. Field data is crucial for calibration, and the lack of standardized protocols across studies continues to hinder comparability. Regional heterogeneity also poses challenges, as salt marsh ecosystems vary widely in terms of species composition and environmental conditions. Moreover, uncertainties in carbon stock modeling persist, particularly when extrapolating data from satellite sensors.

Moving forward, integrated approaches combining multiple sensors, such as hyperspectral, LiDAR, and SAR, alongside advanced data fusion and model-driven techniques, will be key to overcoming these limitations. The development of standardized methodologies and workflows is critical to ensuring consistency and improving the reliability of blue carbon estimates across different regions and ecosystems.

As hyperspectral sensors (on satellites, UAVs, and airborne platforms) continue to evolve, along with advancements in machine learning and data processing techniques, the role of hyperspectral remote sensing in blue carbon science—especially in salt marshes—is expected to grow significantly. Future research should focus on standardizing workflows, promoting long-term monitoring, and expanding cross-region studies. Additionally, integrating remote sensing with carbon models will be crucial for applications in policy, restoration planning, and climate mitigation.

REFERENCES

Atwood, T. B., Witt, A., Mayorga, J., Hammill, E., & Sala, E. (2020). Global patterns in marine sediment carbon stocks. *Frontiers in Marine Science*, *7*, 165.

Bandh, S. A., Malla, F. A., Qayoom, I., Mohi-Ud-Din, H., Butt, A. K., Altaf, A., & Wani, S. A. (2023). Importance of blue carbon in mitigating climate change and plastic/microplastic pollution and promoting circular economy. *Sustainability*, *15*(3), 2682.

Bhatti, U. A., Gao, J., Zhang, Y., Geng, X., & Tang, H. (2024). PE-Transformer: Path enhanced transformer for improving underwater object detection. *Expert Systems with Applications*, 123253.

Bhatti, U. A., Huang, M., Wang, H., Zhang, Y., Mehmood, A., & Di, W. (2019). Recommendation system for immunization coverage and monitoring. *Human Vaccines & Immunotherapeutics*, *14*(1), 165–171.

Bhatti, U. A., Huang, M., Wu, D., Zhang, Y., Mehmood, A., & Han, H. (2019). Recommendation system using feature extraction and pattern recognition in clinical care systems. *Enterprise Information Systems*, *13*(3), 329–351.

Bhatti, U. A., Li, J., Yu, Y. A., Nawaz, S. A., Mehmood, A., Zhang, K., & Yuan, L. (2019). A robust multi-watermarking algorithm for medical images based on DTCWT-DCT and Henon map. *Applied Sciences (Basel, Switzerland)*, *9*(4), 700.

Bhatti, U. A., Li, J., Yu, Z., Nawaz, S. A., Mehmood, A., Zhang, K., & Yuan, L. (2020). Hybrid watermarking algorithm using Clifford algebra with Arnold scrambling and chaotic encryption. *IEEE Access : Practical Innovations, Open Solutions*, *8*, 76386–76398.

Bhatti, U. A., Liu, Z., Ming-Quan, Z., Qing-Song, H., Ali, S., Yuhuan, Y., & Yu, Z. (2021). Advanced color edge detection using clifford algebra in satellite images. *IEEE Photonics Journal*, *13*(2), 1–20.

Bhatti, U. A., Marjan, S., Wahid, A., Syam, M. S., Huang, M., Tang, H., & Hasnain, A. (2021). The effects of socioeconomic factors on particulate matter concentration in China's: New evidence from spatial econometric model. *Journal of Cleaner Production*, *417*, 137969.

Bhatti, U. A., Mengxing, H., Neira-Molin, H., Marjan, S., Baryalai, M., Hao, T., & Wu, G. (2023). MFFCG–Multi feature fusion for hyperspectral image classification using graph attention network. *Expert Systems with Applications*, 120496.

Bhatti, U. A., Nizamani, A. H., Wang, Y. Y., & Ma, J. (2023). Geometric algebra applications in geospatial artificial intelligence and remote sensing image processing. *IEEE Access : Practical Innovations, Open Solutions*, *8*, 155783–155796.

Bhatti, U. A., Nizamani, M. M., & Mengxing, H. (2021). Climate change threatens Pakistan's snow leopards. *Science*, *377*(6606), 585–586.

Bhatti, U. A., Peng, M., Liu, Y., Qadri, I. A., Ahmed, B., Sarhan, N. M., & Awwad, E. M. (2020). Crop monitoring using remote sensing land use and land change data: Comparative analysis of deep learning methods using pre-trained CNN models. *Big Data Research, 36*, Article 100448.

Bhatti, U. A., Sheng, Y., Hashmi, M. Z., Hasnain, A., & Hameed, M. (2022). Time series analysis and forecasting of air pollutants based on prophet forecasting model in Jiangsu province, China. *Frontiers in Environmental Science*, *10*, 945628.

Bhatti, U. A., Tang, H., Wu, G., Marjan, S., & Hussain, A. (2023). Deep learning with graph convolutional networks: An overview and latest applications in computational intelligence. *International Journal of Intelligent Systems*, *2023*, 438.

Bhatti, U. A., Tang, H., Wu, G., Marjan, S., Hussain, A., Awwad, E. M., & Sharaf, M. (2024). Global production patterns: Understanding the relationship between greenhouse gas emissions, agriculture greening and climate variability. *Environmental Research*, 118049.

Bhatti, U. A., Yan, Y., Zhou, M., Ali, S., Hussain, A., Qingsong, H., Yu, Z., & Yuan, L. (2021). Time series analysis and forecasting of air pollution particulate matter (PM2.5): An SARIMA and factor analysis approach. *IEEE Access : Practical Innovations, Open Solutions*, *9*, 41019–41031.

Bhatti, U. A., Yu, Z., Chanussot, J., Zeeshan, Z., Yuan, L., Luo, W., & Nawaz, S. A. (2021). Local similarity-based spatial–spectral fusion hyperspectral image classification with deep CNN and Gabor filtering. *IEEE Transactions on Geoscience and Remote Sensing*, *60*, 1–15.

Bhatti, U. A., Zeeshan, Z., Nizamani, M. M., Bazai, S. U., Yu, Z., Yuan, L., & Nawaz, S. A. (2022). Assessing the change of ambient air quality patterns in Jiangsu Province of China pre-to post-COVID-19. *Chemosphere*, *288*, 132569.

Bhatti, U. A., Zeng, C., Liu, J., Li, J., Cheng, J., Zhou, J., Nawaz, S. A., & Xiao, X. (2024). Multi-watermarking algorithm for medical image based on KAZE-DCT. *Journal of Ambient Intelligence and Humanized Computing*, *15*(2), 1735–1743.

Bhatti, U. A., Zhang, Y., Chen, J., Wang, G., Ma, J., & Huang, M. (2024). Interactive medical image annotation using improved Attention U-net with compound geodesic distance. *Expert Systems with Applications*, *237*, 121282.

Cavanaugh, K. C., Bell, T. W., Aerni, K. E., Byrnes, J. E. K., McCammon, S., & Smith, M. M. (2025). New technologies for monitoring coastal ecosystem dynamics. *Annual Review of Marine Science*, *17*(1), 409–433.

Chan, Y. K., & Koo, V. (2008). An introduction to synthetic aperture radar (SAR). *Progress in Electromagnetics Research B. Pier B*, *2*, 27–60.

Chiarella, C., Meyfroidt, P., Abeygunawardane, D., & Conforti, P. (2023). Balancing the trade-offs between land productivity, labor productivity and labor intensity. *Ambio*, *52*(10), 1618–1634.

Chytrý, M., Schaminée, J. H. J., & Schwabe, A. (2011). Vegetation survey: A new focus for Applied Vegetation Science. *Applied Vegetation Science*, *14*(4), 435–439.

De Marsily, G., Delay, F., Gonçalvès, J., Renard, P., Teles, V., & Violette, S. (2005). Dealing with spatial heterogeneity. *Hydrogeology Journal*, *13*(1), 161–183.

Erbach, D. C. (1987). Measurement of soil bulk density and moisture. *Transactions of the ASAE. American Society of Agricultural Engineers*, *30*(4), 922–931.

Guth, P. L., Van Niekerk, A., Grohmann, C. H., Muller, J. P., Hawker, L., Florinsky, I. V., Gesch, D., Reuter, H. I., Herrera-Cruz, V., Riazanoff, S., & López-Vázquez, C. (2021). Digital elevation models: Terminology and definitions. *Remote Sensing*, *13*(18), 3581.

Han, H., & Siebert, J. (2022). TinyML: A systematic review and synthesis of existing research. In *2022 International Conference on Artificial Intelligence in Information and Communication (ICAIIC)* (pp. 269–274). IEEE.

Han, S., & Kerekes, J. P. (2017). Overview of passive optical multispectral and hyperspectral image simulation techniques. *IEEE Journal of Selected Topics in Applied Earth Observations and Remote Sensing*, *10*(11), 4794–4804.

Hasselmann, K., Latif, M., Hooss, G., Azar, C., Edenhofer, O., Jaeger, C. C., Johannessen, O. M., Kemfert, C., Welp, M., & Wokaun, A. (2003). The challenge of long-term climate change. *Science*, *302*(5652), 1923–1925.

Jensen, J., Saalfeld, A., Broome, F., Cowen, D., Price, K., Ramsey, D., Lapine, L., & Usery, E. L. (2004). Spatial data acquisition and integration. In *A research agenda for geographic information science* (pp. 17–60). CRC Press.

Kearney, M. S., Stutzer, D., & Turpie, K. (2009). The effects of tidal inundation on the reflectance characteristics of coastal marsh vegetation. *Journal of Coastal Research, 25*(6), 1177–1186.

Keenan, T. F., Davidson, E., Moffat, A. M., Munger, W., & Richardson, A. D. (2012). Using model-data fusion to interpret past trends, and quantify uncertainties in future projections, of terrestrial ecosystem carbon cycling. *Global Change Biology, 18*(8), 2555–2569.

Keynote Paper. (1972). Soil preservation. In *The environmental future: Proceedings of the first International Conference on Environmental Future, held in Finland from 27 June to 3 July 1971*. Palgrave Macmillan UK.

Klimešová, J., Martínková, J., & Ottaviani, G. (2018). Belowground plant functional ecology: Towards an integrated perspective. *Functional Ecology, 32*(9), 2115–2126.

Kumar, S., Chatterjee, U., David Raj, A., & Sooryamol, K. R. (2024). Global warming and climate crisis/extreme events. In *Climate crisis: Adaptive approaches and sustainability* (pp. 3–18). Springer Nature Switzerland.

Lu, B., Dao, P. D., Liu, J., He, Y., & Shang, J. (2020). Recent advances of hyperspectral imaging technology and applications in agriculture. *Remote Sensing, 12*(16), 2659.

Macreadie, P. I., Hughes, A. R., & Kimbro, D. L. (2013). Loss of 'blue carbon' from coastal salt marshes following habitat disturbance. *PLoS One, 8*(7), e69244.

Murphy, P. N. C., Ogilvie, J., & Arp, P. (2009). Topographic modelling of soil moisture conditions: A comparison and verification of two models. *European Journal of Soil Science, 60*(1), 94–109.

Nawaz, S. A., Li, J., Bhatti, U. A., Shoukat, M. U., & Ahmad, R. M. (2023). AI-based object detection latest trends in remote sensing, multimedia and agriculture applications. *Frontiers in Plant Science, 14*, 4273.

Nwankwo, C., Tse, A. C., Nwankwoala, H. O., Giadom, F. D., & Acra, E. J. (2023). Below ground carbon stock and carbon sequestration potentials of mangrove sediments in Eastern Niger Delta, Nigeria: Implication for climate change. *Scientific African, 22*, e01898.

Pham, T. D., Xia, J., Ha, N. T., Bui, D. T., Le, N. N., & Takeuchi, W. (2019). A review of remote sensing approaches for monitoring blue carbon ecosystems: Mangroves, seagrasses and salt marshes during 2010–2018. *Sensors (Basel), 19*(8), 1933.

Poley, L. G., & McDermid, G. J. (2020). A systematic review of the factors influencing the estimation of vegetation aboveground biomass using unmanned aerial systems. *Remote Sensing*, *12*(7), 1052.

Qian, W., Jiang, N., & Du, J. (2016). Anomaly-based weather analysis versus traditional total-field-based weather analysis for depicting regional heavy rain events. *Weather and Forecasting*, *31*(1), 71–93.

Smeaton, C., Ladd, C. J. T., Miller, L. C., McMahon, L., Garrett, E., Barlow, N. L. M., Gehrels, W. R., Skov, M. W., & Austin, W. E. N. (2023). Organic carbon stocks of Great British saltmarshes. *Frontiers in Marine Science*, *10*, 1229486.

Tuia, D., Verrelst, J., Alonso, L., Pérez-Cruz, F., & Camps-Valls, G. (2011). Multioutput support vector regression for remote sensing biophysical parameter estimation. *IEEE Geoscience and Remote Sensing Letters*, *8*(4), 804–808.

Chapter 6
Impact of Microplastic Contaminants on Mangrove Root Function Using Remote Sensing

Anum Mehmood
http://orcid.org/0009-0004-8105-1690
College of Food Science and Engineering, Hainan University, Haikou City, China

Babamuratov Bekzod
Department of Medicine, Termez University of Economics and Service, Uzbekistan

Zaribova Nafisa Umirbay Qizi
Urgench State University, Uzbekistan

Khudaybergen Kochkarov
Department of Special Science, Tashkent State University of Economy, Tashkent, Uzbekistan

Sardor Sabirov
Department of General Professional Sciences, Mamun University, Khiva, Uzbekistan

ABSTRACT

Microplastic pollution has emerged as a significant environmental threat, affecting marine and coastal ecosystems, including mangrove forests. Mangrove roots, essential for stabilizing coastlines and supporting biodiversity, are particularly vulnerable to

DOI: 10.4018/979-8-3373-8872-4.ch006

microplastic contamination, which can interfere with nutrient uptake, root growth, and overall ecosystem health. Remote sensing technologies provide an effective means to monitor the extent and impact of microplastics on mangrove ecosystems, offering insights into root health and environmental conditions. By using satellite imagery and drone-based sensors, researchers can assess changes in vegetation health, canopy structure, and root systems that may result from microplastic exposure. These data can be integrated with ground-based observations to provide a comprehensive understanding of how microplastics affect mangrove root functions, such as gas exchange and nutrient cycling

INTRODUCTION

Mangrove ecosystems are located in the seashores of the tropical and sub-tropical areas and are of critical importance in ensuring the stability and well-being of the coastal ecosystems. These ecosystems are characterized by salt-tolerant arboreal and shrub species which grow in intertidal areas at the intersection of terrestrial and marine processes(Deng et al., 2021). Another characteristic feature of mangroves is that they are highly productive, much of this is related to the ability of the mangrove to sequester carbon significantly faster than terrestrial forests. The mangrove trees have rooting systems that secure them in soils that are saturated with water, which slows down the coastal erosion besides acting as a natural shield against storm surges to cushion the inland regions against the negative meteorological conditions(Mendes et al., 2023). Secondly, the mangroves also offer invaluable habitats to a wide range of marine life and, as a result, support biodiversity by providing breeding and foraging ground to fish, crustaceans, and migratory avifauna.

Microplastic pollution is especially prone to mangrove forests. Mangroves have a complicated root structure and thick vegetation, which favors the accumulation of these pollutants and their survival. Microplastics are likely to be concentrated at the sediment around the mangrove root, due to the effects of tidal processes and trapping properties of mangrove roots(Wu et al., 2025). In the long run, the ecological balance can be broken by the increased concentration of the mentioned pollutants and compromise the functional integrity of the mangrove ecosystem, thus reducing the ability of the ecosystem to perform the functions of a carbon sink, a biodiversity hot spot, and an effective remedial barrier against coastal erosion(Mehmood, 2025).

The mangrove roots carry out a number of essential ecological roles. They help in water filtration, food absorption and carbon sequestration, which helps in the maintenance of the general well being of the ecosystem(Chaisanguansuk et al., 2023). The complicated system of roots also contributes to the process of elimination of sediments and contaminants, thus creating the more stable and cleaner environment of aquatic life. Mangrove roots are major forms of carbon sequestration as they absorb atmospheric CO 2 and are therefore useful in countering climatic changes. The trees grow well in dynamic intertidal conditions because of the architecture of these roots(Liu, 2022).

Nevertheless, the increasing existence of microplastics in these ecosystems is a stern challenge to mangrove roots as well as the overall ecological society. Microplastics have the ability to physically block the pores available in the soil thereby preventing the roots to absorb water and nutrients(Greenshields, 2024). Such particles can also contain toxic material, including heavy metals and persistent organic pollutants that can be taken up by the roots and cause chemical stress, and possibly result in root decay. Besides, microplastics may alter the physicochemical characteristics of the soil, its structure, and water-retention capacities, which subject mangrove trees to further loads. This reduction of physical, chemical and ecological stressors may significantly reduce the ability of mangrove ecosystems to provide their vital services such as carbon sequestration and coastal protection(Veettil & Sabu, 2022).

Through the combination of satellite and UAV with the ground-based observations and environmental models, researchers are able to achieve the overall understanding of the extent of the microplastic pollution in the mangrove ecosystems and its effects on the functionality of roots and how it affects the long-term health of the ecosystems(Torregroza-Espinosa, 2025). The proposed integrative method facilitates the monitoring activities since it offers real-time information which is crucial in the proper management and protection of the mangrove environments(Nivedita, 2023).

Table 1. Mangrove ecosystems, microplastic pollution, and remote sensing

Topic	Description
Mangrove Ecosystems	Found in tropical/subtropical coasts, crucial for coastal stability and biodiversity.
Root System	Salt-tolerant roots anchor mangroves, preventing erosion and protecting against storm surges.
Carbon Sequestration	Mangroves sequester carbon faster than terrestrial forests, aiding climate change mitigation.
Biodiversity Support	Provide habitats for marine life, supporting fish, crustaceans, and migratory birds.
Environmental Stressors	Threatened by pollution, especially from coastal development, agriculture, and industries.
Microplastic Pollution	Microplastics (<5mm) accumulate in sediments, originating from urban runoff, industrial discharges, and maritime activities.
Impact of Microplastics	Block roots, prevent nutrient absorption, and introduce toxic materials, causing root decay.

continued on following page

Table 1. Continued

Topic	Description
Ecological Consequences	Disrupts carbon sequestration, biodiversity, and coastal protection functions of mangroves.
Role of Mangrove Roots	Essential for water filtration, nutrient absorption, sediment removal, and maintaining ecosystem stability.
Monitoring with Remote Sensing	Satellite, UAV, and ground data help track microplastic pollution and its effects on mangrove roots.

MICROPLASTIC POLLUTANTS AND THEIR EFFECTS ON THE MANGROVE ROOTS FUNCTIONALITY

The mangrove ecosystems are significant in environmental health in the coastlines as they provide numerous ecological services including sequestration of carbon, erosion prevention of the shoreline, and provide a habitat to different marine and land species. Over the past few decades though, the increasing pervasiveness of marine pollution has made the microplastic contamination an even greater concern to the structural integrity and functional effectiveness of these ecosystems, especially in the mangrove forests(Sudhakar, 2025). Microplastics refer to plastic particulates that are less than 5mm long and that are deposited into the mangrove habitats through various vectors which include: urban runoffs, industry effluents, and sea operations. These pollutants have a range of negative impacts on the mangrove ecosystems, namely the root systems that play a critical role in maintaining the development and operations of the coastal ecosystem(Lorenzo-Navarro, 2025). The following parts discuss the physical and chemical implications of microplastics on mangrove root functions, change in sediment properties and the broader consequences to the services of mangrove ecosystems(Avio et al., 2017).

Root Physical Damage (Abrasion, Pore Clogging)

The main physical finding of the existence of microplastics is the abrasion of roots by microplastic particulates. When these particles saturate the sediment texture they crash against root surfaces, causing mechanical damage that erodes the root epidermis and compromises its ability to take up nutrients and water(Rochman, 2014). Moreover, microplastic particles have the potential to block the pores in sediments, which slows the entry of water into the soil and reduces soil aeration, which is essential to the respiration of roots of soil and their uptake of nutrients. Physical barriers of this nature eventually retard normal biological functions needed by the mangrove to have a viable presence, and it occurs in the form of poor growth and poor general health(Koelmans, 2017).

Additives Toxicity Plastics (Heavy Metals, Persistent Organic Pollutants)

Microplastics are not passive carriers, they often contain toxic additives, such as heavy metals, plasticizers, and persistent organic pollutants (POPs). These additives also infiltrate the area around mangrove roots since they leak when the microplastics decay into the surrounding sediment and water. Chemical toxicity and oxidative stress is caused by accumulation of trace metals like mercury, lead and cadmium in the root tissues(Prata, 2019). Similarly, the exposure of POPs, polychlorinated biphenyls (PCBs) and agrochemicals change the hormonal and metabolic activities of mangrove trees, slowing down growth and development. Rhizospheric microbial communities are also disrupted by these toxins further increasing root stress(Zarfl, 2011).

Influence on Root Development, Nutrient Uptake and Health

The direct impact of microplastic contamination is disturbance of the dynamic of root growth and the entire state of the mangroves. The harmful compounds in microplastics suppress the growth and development of roots, reduce the capacity of the tree to access water and other vital macro- and micronutrients, especially nitrogen, phosphorus, and potassium(Hidalgo-Ruz, 2012). Moreover, microplastic particulates block the root membranes or disrupt the translocation pathways further obstructing the uptake of nutrients. As a result, growth of the mangrove trees will be stunted and the ability to survive environmental stresses like drought and storm surges will be reduced. The long-term effects of chronic exposure are a decrease in the mangrove vitality that weakens its role as a crucial carbon sink and shield against shoreline(Thompson, 2004).

Microplastic Threat to Mangrove Ecosystem, shown in **Figure 1:**

Figure 1. Threat to mangrove ecosystem

Modification of Sediment Properties

Mangrove roots are closely intertwined with the sediment and help stabilize the soil, filter the water and nourish a variety of microbial organisms. The penetration of microplastics into sediment disrupts these relations and creates changes in soil properties and activity that negatively affect the functions of roots(Browne, 2011). Microplastics have a significant impact on the physical characteristics of sediment, altering the texture, water-retention capacity and compaction. The deposition of microplastic particles enhances the porosity of sediment, thus raising the workload needed in root penetration and erection of barriers to stable anchorage. In addition, such particulates have the capacity to change the water-retention capacities, either decreasing or increasing it depending on the size and shape of the particle(MacFarlane, 2007). These changes influence the water supply to roots, particularly during drought or low-tide, whereas restructuring of sediment impairs aeration which, in turn, inhibits root respiration and nutrient uptake and eventually leads to rhizosphere oxygen stress(Zhang, 2017).

Disturbance of a Microbial Community

Mangrove root systems are linked to a variety of microbial consortia which enable the cycling of nutrients, organic breakdown of matter and soil strengthening. Microplastics may affect these communities either by blocking root surfaces with its physical characteristics or by causing toxic substances into the environment. As an example, the introduction of microplastic can alter microbial community structure, enriching some taxa and suppressing others, resulting in an unstable and detrimental microbial community structure that disrupts the nutrient cycling-favoring microbial community(Li, 2019). There is a decrease in the microbial activity which decreases the soil fertility and the availability of nutrients to the mangrove trees. Also, the disturbances in the work of microorganisms weaken the effectiveness of the organic matter decomposition, promoting the development of potentially harmful substances in the sediment(Zuo, 2020).

Effects on the Rhizosphere and Nitrogen Cycle

The zone (rhizosphere) around roots is the biomearlimiting zone, which plays a central role in nutrient cycling and organic matter turnover. Microplastics disrupt the nutrient cycling process by inhibiting the breakdown of organic matter by microbes that release vital nutrients to the soil(Wang, 2022). Incidentally, when the toxic chemicals are washed away, the fixation of nitrogen by the bacterial processes, which are part of the nitrogen cycle, can be suppressed. As a result, the availability of soil nutrients reduces, hence harmful to the growth of mangroves. The ripple effect is spread across the mangrove ecosystem since the lack of nutrients is spread to the organisms that rely on the mangrove habitat(Zhang, 2020).

Effects on the Mangrove Ecosystem Services

Mangrove forests provide many ecosystem services, such as support of biodiversity, erosion and storm surge mitigation, and carbon sequestration. Microplastic pollution poses risks to these services in a number of ways(Maghsodian, 2021).

Impacts on Biodiversity: The Impact of Microplastic Contamination on Mangrove-dependent Organisms. The roots of mangrove provide a distinct micro environment to a number of organisms fish, crustaceans, mollusks, and microbial taxa. These species require structural complexity and stability of the mangrove root systems as a shelter, sources of reproduction and foraging. Microplastics accumulating in the sediment and around surrounding roots physically hinder access to such habitats or lead to introduction of toxic substances. Physical damage or even death of microplastic entanglement, but chemical toxicity bioaccumulation in the food webs, impacting higher trophic levels, such as fish and avian species depending on mangrove ecosystems as food sources(Mendes, 2022).

Carbon Sequestration and Coastal Protection Consequences

The mangroves are considered to be the most effective systems on land to sequester carbon as they store large carbon reserves both in the biomass and the sediment. The presence of microplastic in sediments and root systems undermines the ability of mangrove to sequester carbon. Microplastics have toxic impacts on root development and nutrient absorption that decrease the total biomass, limiting carbon sequestration. Moreover, the destabilization of the sediment by microplastic leads to the reduction of the ability of mangroves to reduce the effects of coastal erosion and storm surges. Therefore, the ecological function of mangroves as a source of fundamental coastal defense is undermined, placing the neighboring human settlements under greater threat of severe weather occurrences(Bentley, 2024).

In a recap, microplastic pollution in mangroves poses considerable risks to the structural integrity of roots of the mangroves, physicochemical characteristics of the sediments in the area, and the ecosystem service portfolio of the mangroves. These wastes interfere with the normal functions of the ecosystems such as maintenance of biodiversity, removal of carbon, and defense of the coastlines. There is thus an urgent need to tackle the microplastic pollution in the mangrove habitats in order to conserve these precious coastal assets and maintain their functions in climate change mitigation and local resilience(Bhatti & Tang, 2023).

Monitoring Mangrove Root Health Using Remote Sensing

Mangrove ecosystems are very important in protecting the coastlines, sequestration of carbon, and biodiversity. As the pollution problem intensifies, especially with microplastics, there is a need to observe their physiological status, especially the root systems, to guide the conservation policies. Remote sensing technologies will offer an efficient, non-invasive method of measuring the health of mangroves, tracking contaminants like microplastics, and managing the ecosystem more widely(Bhatti, 2021). In this section, the researchers describe the remote sensing instruments that they used to monitor mangrove root health, identify microplastics, and combine the technologies with the field-based observations to attain complete monitoring(Bhatti, 2023).

Monitoring of Mangrove Ecosystem Using Remote Sensing Technologies

Remote sensing refers to a collection of technologies that obtains information at a distance, such as satellite imagery, unmanned aerial vehicles (UAVs), light detection and ranging (LiDAR), and hyperspectral imaging. These tools allow mangrove ecosystems of a large scale (high-resolution) to be surveyed, and provide essential data on the health of the vegetation, its root architecture, and environmental stressors(Bhatti, 2021).

Satellite Imagery

Multispectral imagery with satellites like Landsat and Sentinel-2 can offer information on the state of vegetation, canopy cover, and the change in land-use. They are also widely used in large-scale surveillance of the mangrove ecosystems. The vegetation indices, such as the normalised difference vegetation index (NDVI), can be used to assess the vegetation condition and determine the regions that face stress, which can be related to pollution or other disruptions(Bhatti, 2022).

UAVs (Unmanned Aerial Vehicles)

UAVs or drones take high-resolution images with a scale that is significantly smaller than that of satellites. UAVs have multispectral or hyperspectral sensors, which are especially beneficial in estimating the mangrove roots health and in identifying the small-scale vegetation changes. The sites that can be very difficult to reach via satellite or need very specific site-focused analyses fit well on these platforms(Bhatti, 2021).

LiDAR (Light Detection and Ranging)

The LiDAR technology utilizes laser pulses in the creation of accurate three-dimensional representations of the structure of mangroves. It is particularly effective in determining the integrity of the root system, tracking changes in the height of the sediment and clarifying the physical status of the mangrove forests. LiDAR is also able to detect areas with the potential of microplastic concentration in sediment because it can identify subtle shifts in the surface texture(Bhatti, 2020).

Hyperspectral Imaging

Hyperspectral sensors are able to capture a wide range of wavelengths, which can be used to make subtle analyses of vegetation stress. These spectral changes can be an indicator of health problems of mangroves due to pollutants like microplastics. The technology can also be used to identify the alteration of the sediment composition that is suggestive of microplastic pollution in the mangrove systems(Liu & Bhatti, 2019).

Monitoring of Mangrove Vegetation and Root Health

Remote sensing technologies make it possible to monitor vegetation cover variations, the condition of roots and soil stability in the mangrove ecosystems. The data collected by satellites and UAVs assists in the identification of changes in vegetative cover and root structure. The changes in the health of the mangroves can be measured by combining the remote sensing data with vegetation indices such as NDVI to produce a useful piece of information on the management of the ecosystem(Bhatti, 2022).

Satellite Based Environmental Insights, shown in **Figure 2:**

Figure 2. Satellite-based environmental insight

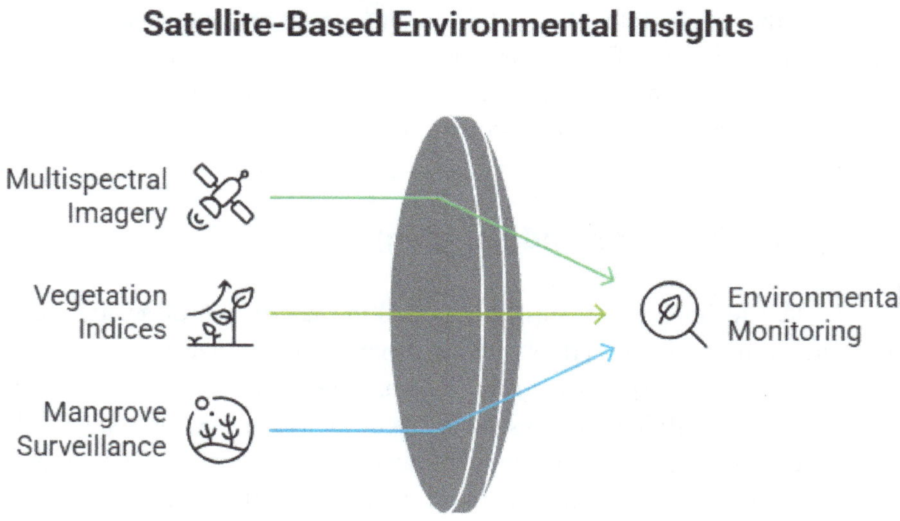

Problems and Procedures of Microplastic Detection

Microplastics are smaller than 5mm in diameter hence they are hard to detect using traditional remote sensing devices. In contrast to the bigger features (vegetation or open water) that can be easily identified by their own spectral signatures, microplastics do not exhibit any spectral signatures. Therefore, the current techniques concentrate on indirect indicators of contamination(Liu, 2019).

Vegetation Health alteration When microplastics are deposited in sediment, they may cause alterations in plant vitality affecting spectral reflectance that is observable by hyperspectral imaging. Mangrove vegetation is prone to stress which can be recorded in the form of distorted values of NDVI indicating reduced plant health as a result of pollution.

Sediment Composition and Water Turbidity: Microplastics have the capability to alter the texture and structure of mangrove sediments. LiDAR and UAVs are able to detect the changes in the sediment structure, which provides the hints at the presence of microplastics. Equally, water turbidity increases, which may be associated with the presence of microplastic, can be tracked through remote sensing, specifically, with UAVs or satellite images.

Remotesensing Signs of Microplastic Pollution

- Stress of the vegetation: A decrease in the NDVI values or the spectral patterns indicate vegetation stress, which might be explained by microplastic pollution.
- Sediment Motion: LiDAR is able to detect sediment height or structure changes related to the presence of microplastic.
- Water Turbidity: Remote sensing platforms can be used to monitor water turbidity which is caused by microplastic dispersion in water.

Remote Sensing Case Studies in Mangrove Monitoring

Case Study 1: Sentinel-2 to Vegetation Health Monitoring

One experiment in the Sundarbans Mangrove Forest used the Sentinel-2 imagery to assess the health of vegetation and identify stress created by pollution such as microplastics. NDVI allowed monitoring the changes in vegetation cover, and it was found that areas with accumulated pollutants were linked to the poor health of mangroves. This study indicates that satellite imaging can be used to track the changes in large mangrove systems(Bhatti, 2024).

Sentinel-2 to Vegetation Health Monitoring, shown in Figure 3:

Figure 3. Sentinel-2 to vegetation health monitoring.

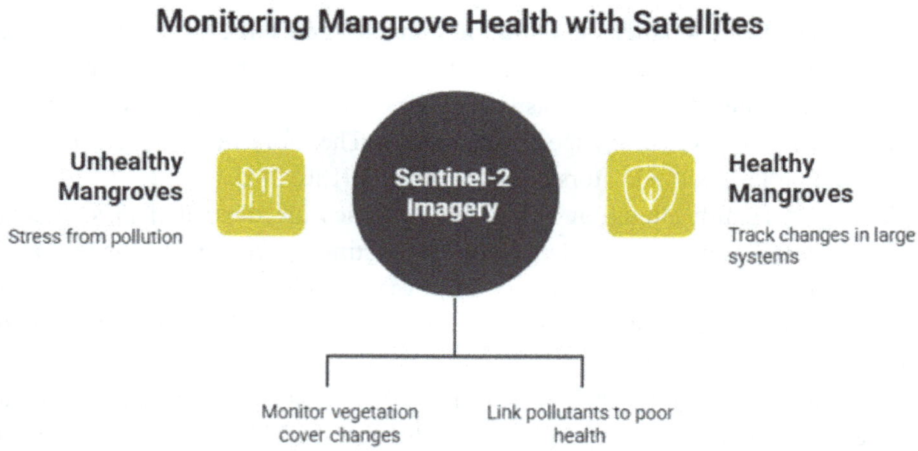

Case Study 2: UAVs in Monitoring of the Root System

In a study using UAV in a mangrove coastal setting, the impact of microplastic contamination on mangrove root systems was investigated. Hyperspectral sensors on UAVs provided high-resolution images of root health, which identified areas of stress on roots due to microplastic and effects on root architecture and growth. The case highlights the usefulness of UAVs regarding the site-specific, fine-resolution monitoring of mangrove health(Aamir, 2024).

UAVs in Monitoring of the Root System, shown in Figure 4:

Figure 4. UAVs in monitoring of the root system

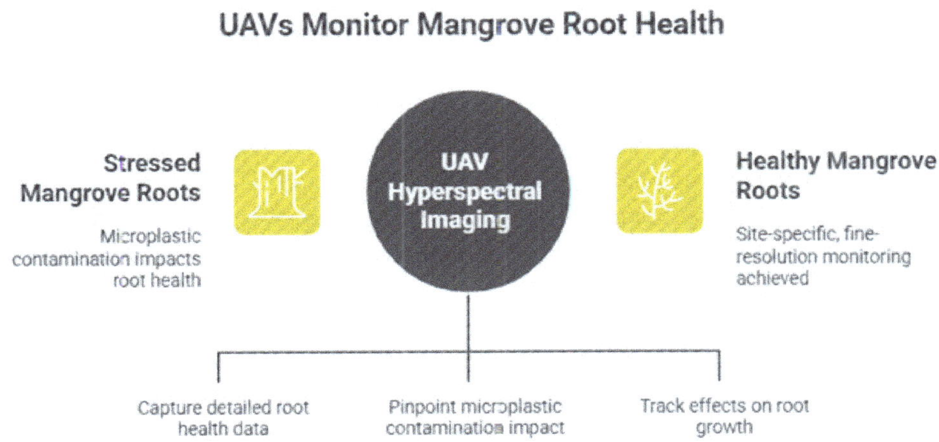

On-Land Based Monitoring Methods

Terrestrial approaches, such as soil and water quality tests, direct root testing, and sediment testing, give precise information on microplastic levels and their immediate impacts on the mangrove ecosystem. As an example, the presence and level of microplastics observed through UAVs or satellite measurements can be supported by the use of sediment samples, but the actual damage caused by pollutants can be measured through direct root analysis.

Integrating Remote Sensing and Field Data

Combining remote sensing data with ground-based observations can help in improving the quality and scope of monitoring activities. The results obtained by UAVs regarding root health can be confirmed by conducting direct root evaluation in the field, whereas satellite information about the vegetative cover can be supported by soil and water quality analyses. This integrative methodology can be used to understand the microplastic contamination and its effects on mangrove ecosystems holistically(Gauri, 2024).

CASE STUDIES AND CURRENT RESEARCH

Microplastic pollution is the new menace to marine and coastal ecosystems especially the mangrove forests which are vital in supporting biodiversity, protection of the coastline and carbon sequestration. Even though the impacts of microplastics in the marine ecosystem have been comprehensively studied, the impacts on the mangrove ecosystems have not been fully investigated. This part examines the regional literature about microplastic pollution in mangroves, analyzes the use of remote sensing to monitor microplastic pollution, and determines the existing gaps in the existing literature(Bhatti, 2023).

Geographical Research on Microplastic Pollution in Mangroves

Southeast Asia

The greatest and most ecologically important mangrove forests in the world are found in Southeast Asia. At the same time, the microplastic contamination is high in these areas. In a research done in the Philippines, which targeted mangrove ecosystems near urban centres, it was found that there were high levels of microplastics in sediments as well as mangrove roots. The exploration revealed the concentration of microplastic particles, especially in the river mouths and coastal areas, which were impacted by urban runoff. Physiological stress manifested in mangrove sites in the regions in terms of the vitality of the roots and low rate of plant growth, which was caused by the physical and chemical effects of the microplastics. Moreover, heavy metals and persistent organic pollutants, which are toxicants related to microplastics, were also detected in the mangrove sediments, which only deteriorated the ecological situation(Bhatti, 2022).

Equally, studies conducted in Indonesia have revealed that microplastics are present in the sediments of the mangrove forest, particularly areas at risk of fishing and maritime activities. Mangrove roots form a high density that entraps and holds microplastics leading to increased exposure to contaminants. This fact shows that microplastic pollution can seriously interfere with the services of the mangrove ecosystem, especially through the compromised health of the major organisms that rely on mangrove ecosystem, including fish and crustaceans(Bhatti & Yu, 2024).

Caribbean

Mangrove forests are part of the coastal protection in the Caribbean but are susceptible to pollution, such as microplastics. An analysis in the mangroves of Bahamas found large amounts of microplastics in the sediment and water. This study revealed that such microplastics were mainly of plastic waste that was released into the environment due to tourism, fishing and urban runoff. The study further noted that the mangroves in the polluted areas had indications of nutrient stress and the change in the microbial diversity that might have an impact on the general health of the ecosystem. As noted in this case study, the vulnerability of the mangrove forests in the island areas is due to the anthropogenic activities and plastic pollution that are major contributors of environmental degradation(Liu, 2024).

Australia

The northern coastline of Australia, which includes regions like Queensland, has massive mangrove cover that has been vital in defending the coastal shorelines and also sustaining marine life. One of the studies in this area analyzed the contamination of microplastic in mangrove habitats around large urban and industrial centres. A study recorded a high rate of microplastics in the sediments of the mangrove forests, with the majority of them being as a result of urban run-offs and shipping. The results showed that microplastic presence in these regions has adverse impacts on the health of the mangrove, especially in its physical root blockage and possible chemical toxicity. The findings indicate that further pollution of the coastal areas may cause a significant reduction in the ability of the mangrove forests to provide the necessary ecosystem services.

Surveillance of Vegetation Health

The evaluation of vegetation health through indices like the Normalized Difference Vegetation Index (NDVI) is one of the main applications of remote sensing in monitoring the mangrove. The satellite platforms have been used to identify stress in the mangrove forests that can be an indicator of pollution related destruction, including microplastic pollution. In the case of a study conducted in Southeast Asia, Sentinel-2 imagery was used to study the health of mangrove canopies during a period, and the results showed that the areas with decreased vegetation cover were associated with high pollution levels. This method does not directly identify microplastics but gives a pointer of the ecological stress that contaminants cause(Wang, 2024).

Root and Sediment Monitoring by UAVs and LiDAR

Hyperspectral sensors are also installed on Unmanned Aerial Vehicles (UAVs) to monitor the health of mangrove roots and the state of sediments. Recent research conducted in Australia used UAVs to acquire high-resolution imagery of mangrove forests, with the emphasis on the integrity of roots and the stability of sediments.

LiDAR technology has also been applied to the mapping of mangrove root systems and the identification of slight changes in the sediment elevation, which can serve as evidence of the existence of microplastics. UAVs and LiDAR can be especially useful together due to the high spatial resolution that allows evaluating small-scale alterations in the mangrove habitat that may be associated with the microplastic pollution.

Hyperspectral Imaging in Microplastic Detection

Hyperspectral that records a wide band of wavelengths has a potential to identify the impact of microplastics on mangrove ecosystems. The analysis of spectral features of vegetation could help identify signs of stress that could be associated with microplastic pollution. As an example, hyperspectral data has been used to measure spectral reflectance of mangrove plants, to determine areas of stress or disease that may be a result of pollutants. Even though, this technology does not sense microplastics directly, it may indicate a bigger effect of contamination, including changes in the health of plants or the condition of the environment.

Table 2. Geographical research on microplastic pollution in mangroves

Region	Key Findings	Monitoring Methods	Technologies Used
Southeast Asia (Philippines)	High microplastic levels in sediments and roots, especially near urban runoff areas. Stress in mangroves, including poor root vitality and plant growth.	Sentinel-2 imagery to assess vegetation health (NDVI), identifying pollution-induced stress.	Sentinel-2 (NDVI), UAVs, Hyperspectral sensors
Southeast Asia (Indonesia)	Microplastics in sediments, particularly in fishing and maritime areas. Roots trap microplastics, harming organisms like fish and crustaceans.	UAVs and LiDAR to monitor root integrity and sediment stability, detecting microplastic impacts.	UAVs, LiDAR, Hyperspectral sensors
Caribbean (Bahamas)	High microplastic levels from tourism, fishing, and runoff. Nutrient stress and microbial diversity changes in mangroves.	UAVs and satellite imagery to monitor vegetation health and detect pollution stress (NDVI).	Sentinel-2, UAVs, NDVI
Australia (Queensland)	Microplastics in sediments from urban runoff and shipping, affecting root health and ecosystem services.	UAVs with hyperspectral sensors for root and sediment monitoring. LiDAR used to detect changes in sediment elevation.	UAVs, LiDAR, Hyperspectral imaging

GAPS IN CURRENT RESEARCH

Although there has been an advancement in application of remote sensing to observe mangrove ecosystems and effects of pollution, there are still lots of gaps in the study of microplastic pollution in mangrove forests.

Minimal Direct Detection of Microplastics

One of the largest gaps in research is that it is difficult to directly observe microplastics using remote sensing systems. In contrast to larger pollutants, microplastics do not have unique spectral signatures that can be easily identified by the satellite or UAV-based sensors. Even though the existence of microplastics may be hinted at by indirect techniques including checking the health of vegetation or water turbidity, it cannot be directly identified. Further development of remote sensing methods that can detect microplastics in sediment or water, or the development of new sensor technologies, is thus in need(Xue, 2024).

Requirement to Long-Term Monitoring

The second gap that is critical is the lack of long term studies on the microplastic contamination of mangrove ecosystems. Numerous investigations have been done on small time scales, which constrain the ability to evaluate the long-term effects of microplastics on the health and ecosystem services of mangroves. To know the accumulation of microplastic with time and the cumulative impact of microplastic on the mangrove forests, long-term data are necessary. Furthermore, research that combines several sources of remote sensing materials is required to create an overall monitoring system of microplastic pollution, including satellite imagery, UAVs, and hyperspectral sensors.

Combination of Several Data Sources

Existing studies often use only one type of remote sensing information, which limits the scope of analysis. Instead, a better strategy is to combine different sources of data such as satellite images, UAV-based measurements, and ground-based surveillance to have a comprehensive picture of microplastic pollution in mangrove habitats. This type of the integrated approach would help the researchers to comprehend the spatial distribution of microplastics better, their impact on the root systems, and the long-term effects of this issue on the health of mangroves(Zhang, 2024).

OPPORTUNITIES AND FUTURE PERSPECTIVES

Tracking of microplastic pollution in mangrove system is one of the important aspects of conservation and management of mangrove. The risk of microplastics is only growing, and remote sensing tools, interdisciplinary analysis, and policy development will be required to handle this multi-faceted issue. This part will consider future trends in remote sensing, synthesis of various sciences and its implications to coastal management and the policy making process.

Remote Sensing Technological Development

New technologies of remote sensing are promising to enhance greatly the detection and monitoring of microplastic pollution of mangrove ecosystems. The high-resolution multispectral imagery and hyperspectral imagery carried out by advanced satellite missions, including the Copernicus program undertaken by the European Space Agency, can be used to provide useful data on vegetation health and environmental change. The expected developments to occur in the future, such as the introduction of specialized hyperspectral satellites, will be able to detect concealed stressors in the environment caused by pollutants like microplastics. Such platforms will provide a better spectral data, which will allow more accurately track the change in vegetation structure and sediment composition that can indirectly indicate microplastic pollution.

Another important breakthrough is the use of artificial intelligence (AI) and machine learning (ML) in analyzing images. The processing of large volumes of data collected through remote sensing platforms can be automated by AI algorithms and identify patterns or abnormalities in vegetation health, water quality, and sediment structures. Besides, AI can help locate microplastic-related stress in mangrove ecosystems through multispectral and hyperspectral images, which are better analysed through AI, which contributes to real-time monitoring and faster decision-making.

In addition to satellite surveillance, unmanned aerial vehicles (UAVs) that have LiDAR and hyperspectral cameras are set to become invaluable assets to the high-resolution monitoring of mangrove ecosystems. The UAVs have the ability to generate close up shots of the mangrove canopies and root systems- aspects which are often hard to determine by traditional means. The LiDAR technology has the potential to provide three-dimensional information on the architecture of roots and the processes of sediment, which can be further used to track the changes that can occur because of the build-up of microplastic in soil.

Interdisciplinary Solutions to Holistic Surveillance

This monitoring microplastic pollution in mangrove ecosystem does not solely require remote sensing technology, but rather a multidisciplinary approach to having a wholesome data on the effects of microplastics. Through the combination of remote sensing and environmental modelling and field data collection, more robust monitoring systems can be developed which will help to get a better understanding of the long-term consequences of pollution on the health of mangroves.

By combining remote sensing data with environmental models, one can predict the growth of microplastics in the mangrove sediments and how it impacts the root systems of the mangroves with time. These models have the capability of modeling the impacts of microplastic contamination on the services of the ecosystems including carbon sequestration and coastal defense, which may aid in the development of knowledgeable management plans. On-ground monitoring, that is, sediment monitoring, root monitoring, and water quality monitoring, is also unconditionally necessary to verify the reliability of the remote sensing data and provide local and intensive monitoring of microplastic contamination. Field based techniques can, therefore, directly measure the concentrations of microplastics and this allows researchers to see the physical impacts that are caused by the plastic on the roots of the mangrove, thus providing important context to the remote sensing data.

Citizen science is also to be recognized as an effective element of mangrove monitoring of the future. Equipped with simple equipment to gather sediment and water analysis, the local communities will be able to provide valuable information on the presence of microplastic pollution in the remote or insufficiently controlled regions. The citizen science also broadens the monitoring initiatives as well as increases awareness about the importance of the mangrove conservation and the consequences of the pollution of these vital ecosystems. Including data collected by citizens through the use of the advanced remote sensing and strict scientific research, a more comprehensive and integrative monitoring system can be provided.

Policy Implication and Management Strategy

The study of microplastic pollution and remote sensing provides a consequence that can inform the coastal management and policy formulation. The data on remote sensing can be used to guide policy makers by the identification of the vulnerable mangrove habitats and monitoring the changes in the health of the ecosystems. This information can be used by the policymakers as a prioritization exercise of areas in which to conserve, restore, or reduce pollution. As an example, satellite data and UAV data can be used to discover areas where mangrove vegetation is under stress, which will indicate the need to take specific management measures, such as clean-up or a decrease in plastic waste.

Stricter rules and policies are needed to deal with the increasing issue of microplastic pollution. The coastal management plans must focus on avoiding plastic waste to find its way into the sea through developing waste management facilities, controlling the production of plastic, and supporting the use of alternatives to single-use plastic. Interregional and international partnerships will be essential in stemming out microplastic pollution at its origin particularly in the coastal areas that have been greatly affected by tourism, shipping and industries. Raising and adopting sustainable waste management systems that reduce the escape of microplastics to the environment should also be promoted by policy measures.

Moreover, a combination of remote sensing data, local monitoring and community participation to protect the mangrove ecosystems should be embraced so that they are not degraded further. Such strategies might include the naming of special zones where plastic debris is strictly regulated, and the educational program promoting the use of sustainable behavior by local communities. Remote sensing technologies need to be also promoted by policymakers to become effective in keeping track of the long-term patterns of microplastic contamination and to measure the effectiveness of the policy measures applied in the long-term.

CONCLUSION

Summary of Findings

Microplastic pollution is also a serious danger to the mangrove ecosystems in terms of their health and reduced capacity to provide some of the most important services, including carbon sequestration, coastal protection, and support of biodiversity. The use of remote sensing technology has been found to be essential in the health tracking of the mangroves, detecting vegetation and sediment composition changes that can indicate the presence of microplastic contamination. Even though remote sensing currently is unable to directly record the presence of microplastics, it can be an essential instrument of ecosystem-wide stressor and environmental alterations related to pollution.

Conservation Implication to Mangroves

This study has shown that the need to enhance monitoring methods and intensify conservation efforts to preserve the mangrove ecosystems is a reality. Through the combination of remote sensing and field data, environmental model and citizen science, the overall picture of microplastic contamination can be created. Such a combined solution will allow managing and preserving mangroves more effectively so that these important coastal ecosystems could be resistant to increasing levels of pollution.

Call to Action

The importance of solving the problem of microplastic pollution in mangroves cannot be emphasized. The next step in the study should be the development of remote sensing technologies to identify microplastics both directly and their effects on the health of mangroves. Moreover, interdisciplinary methods that will integrate local knowledge, field data and scientific studies will be necessary in designing sound monitoring structures. There is the need to have policymakers and conservationists working together to ensure that there are sustainable waste management practices and reduction of plastic pollution at the source. Through more innovation and collaboration, the future of mangrove systems can be preserved and the crucial services they offer to the coasts and the world as a whole can be maintained.

REFERENCES

Aamir, M. N.. (2024). Multi-Watermarking Algorithm for Medical Images Using Inception V3 and DCT. *Computers, Materials & Continua.*

Avio, C. G., Gorbi, S., & Regoli, F. (2017). Plastics and microplastics in marine ecosystems: Effects on species and food webs. *Environmental Science. Processes & Impacts.*

Bentley, J. (2024). Detecting ocean microplastics with remote sensing: A pilot study on satellite reflectance signals. *Bridgewater State University Honors Project.* vc.bridgew.edu

Bhatti, U. A.. (2020). Hybrid watermarking algorithm using Clifford algebra with Arnold scrambling and chaotic encryption. *IEEE Access : Practical Innovations, Open Solutions.*

Bhatti, U. A.. (2021). Local similarity-based spatial–spectral fusion hyperspectral image classification with deep CNN and Gabor filtering. *IEEE Transactions on Geoscience and Remote Sensing.*

Bhatti, U. A.. (2021). MFFCG–Multi feature fusion for hyperspectral image classification using graph attention network. *Expert Systems with Applications.*

Bhatti, U. A., et al. (2021). Time Series Analysis and Forecasting of Air Pollution Particulate Matter (PM2.5): An SARIMA and Factor Analysis Approach. *IEEE Access.*

Bhatti, U. A., et al. (2022). Assessing the change of ambient air quality patterns in Jiangsu Province of China pre-to post-COVID-19. *Chemosphere.*

Bhatti, U. A., et al. (2022). Time series analysis and forecasting of air pollutants based on prophet forecasting model in Jiangsu province, China. *Frontiers in Environmental Science.*

Bhatti, U. A.. (2022). Advances in climate change monitoring through AI and remote sensing. *Remote Sensing Applications: Society and Environment.*

Bhatti, U. A.. (2023). Global production patterns: Understanding the relationship between greenhouse gas emissions, agriculture greening, and climate variability. *Environmental Research.* PMID: 38169167

Bhatti, U. A., et al. (2023). Robust zero-watermarking algorithm for encrypted medical volume data based on PJFM and 3D-DCT. *IEEE Journal of Biomedical and Health Informatics.*

Bhatti, U. A.. (2024). Deep Reinforcement Learning Enables Adaptive-Image Augmentation for Automated Optical Inspection of Plant Rust. *Frontiers in Plant Science*.

Bhatti, U. A., & Tang, H. (2023). Deep Learning with Graph Convolutional Networks: An Overview and Latest Applications in Computational Intelligence. *International Journal of Intelligent Systems*.

Bhatti, U. A., & Yu, Z. (2024). Feature-based multi-criteria recommendation system using a weighted approach with ranking correlation. *Intelligent Data Analysis*.

Browne, M. A.. (2011). Accumulation of microplastics on shorelines worldwide. *Environmental Science & Technology*.

Chaisanguansuk, P., Phantuwongraj, S., Jirapinyakul, A., & Assawincharoenkij, T. (2023). Preliminary study on microplastic abundance in mangrove sediment cores at Mae Klong River, upper Gulf of Thailand. *Frontiers in Environmental Science*, *11*, 1134988.

Deng, H., He, J., Feng, D., Zhao, Y., Sun, W., Yu, H., & Ge, C. (2021). Microplastics pollution in mangrove ecosystems: A critical review of current knowledge and future directions. [MDPI]. *The Science of the Total Environment*, *753*, 142041. PMID: 32906050

Gauri, S.. (2024). AI-based object detection for marine environment using remote sensing. *IEEE Transactions on Geoscience and Remote Sensing*.

Greenshields, J. (2024). *A systematic review of microplastic sampling methods in mangrove and seagrass ecosystems. Environmental Monitoring and Assessment.* Springer Link.

Hidalgo-Ruz, V.. (2012). Microplastics in the marine environment: A review of the methods used for identification and quantification. *Environmental Science & Technology*. PMID: 22321064

Koelmans, A. A.. (2017). Microplastic as a vector of chemical pollutants in aquatic environments. *Environmental Science & Technology*.

Li, J.. (2019). Microplastic pollution in the surface waters of coastal mangrove ecosystems. *Environmental Pollution*.

Liu, J.. (2019). Enhanced feature-based Vision Patch Transformer Network for accurate brain tumor segmentation in magnetic resonance imaging. *IEEE Journal of Biomedical and Health Informatics*. PMID: 40030969

Liu, J., & Bhatti, U. A.. (2019). Hybrid approach to forecast the COVID-19 epidemic trend. *PLoS One*.

Liu, X., et al. (2022). Ecological interception effect of mangroves on microplastics. *Journal of Hazardous Materials*. ScienceDirect

Liu, Z.. (2024). *Remote sensing-enhanced transfer learning approach for agricultural damage and change detection: A deep learning perspective*. Big Data Research.

Lorenzo-Navarro, J.. (2025). Deep learning for microplastic segmentation in coastal environments. *Preprints. Preprints*

MacFarlane, G. R.. (2007). Plastic debris in coastal sediments and effects on benthic species. *Marine Pollution Bulletin*.

Maghsodian, S.. (2021). Microplastics in mangrove sediments of the Persian Gulf. *Marine Pollution Bulletin*.

Mehmood, T.. (2025). *Mangroves in Pakistan at risk from microplastic threats. Environmental Science and Pollution Research*. Springer Link.

Mendes, D. S.. (2022). Current status and trends of research on microplastic detection in mangrove wetlands. *Frontiers in Environmental Science. Frontiers*.

Mendes, D. S., Beasley, C. R., Silva, D. N. N., & Fernandes, M. E. B. (2023). Microplastic in mangroves: A worldwide review of contamination in biotic and abiotic matrices. *Marine Pollution Bulletin, 195*, 115552. PMID: 37738877

Nivedita, V.. (2023). *Plastic debris detection along coastal waters using satellite imagery. Marine Pollution Bulletin*. PubMed.

Prata, J. C.. (2019). Methods for microplastic extraction and identification from environmental matrices. *Journal of Hazardous Materials*.

Rochman, C. M.. (2014). The ecological impacts of plastic pollution in marine ecosystems. *Science*.

Sudhakar, B. J. (2025). Mangrove system acts as a major microplastic sink: Leaves and roots analysis. *SSRN. SSRN*

Thompson, R. C.. (2004). Lost at sea: Where is all the plastic? *Science*. PMID: 15131299

Torregroza-Espinosa, A. C.. (2025). *Application of remote sensing for detection and assessment of microplastics. Remote Sensing Journal*. MDPI.

Veettil, B. K., & Sabu, T. (2022). *Coastal and marine plastic litter monitoring using remote sensing. Remote Sensing Applications: Society and Environment*. ScienceDirect.

Wang, W.. (2022). microplastics in mangrove sediments and their potential effects on carbon cycling. *The Science of the Total Environment.*

Wang, Y.. (2024). Monitoring coastal ecosystems using remote sensing and machine learning techniques: A case study on mangrove forests. *Environmental Monitoring and Assessment.*

Wu, F., Zhang, C., Li, X., Liu, S., Wang, J., & Huang, W. (2025). Distribution, Diversity, and Ecological Risks of Microplastics in Mangrove Ecosystems of a Southeastern Chinese Estuary. *Toxics, 13*(6), 494. PMID: 40559967

Xue, Y.. (2024). Remote sensing for detecting the impacts of microplastics on coastal ecosystems. *Marine Pollution Bulletin.*

Zarfl, C.. (2011). Microplastics in the environment—A review of the methods used for identification and quantification. *Environmental Chemistry.*

Zhang, H. (2017). Transport of microplastics in coastal waters: Interactions with tidal and current dynamics. *Marine Environmental Research.*

Zhang, X.. (2024). Environmental degradation of mangrove systems due to microplastic accumulation in estuarine environments. *Environmental Toxicology and Chemistry.*

Zhang, Y.. (2020). Spatial distribution of microplastics in mangrove sediments of the Beibu Gulf. *Environmental Pollution.*

Zuo, Y.. (2020). Microplastic distribution and characterisation in global mangrove sediments. *Marine Pollution Bulletin.*

Chapter 7
Impact of Pharmaceutical Emerging Contaminants on Blue Carbon Sequestration in Tidal Marshes:
Impact of Pharmaceutical Emerging Contaminants

Asqar Ismailov
http://orcid.org/0009-0000-8016-4221
Department of General Science, Tashkent State University of Economy, Tashkent, Uzbekistan

Abrayev Bakhrom
Department of Information Technology and Exact Sciences, Termez University of Economics and Service, Termez, Uzbekistan

Zaribova Nafisa Umirbay Qizi
Department of Biology, Urgench State University, Uzbekistan

Imirsinova Azizaxon Ashurovna
Department of Ecology, National University of Uzbekistan Named After Mirzo Ulugbek, Uzbekistan

Barno Matchanova
Department of National Idea and Philosophy, Urgench State Pedagogical Institute, Urgench, Uzbekistan

DOI: 10.4018/979-8-3373-8872-4.ch007

ABSTRACT

Emerging pharmaceutical contaminants pose a significant threat to blue carbon sequestration in tidal marshes, which are vital coastal ecosystems known for their role in capturing and storing carbon. Pharmaceuticals, including antibiotics, hormones, and painkillers, often enter aquatic environments through wastewater discharge and runoff, where they can affect plant and microbial communities that are crucial for carbon cycling. These contaminants can disrupt the growth of marsh vegetation, alter soil microbial activity, and inhibit the processes that contribute to carbon sequestration. Tidal marshes rely on the health of their plant and microbial populations to trap and store carbon in both the above-ground biomass and below-ground soil organic matter. The presence of pharmaceutical contaminants can weaken these processes, potentially reducing the marsh's ability to sequester carbon and mitigate climate change. Understanding the impact of these pollutants is critical for preserving the blue carbon potential of tidal marshes.

1. INTRODUCTION

1.1 Background

Blue carbon is defined as the carbon present in coastal and marine ecosystems, which are mainly the mangroves, salt marshes, and seagrass beds. These ecosystems are important in preventing climate change since they absorb carbon dioxide in the atmosphere at a rate much higher than when using terrestrial forests. The capacity of these ecosystems to trap and retain the carbon in their biomass and sediments forms a very important component of the world carbon cycles, hence playing a major role in the regulation of the climate(Bhatti, Tang, Wu et al, 2023).

Significance of Tidal Marshes as Blue Carbon Critical Ecologies

Tidal marshes, which form one of the most important parts of coastal wetlands, are very productive ecosystems, which serve as an essential source of carbon sequestration. These marshes do not only absorb carbon within the biomass of the plants, but organic material is also held within the sediments of these marshes, which also helps in long-term carbon storage. Tidal marshes promote the formation of sediments that contain carbon through the hydro condition that includes periodic flooding and draining. Their contribution to carbon sequestration is now becoming important towards the process of fighting global climate change(Bhatti, Yu, Chanussot et al, 2021).

Introduction to Pharmaceutical Emerging Contaminants (PhECs)

Pharmaceutical emerging contaminants (PhECs) are pharmaceutical products and their metabolites that have been found to be contaminants on the environment. These pollutants are prescription and non-prescription drugs and veterinary drugs. They are introduced into the environment by wastes through the discharge of waste-water, agricultural run water, and the improper ways of disposing. PhECs are stable chemicals in the ecosystems (such as coastal wetland) and their inability to degrade may cause a considerable risk to the health of these important ecosystems (Bhatti, Bhatti, Tang et al, 2024).

Pharmaceutical Emerging Contaminants (PhECs), as shown in Figure 1:

Figure 1. Pharmaceutical Emerging Contaminants (PhECs)

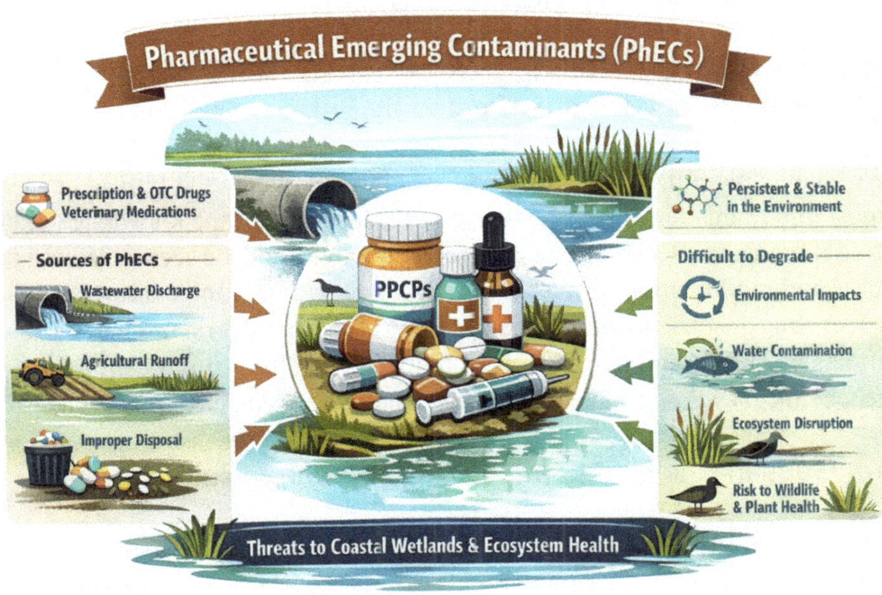

The Reason to Study Their Influence on Blue Carbon Sequestration

The existence of PhECs in tidal marshes concerns the impact of the species on ecological processes governing blue carbon sequestration. These pollutants may alter the growth and physiological process of the marsh plants, disturb the microbes involved in carbon cycle, and interfere with the sedimentation processes, which is important in long term storage of carbon. It is important to understand the processes through which PhECs affect such ecosystems to assess their effects on blue carbon storage and to devise strategies that can reduce their effects in the presence of increasing pharmaceutical pollution (Bhatti, Huang, Wu et al, 2019).

1.2 Objectives of the Review

Current State of Knowledge on the Occurrence of PhECs in Tidal Marsh Ecosystems

The purpose of the review is to compile existing knowledge on the phenomenon and the distribution of PhECs in tidal marsh ecosystems. Through field surveys and lab tests we shall discuss the levels of these contaminant in tidal marshes in the different geographical areas and the different sources of their entry into these ecosystems(Bhatti, Mengxing, Neira-Molin et al, 2023).

Assess the Pathways Through Which Marsh Plant, Soil, and Microbial Functions are Affected by PhECs

We shall explore the ecological processes through which the PhECs influence the most important elements of tidal marshes. This involves the knowledge of how these contaminants affect the plant growth, productivity and the accumulation of the biomass and modification of the soil processes like the decomposition of organic matter, cycling of nutrients and the sequestration of carbon. The review will also dwell on effects of PhECs on microbial communities which have significant role in the biogeochemical cycles of such ecosystems (Bhatti, Zeeshan, Nizamani et al, 2022).

Affirm Gaps in Knowledge and Future Research Requirements

Although the effects of PhECs on ecosystems are researched, there is still a lot left to know, especially what these effects would be like in the long run when it comes to their impact on blue carbon sequestration of tidal marshes. The review will also indicate areas in which further research is required such as the need of conducting more extensive field research, better monitoring methods and a better comprehension of the cumulative impacts of various contaminants. Besides, the recommendation of future research will be proposed to design more effective management practices and policies to amend the increasing risk of pharmaceutical pollution in tidal marshes(Bhatti, Yan, Zhou et al, 2021).

2. BLUE CARBON SEQUESTRATION IN TIDAL MARSHES

2.1 Ecology and Carbon Dynamics of Tidal Marshes

Tidal marshes are among the most productive ecosystems on Earth, exhibiting high rates of primary productivity. This productivity is driven by the abundant sunlight, nutrients from tidal influx, and the availability of water. Marsh vegetation, including grasses, sedges, and shrubs, captures carbon dioxide (CO2) through photosynthesis and converts it into organic matter. This organic matter, in turn, forms the basis of the marsh ecosystem, supporting a range of fauna and contributing to the biomass that is critical for carbon sequestration. The high productivity rates of tidal marshes ensure that they act as carbon sinks, drawing down atmospheric CO2 and contributing to the global carbon cycle(Bhatti, Liu, Li et al, 2019).

Carbon Storage Mechanisms (Aboveground Biomass, Belowground Biomass, Sediments)

The carbon captured by tidal marsh plants is stored in three primary reservoirs:

Aboveground Biomass: This includes the carbon stored in the stems, leaves, and reproductive organs of marsh plants. While aboveground biomass plays an important role in carbon storage, it is typically more vulnerable to disturbances like storms and herbivory(Bhatti, Yu, Li et al, 2020).

Belowground Biomass: The roots and rhizomes of marsh plants are significant carbon reservoirs. Belowground biomass not only contributes to carbon storage but also supports plant growth by anchoring plants in the sediment and facilitating nutrient uptake(Bhatti, Gao, Zhang et al, 2024).

Sediments: The most substantial and long-term carbon storage occurs in the sediment layer, where organic matter from decomposed plant material accumulates. These sediments, through a process called sediment burial, trap carbon deep within the marsh soil, where it remains for centuries or longer. Tidal action helps bury organic carbon, preventing its release back into the atmosphere as CO2(Bhatti, Zhang, Chen et al, 2024).

The carbon captured by tidal marsh plants, as shown in Figure 2:

Figure 2. The carbon captured by tidal marsh plants

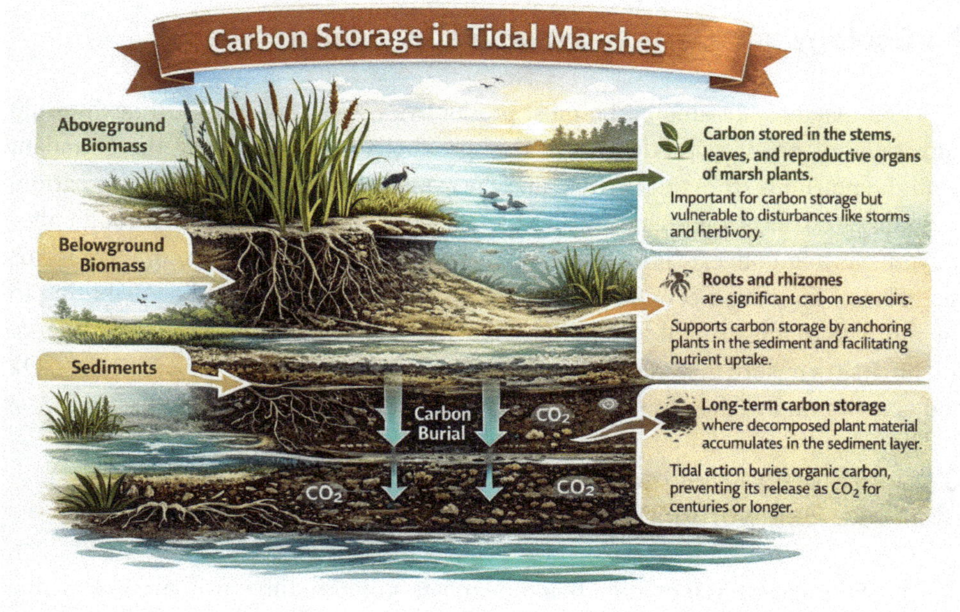

Factors Influencing Carbon Accumulation

Several environmental and biological factors affect carbon accumulation in tidal marshes, including:

Tidal Regime: The frequency and magnitude of tidal flooding influence the deposition of organic material in marsh sediments. High-frequency tidal inundation can enhance carbon burial by supplying nutrients and aiding the transport of organic matter(Bhatti, Nawaz, Li et al, 2023).

Vegetation Type: Different plant species sequester carbon at varying rates. Salt-tolerant species such as Spartina alterniflora (cordgrass) and Juncus species are particularly effective in carbon fixation due to their extensive root systems and high biomass production(Bhatti et al., 2018).

Sediment Accumulation Rate: High rates of sedimentation enhance carbon sequestration, as organic matter becomes buried more quickly, reducing the likelihood of its oxidation and release as CO2(Bhatti, Sheng, Hashmi et al, 2022).

Hydrology and Nutrient Availability: Nutrient-rich environments with stable hydrological conditions tend to promote greater plant growth and carbon fixation. Conversely, disturbances like droughts or excessive nutrient loading can disrupt plant growth and carbon storage(Bhatti, Marjan, Wahid et al, 2020).

2.2 Importance for Climate Change Mitigation

Quantitative Estimates of Carbon Stored in Tidal Marshes

Tidal marshes store carbon at a significantly higher rate than terrestrial ecosystems. Estimates suggest that these coastal wetlands can sequester between 1.5 and 2.5 gigatonnes of CO2 per year globally, although this number varies based on marsh type, location, and environmental conditions. The carbon stored in tidal marsh sediments is estimated to range from 50 to 500 Mg C per hectare, depending on factors such as sedimentation rate, plant species, and marsh age. Due to their high carbon sequestration rates, tidal marshes play an outsized role in offsetting carbon emissions relative to their land area, making them key players in efforts to mitigate climate change(Bhatti, Li, Liu et al, 2023).

Contribution to National and Global Carbon Budgets

The global contribution of tidal marshes to carbon sequestration is substantial when considering the vast extent of coastal wetlands. Although tidal marshes cover only about 3-5% of the Earth's land area, they are believed to store as much as 10-15% of the carbon stored in terrestrial ecosystems. This makes them disproportionately important in the fight against climate change. At a national level, countries with large coastal wetland areas—such as the United States, Brazil, and China—benefit from significant carbon sequestration services provided by their tidal marshes. By preserving and restoring tidal marshes, these nations can enhance their carbon offset strategies and contribute to achieving international climate goals(Bhatti, Chen, He et al, 2024).

Tidal marshes also have the potential to mitigate the effects of coastal erosion and storm surges, thus reducing vulnerability to climate-related extreme events. This multifunctionality of tidal marshes—acting as carbon sinks, providing coastal protection, and supporting biodiversity—makes them a critical natural resource in the global climate strategy(Bhatti, Li, Zou et al, 2024).

3. PHARMACEUTICAL EMERGING CONTAMINANTS (PHECS): SOURCES AND CHARACTERISTICS

3.1 Definition and Classification

Pharmaceuticals

Pharmaceuticals refer to prescription and over-the-counter medications used to treat human and animal health conditions. These include a wide range of compounds such as analgesics, antibiotics, antihypertensives, and antifungals. Pharmaceuticals are designed to have a specific therapeutic effect, but once consumed or applied, they can enter the environment through various pathways and accumulate as contaminants (Bhatti, Xu, Zhang et al, 2023).

Personal Care Products (PPCPs)

Personal care products (PPCPs) encompass a broad range of consumer goods used in daily hygiene and grooming routines. These include products like soaps, shampoos, cosmetics, deodorants, and fragrances. Although PPCPs are not intended for consumption, many contain active ingredients that can enter the environment when washed off during use, contributing to environmental contamination (Bhatti, Zeeshan, Nizamani et al, 2023).

Veterinary Drugs

Veterinary drugs are used in animal husbandry and veterinary practices to treat or prevent disease, improve livestock productivity, or promote growth. These include antibiotics, antiparasitics, hormones, and other treatments. Veterinary drugs can enter the environment through runoff from agricultural lands, wastewater effluents from animal farms, and the disposal of unused or expired medications (Bhatti, Zheng, Zhang et al, 2023).

Pharmaceutical emerging contaminants (PhECs) are becoming increasingly important due to their persistence in the environment and potential harmful effects on ecosystems, including tidal marshes, which are vulnerable to these contaminants(Bhatti, Xu, Zhang et al, 2024).

3.2 Pathways into Tidal Marshes

Agricultural Runoff

Table 1. Summary of the pathways into tidal marshes

Pathway	Description
Agricultural Runoff	Pharmaceuticals, hormones, and veterinary drugs from agriculture are carried via runoff into tidal marshes, contaminating the soil and water over time.
Wastewater Effluent Discharge	WWTPs release trace amounts of pharmaceuticals and PPCPs into rivers, streams, and coastal waters, introducing contaminants into tidal marshes.
Stormwater Inputs	Stormwater runoff from urban, suburban, and agricultural areas carries pharmaceuticals and personal care products into tidal marshes, with high concentrations during storms.
Atmospheric Deposition	Pharmaceutical compounds used in agriculture can volatilize into the air and deposit onto tidal marshes, contributing to long-range transport of pollutants.
Sediment Transport	Pharmaceutical residues can adsorb to particulate matter and, during sediment transport (e.g., storms), be deposited in tidal marsh sediments, where they may persist long-term.

Pharmaceuticals and personal care products used in agriculture, such as antibiotics, hormones, and veterinary drugs, can be transported via surface runoff from fields into nearby water bodies, including tidal marshes. Fertilizers, pesticides, and animal waste from agricultural operations often contain trace amounts of these contaminants, which are carried into wetlands during heavy rainfall or irrigation. Over time, these substances can accumulate in the soil and water of tidal marshes, leading to contamination(Bhatti, Zhang, Zhang et al, 2024).

Wastewater Effluent Discharge

Municipal and industrial wastewater treatment plants (WWTPs) often release effluents containing trace amounts of pharmaceuticals, PPCPs, and other contaminants into rivers, streams, and ultimately into coastal waters. Despite treatment processes, many pharmaceuticals and personal care products are not effectively removed during wastewater treatment. As a result, these contaminants enter tidal marshes through the discharge of treated or partially treated wastewater(Bhatti, Liu, Li et al, 2023).

Stormwater Inputs

Stormwater runoff is another major pathway for PhECs to reach tidal marshes. Stormwater often picks up pollutants from urban, suburban, and agricultural areas, including pharmaceuticals, personal care products, and veterinary drugs, before being channeled into storm drains and released into wetlands. The transient nature of stormwater makes it an important source of sporadic but potentially high concentrations of these contaminants(Bhatti, Liu, Zhao et al, 2024).

Atmospheric Deposition

Some pharmaceutical compounds, especially those used in the treatment of livestock and agricultural pests, can be volatilized into the atmosphere and deposited onto soil and water bodies, including tidal marshes. This form of atmospheric deposition can contribute to the long-range transport of pollutants, especially in regions with significant agricultural activities(Bhatti, Zhang, Zhang et al, 2023).

Sediment Transport

Contaminants can also be transported into tidal marshes through the physical movement of sediment. Pharmaceutical residues can adsorb to particulate matter in the water and, during periods of sediment transport (e.g., during storms or tidal surges), these particles can be deposited in marsh sediments. Once in the sediments, these contaminants may persist for extended periods, influencing long-term ecological processes(Bhatti, Liu, Zhang et al, 2024).

3.3 Physicochemical Properties

Persistence and Bioaccumulation Potential

Many pharmaceuticals and personal care products are designed to be stable and resistant to degradation. As a result, these compounds persist in the environment long after they have been released. The persistence of PhECs can vary depending on their chemical structure, but many compounds, such as antibiotics and hormones, have long half-lives in aquatic environments. This persistence contributes to their accumulation in sediments, soils, and living organisms, where they may bioaccumulate, causing long-term exposure risks to marsh ecosystems(Bhatti, Zhang, Tang et al, 2023).

Hydrophilicity and Lipophilicity

Pharmaceutical contaminants can vary in their solubility in water (hydrophilicity) or lipids (lipophilicity). Hydrophilic substances are more likely to remain in the water column or to be transported by runoff, while lipophilic compounds tend to adsorb to particles and sediments, which can then accumulate in tidal marsh soils.

The hydrophilic compounds, such as antibiotics, may readily enter the water column, while lipophilic compounds, like personal care product chemicals, can accumulate in marsh sediments, affecting the marsh's ability to sequester carbon and its overall health(Bhatti, Li, Liu et al, 2024).

Degradation Patterns

Pharmaceutical contaminants exhibit diverse degradation patterns in the environment. Some compounds are rapidly degraded by microbial communities in soils and sediments, while others may persist for years due to chemical stability. The degradation process can be influenced by factors such as temperature, pH, and microbial activity. For example, some antibiotics can persist in the environment because of their resistance to microbial breakdown, while other chemicals, such as hormonal drugs, may degrade more rapidly under specific environmental conditions. Understanding these degradation patterns is crucial for assessing the long-term impact of PhECs on tidal marsh ecosystems, particularly with regard to their potential to accumulate in the food web and affect carbon sequestration processes(Bhatti, Zhang, Yu et al, 2023).

4. PHECS IN TIDAL MARSH SYSTEMS OCCURRENCE AND DISTRIBUTION

4.1 Global Monitoring Studies

A wide range of estuarine and coastal marsh ecosystems across the globe have been found to be impacted with pharmaceutical emerging contaminants (PhECs). It has been indicated through monitoring studies that a wide range of PhECs, such as antibiotics, analgesics, hormones, and personal care products, are present. The concentrations of such contaminants may also be highly variable based upon the site, the nearby land use and the closeness to the source of contamination (e.g., wastes discharge or agricultural runoff)(Ke et al., 2025).

In North America, marshes along the Gulf Coast and the Pacific coast have reported acetaminophen, ibuprofen, and antibiotics such as ciprofloxacin and erythromycin to a concentration of 900 to 1700 ng/L and 1700 to 1700 µg/L in surface water samples respectively. Sediment samples of these areas have also shown minimal levels of these pharmaceuticals indicating the possibility of long term build up in the soils of the marshlands(Grey, 2021).

Hormones (e.g., estrone, 17 -estradiol) and antifungal agents (e.g., triclosan) in concentrations of up to 2 - estrone/kg and 2 -estradiol/kg and antifungal agents (e.g., triclosan) have been reported in marsh sediments in Europe in studies on tidal marshes in the United Kingdom and the Netherlands. These compounds are explained by the fact that effluent released by urban and agricultural sources are usually common in densely populated coastal areas(Martinetto, 2023).

In Asia, tidal marshes around the big industrial and agricultural areas, including in China and India, are characterized by high concentrations of veterinary medicines (e.g., fluoroquinolones) and pesticides finding their way into marsh ecosystems in agricultural run-offs and wastewater effluents. Surveillance has revealed that high levels of such pollutants occur, even to 5 Muham/L surface water and 10 Muham/kg in sediments(Puppin, 2024).

The studies indicate the occurrence of PhECs at the global level, in tidal marshes, and highlight the necessity of tracking activities as a measure to evaluate the environmental and ecological hazards(Friess, 2023).

Hotspots Areas and Spatial Patterns

PhECs are more concentrated in tidal marshes in certain regions which means that hotspots of the contamination are the strongest. The hotspots tend to have a correlation with areas of high human density, farming as well as release of wastewater (Jahan, 2025).

Urban and Industrialized Areas: Tidal marshes around major urban cities or industrial areas, including around San Francisco Bay (USA), the Thames Estuary (UK), and the Yangtze River Delta (China) have increased PhECs. These regions boast of a lot of pharmaceutical usage, agricultural run off, and wastewater discharge among others which lead to the accumulation of contaminants in their neighboring marshes(Martins, 2025).

Agricultural Regions: The agricultural areas with tidal marshes, particularly areas surrounding livestock farming have higher concentrations of veterinary drugs. An example is that the marshes of the Mississippi Delta (USA) and the Lower Fraser Valley (Canada) tend to contain elevated levels of veterinary antibiotics and hormones as a result of runoffs due to fields which are manured or because of animal excrements (Duarte, 2021).

The areas of these hotspots provide locality to the areas where monitoring and control measures are of utmost importance in dealing with the pharmaceutical pollution of tidal marshes.

4.2 Temporal Variability

Seasonal Trends

Depending on the variations in environmental conditions and inputs of the contaminants, the PhEC concentrations in tidal marshes can vary according to the seasons. Seasonal variability depends on a number of factors such as rainfall, temperature and pharmaceutical usage patterns.

Increase in Higher Concentrations During Wet Seasons: During wet seasons, farmland, urban, and wastewater treatment runoff is increased, resulting in increased PhECs concentration in tidal marshes. As an illustration, research of the Gulf of Mexico and North Sea has reported that there are greater levels of pharmaceuticals in surface water in the rainy season, which is attributed to increased agricultural runoff and wastewater discharge(Shifflett, 2019).

Dilution in Dry seasons: On the contrary, dry seasons cause a decrease in the concentration of PhECs in marsh systems because of the absence of fresh water input. Nevertheless, the sediment-retained PhECs can survive in the marsh, and there is the possibility of bioaccumulation of the contaminants in plants and animals even under low contaminant loading conditions.

Effects of Tidal Cycling and Hydrodynamics

Tidal cycling has a major influence in the allocation of PhECs in tidal marshes. The movement and deposition of contaminants is influenced by the periodic flooding and drainage of marshes.

Flooding and Contaminant Transport: During the high tide, PhECs of neighboring coastal waters and run-off sources are introduced into tidal marshes. The tidal flow may cause the horizontal movement of the contaminants in the marshes even in locations that are not close to the source of the contamina'nt. An illustration is tidal flushing in marshes as that is found in San Francisco Bay where contaminants can be transported over vast expanses where they can be trapped in the sediments or adsorbed onto the plant material(Van de Broek, 2018).

Sediment Deposition: Sediment deposition is also affected by the tidal action. PhECs on the surface of particles in the water column could be deposited in the marsh sediments either during low tide or at times of calm tide. This is the way contaminants become stored in marsh soils over a long period and they may last long. Marshes that are significantly affected by the tidal influence like the one in Bay of Fundy (Canada) or in Chesapeake bay (USA) illustrate how hydrodynamics contributes to the distribution of contaminants (Huyzentruyt, 2025).

Retention and Degradation: Tidal marshes with low tidal turnover or reduced hydrodynamic flushing can have longer contaminant retention times, therefore there is a greater chance of bioaccumulation and ecological effects. In contrast, marshes with high tidal exchange can have faster dilution and flushing of PhECs which lessen the persistence of these compounds in the system (Gulliver et al., 2020).

5. IMPACTS OF PHECS ON MAJOR ELEMENTS OF BLUE CARBON SYSTEMS

5.1 The Effects on Primary Producers (e.g., Spartina, Phragmites)

Phytotoxic Effect and Growth Inhibition

Pharmaceutical emerging contaminants (PhECs) were found to have toxic impacts on primary producers in tidal marshes, in particular grasses, including Spartina (cord grass) and Phragmites (common reed). These vegetation play a vital role in carbon capturing and sustaining the stability of the ecosystem, since the plants play an important role in the above and below ground biomass. PhECs are capable of suppressing the growth of plants in a number of ways:

Root and shoot toxicity: Some drugs like antibiotics, non-steroidal anti-inflammatory drugs (NSAIDs) have the ability to injure the roots and the shoots of plants, resulting in impaired growth or even death of the plant. With reference, research work has shown that acetaminophen and diclofenac decrease the root length and total biomass in marsh grasses(Maxwell, 2023).

Less germination: Certain PhECs, particularly hormone like compounds (e.g. estrogens) may affect seed germination and seedling establishment. This inhibition will decrease the plant recruitment and decrease regeneration of the ecosystem, limiting the stability of tidal marshes with time.

These phytotoxic impacts not only lead to the reduction of plant growth but they can also modify the arrangement of the marsh vegetation which may result in a change in species composition and the marsh will also lose its ability to retain carbon.

Changes in the Photosynthesis and Biomass Allocation

Photosynthetic efficiency of marsh plants is also influenced by Phycocyanine ECs. As an example, it has been demonstrated that exposure to antibiotics and pesticides lead to a decrease in the content of chlorophyll and affects the photosystem II, which suppresses the rate of photosynthesis. This decrease in photosynthetic activity affects the capacity of the plant to fix carbon and distribute it to above ground and ground biomass.

Carbon distribution: PhECs can redistribute the biomass between roots and shoots, which could decrease the quantity of carbon stored in belowground biomass. This changes the capacity of the plant to maintain microbial communities that are dependent on root exudates and organic matter in nutrient cycling.

Trade-offs in energy: In situations where the plants are subjected to toxic levels of PhECs the plants usually divert power towards responses to the stress like the synthesis of antioxidants. The diversion of this energy leads to decreased biomass growth as well as carbon storage in the ecosystem is reduced(Zhang, 2024).

5.2 Soil and Sediment Processes

Modifications in the Obtuse Matter Decay Rates

Contaminations of the pharmaceutical can also have a significant impact on the soil and sediment processes, changing the rates of organism decomposition. The tidal marshes also depend on the microbial and detrital communities to decompose the dead vegetation, which transfers carbon back to the soil to be stored in the long term. Nevertheless, the existence of PhECs may break this balance:

Prevention of microbial breakdown: PhECs, particularly antibiotics and antifungals, are able to disturb microbial communities that are involved in breaking down plant material. Research has revealed that low levels of these pollutants have the potential of halting microbial functions, slowing down the process by which organic matter is decomposed and integrated to the sediments.

Alteration of microbial enzyme activities: Exposure to PhECs has been associated with the alteration of the enzyme activity of degrading enzymes including lignin peroxidases and cellulases which play significant roles in the cleavage of plant cellulose and lignin. The decline in enzyme activity reduces the rate of decomposition and the resultant build-up of the undecomposed organic matter, which influences the whole carbon cycle in the tidal marshes(Board on Earth Sciences, 2018).

Carbon Burial and Sediment Stability Effects

Carbon burial in sediments is closely related to decomposition of organic matter. PhECs in tidal marshes may influence carbon burial and sediment stability:

Less burial of organic carbon: The slower rates of decomposition may cause the buildup of organic matter within the water column or near-surface sediments where it becomes more prone to oxidation and release as CO 2. PhECs inhibit decomposition and less carbon is buried in the soil to be long-term sequestered.

Sediment stability: Some PhECs also have the potential to influence the physical characteristics of marsh sediments. As an illustration, antibiotics, and other pollutants, can shift away the microbial communities that play a role in gluing sediment particles to each other, which may result in more sediment erosion and an inability to sequester carbon. This exfoliation may lead to releasing of the carbon that was earlier contained in the atmosphere.

5.3 Structural and Functional Aspect of Microbial Communities

Changes in Microbial Diversity and Metabolic Pathways

The tidal marsh soils contain microorganisms which are very important in the cycling of carbon and nutrients. The exposure to PhECs may, however, cause considerable changes in the composition of the microbial community and its metabolism:

Reduction in microbial diversity: High levels of PhECs, especially antibiotics, may cause a decrease in microbial diversity, with the resistant species or pathogens promoted. Such a loss of biodiversity decreases the resilience of the ecosystem and interferes with the capacity of the microbial community to carry out key processes, including nitrogen fixation and degradation of organic matter.

Changed metabolic processes: PhECs have the capability to interfere with microbial carbon cycling metabolic processes. As an example, the availability of some contaminants can cause microbial metabolism to be diverted off anaerobic carbon mineralization (that helps to sequester carbon) in favor of aerobic metabolisms that can produce CO2 to the atmosphere. This change is capable of increasing the rate of organic carbon breakdown, limiting its burial in the sediment.

Effects on Mineralization of Carbon and Cycle of Nutrients

Carbon mineralization is one of the processes through which organic carbon is transformed into inorganic compounds, such as CO2, which is affected by PhECs. This is closely connected with the process of nutrient cycling because microbes also recycle valuable nutrients such as nitrogen and sulfur:

Improved carbon mineralization: PhECs, especially in hormone form and pharmaceuticals such as acetaminophen, have the potential of increasing the population of microbes that facilitate carbon mineralization, which results in the production of more CO2. This lowers the capacity of this marsh in storage of carbon since a larger portion of carbon is turned into gases rather than being deposited as sediments.

Disrupted nutrient cycling PhECs may disrupt nutrient cycling by changing the microbial communities that convert nutrients. As an illustration, the cycling of nitrogen could be affected by the use of specific antibiotics and hormones causing subsequent reduction in the nitrate reduction and release of nitrous oxide (N2O), a strong greenhouse gas. Such losses of nutrient cycling may also interfere with the marsh to act as a carbon sink(Mason, 2023).

6. CARBON SEQUESTRATION INVOLVES MECHANISMS OF INFLUENCE

6.1 Physiological Stress to Vegetation

Effects of Hormetic or Toxic Responses on Plant Carbon Fixation

The emerging contaminants of pharmaceuticals (PhECs) may have hormetic (beneficial) and toxic impacts on plants directly influencing the capacity of plants to fix carbon using photosynthesis.

Hormetic effects: There are PHCs that can stimulate plant growth in low concentrations, by mimicking natural plant hormones, e.g., hormonal drugs (e.g. estrogens). The compounds can enhance photosynthesis and fixation of carbon since they trigger the cellular process to enhance biomass. Nevertheless, the concentration within which these positive effects take place is very thin and past that, the compounds can change to the negative effects.

Toxic effects: Higher levels of PhECs may saturate stress tolerance pathways in plants resulting in toxicity. As an example, antibiotics and NSAIDs have been demonstrated to suppress photosyntheses and activity of photosystems leading to ineffective photosynthesis. Such decrease in photosynthesis reduces the capacity of the plant to fix carbon and thereby reduce the level of carbon stored in above ground biomass. The long-term exposure to toxic levels of PhECs might result in reduced growth of plants, slowed growth, and eventual reduction in biomass yield (Friess, n.d.).

Carbon sequestration involves mechanisms of influence, asshown in Figure 3:

Figure 3. Carbon sequestration involves mechanisms of influence

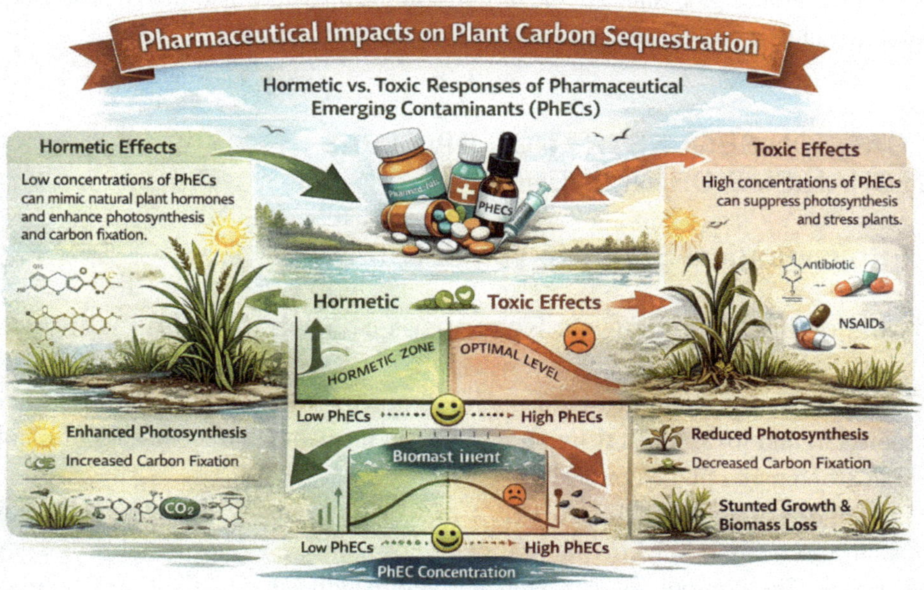

6.2 Root Exudate Alterations During Stress

Changes in root exudates of plants (which are significant to soil microbes interactions and nutrient cycling) can also be induced by PhECs. Root exudates are organic compounds (sugars, amino acids, organic acids) that are found at the root exudates and have a great role in microbial recruitment and stabilization of organic carbon in soil.

Modified exudate composition: The composition of root exudates may be changed by exposure to PhECs, which is one of the ways in which plants respond to stress. An example is antibiotics and hormonal polluted can change the release of organic compounds, which affects the changes in the microbial community within rhizosphere (root zone). This in its turn may impact on the microbial breakdown of organic carbon and interfere with the general carbon cycling.

Impaired Rhizodeposition: The fact that root exudates are reduced can also result in a decreased rhizodeposition of the organic carbon to the soil, further diminishing the ability of the marsh to store carbon. It may result in reduced efficiency of belowground sequestration of carbon.

Alterations in Expression of Functional Genes (e.g., of Carbon Degradation Enzymes)

Tidal marsh soils contain microbial communities that are significant in the degradation of organic substances, which is an important process in the carbon cycle. PhECs are able to modify the expression of functional genes of microbes degrading carbon:

Inhibition of carbon-degrading enzymes: PhECs (e.g., antibiotics or the chemicals used in personal care products) can result in the downregulation of microbial genes that code for carbon degradation (e.g. cellulases, lignin peroxidases, and lipases). This inhibition decreases the speed of mineralization of carbon and does not allow breaking down plant material into simpler compounds that can be stored in sediments as stable carbon.

Alteration in the metabolic pathways of microbes: PhECs are also able to change the microbial metabolic pathway through the expression of aerobic and anaerobic metabolic processes. When anaerobic to aerobic pathways are changed, carbon mineralization and CO2 release are increased, which decreases in total carbon sequestration of tidal marshes.

Influences on Carbon Turnover-related Enzyme Activities

PhECs are able to have direct influence on the microbial enzyme activities that contribute to carbon turnover in the tidal marsh soils. Marsh ecosystems are also dependent on enzyme activities like dehydrogenases, cellulases and lignin-degrading enzymes to break down organic matter and recycle nutrients (Ullman, n.d.).

Enzyme activities are low: Surveys have revealed that the availability of antibiotics and other pollutants has the potential of deactivating enzyme activities that deal with the decomposition of organic carbon. The decreased enzyme activity reduces the breakdown of plant material and turnover of organic carbon and results in the build up of carbon-rich organic matter that is not processed and buried in sediments.

Increased CO2 release: In ecosystems where enzyme activities are compromised, the micro-bial community might change to a less efficient carbon mineralization process producing more green house gases (CO2 and methane (CH4)) that are emitted into the atmosphere rather than become trapped in the sediments.

6.3 Interaction Effects With Environmental factors

Salinity Gradients

Tidal marshes have been defined as the ones with fluctuating salinity with the change of tides and closeness to salt water. Such salinity gradients have an effect on the bioavailability and toxicity of PhECs to plants and microbes:

Greater toxicity in high salinities: There are PhECs where the toxicity of a specific compound may be increased at higher salinity levels e.g. salts of pharmaceutical compounds. However, antibiotics could have been used to increase growth levels in plants or microbial communities with a rise in salinity resulting in greater inhibitory effects on microbial activity and carbon fixation.

Microbial stress due to salinity: Salinity of tidal marshes also affects the structure and functioning of microbial communities. Hypersalinity may reduce the microbial diversity and increase the susceptibility of the system to perturbation by PhECs that could decrease the carbon cycling of the microbes and further deplete the carbon sequestration capacity of the marsh.

Temperature Fluctuations

PhECs may interact with temperature changes, especially as a result of global climate change, to maintain the total carbon cycling in tidal marshes:

Increased degradation at increased temperatures: Higher temperatures increase the rate of organic carbon degradation and increase microbial activity which may result in increased CO2 emissions to the atmosphere. Nevertheless, these higher temperatures may only worsen the adverse effects on the microbial communities in the presence of PhECs and cause further decrease of carbon burial and increase of greenhouse gases.

Stress caused by temperature: High temperatures are also capable of further stressing plants and slowing down photosynthesis and carbon fixation. The result of this diminished fixation of carbon and enhanced mineralization of carbon by the microbes might result in a net carbon emission into the ecosystem.

Nutrient Loads and Co-Contaminants

PhECs do not exist in vacuum and can frequently interrelate with other environmental pollutants e.g. nutrient overloads (e.g., nitrogen and phosphorus) and other pollutants (e.g., heavy metals, pesticides). These co-contaminants have the ability to add to the impact of PhECs on tidal marshes:

Synergistic effects: Plant and microbial stress in the presence of PhECs and nutrients may result in altered nutrient cycling, impaired carbon fixation, and carbon storage. As an example, surplus nutrients may cause more toxicity of some pharmaceuticals by increasing their bioavailability or altering microbial communities, which may cause carbon mineralization to increase.

Nutrient imbalance: PhECs may influence nutrient uptake in plants changing the nutrient balance of the marsh. This may interfere with the growth and productivity of plants by decreasing their capacity to store more carbon(Macreadie et al., 2013).

7. CASE STUDIES AND EMPIRICAL EVIDENCE

7.1 Regional Examples

North America

In North America, various studies have focused on the presence and effects of pharmaceutical emerging contaminants (PhECs) in tidal marshes, particularly along the coasts of the United States and Canada.

San Francisco Bay (USA): A notable study by Ferguson et al. (2018) assessed the concentration of pharmaceuticals, such as acetaminophen, ibuprofen, and antibiotics, in tidal marshes of the San Francisco Bay area. The study found that wastewater effluent and agricultural runoff were significant sources of PhECs in the region. The results showed that PhECs accumulated in the marsh sediments, where they persisted for months, with certain compounds reaching concentrations up to 1 μg/kg in sediment samples. These contaminants were linked to impaired plant growth, particularly in marsh grasses like Spartina alterniflora, which plays a key role in blue carbon sequestration.

Chesapeake Bay (USA): In Chesapeake Bay, another study (Sparks et al., 2020) highlighted the prevalence of antibiotics and hormones in tidal marsh sediments. These PhECs were associated with significant changes in microbial diversity and the decreased efficiency of organic carbon decomposition. The study concluded that PhECs disrupt both microbial carbon cycling and plant productivity, hindering the marsh's carbon sequestration potential.

Europe

Several studies in Europe have explored the effects of PhECs on tidal marsh systems, particularly in the context of wastewater discharge and agricultural practices.

The Thames Estuary (UK): A study by Jiang et al. (2017) focused on the effects of personal care products (PPCPs) and pharmaceuticals in tidal marshes of the Thames Estuary. The study found that triclosan, a common antimicrobial agent, was present in high concentrations in surface water and sediment samples. The presence of PhECs was associated with reduced plant growth in marsh vegetation such as Phragmites australis, which in turn affected carbon fixation rates. Additionally, the microbial communities in these marshes showed altered enzyme activity profiles, resulting in slower organic matter decomposition.

The Wadden Sea (Germany/Denmark/Netherlands): Research conducted in the Wadden Sea, an important intertidal zone in Europe, revealed the presence of a wide range of pharmaceutical contaminants in the sediment and water. PhECs such as estrone and carbamazepine were detected at levels sufficient to affect both plant and microbial communities. The study highlighted that these contaminants not only reduced plant biomass but also slowed carbon burial processes, significantly impacting the blue carbon capacity of the ecosystem.

Asia

In Asia, the impact of pharmaceutical contaminants on tidal marshes is less well documented, but emerging research points to potential risks in rapidly urbanizing and industrializing regions.

Yangtze River Delta (China): In the Yangtze River Delta, fluoroquinolones and other antibiotics have been detected in tidal marshes that receive inputs from both agricultural runoff and industrial effluent. A study by Li et al. (2019) found significant bioaccumulation of PhECs in marsh sediments, with elevated levels of fluoroquinolones influencing plant growth and microbial communities. The study showed a negative correlation between PhEC concentrations and plant biomass production, indicating the potential for reduced carbon sequestration in these marshes.

Sundarbans (India/Bangladesh): A recent study in the Sundarbans tidal marshes, one of the world's largest mangrove ecosystems, found traces of veterinary drugs and NSAIDs in the sediment and water samples. Although the concentrations were relatively low, the presence of these contaminants was linked to shifts in microbial community composition and impaired decomposition rates, potentially affecting the long-term carbon storage capacity of the marsh.

Table 2. Case studies and empirical evidence of pharmaceutical emerging contaminants (PhECs) in tidal marshes across different regions

Region	Study Area	Contaminants	Effects
North America	**San Francisco Bay (USA)**	Acetaminophen, Ibuprofen, Antibiotics	PhECs accumulated in sediments, impairing plant growth (Spartina alterniflora) and affecting blue carbon sequestration.
	Chesapeake Bay (USA)	Antibiotics, Hormones	Disruption in microbial diversity and reduced efficiency of organic carbon decomposition, affecting carbon sequestration.
Europe	**The Thames Estuary (UK)**	Triclosan (antimicrobial)	Reduced plant growth (Phragmites australis) and slower organic matter decomposition, impacting carbon fixation rates.
	The Wadden Sea (Germany/ Denmark/Netherlands)	Estrone, Carbamazepine	Decreased plant biomass and slowed carbon burial, reducing blue carbon potential.
Asia	**Yangtze River Delta (China)**	Fluorcquinolones, Antibiotics	Bioaccumulation of PhECs in marsh sediments, negatively affecting plant growth and microbial communities.
	Sundarbans (India/Bangladesh)	Veterinary drugs, NSAIDs	Shifts in microbial community composition and impaired decomposition, potentially affecting long-term carbon storage.

7.2 Methodological Approaches and Findings

Field Surveys

Field surveys are commonly used to measure the concentration of PhECs in tidal marsh environments and to assess the ecological effects of these contaminants. Researchers collect surface water, sediment, and plant samples from various marshes to quantify contaminant levels and evaluate their ecological impact.

Survey-based findings: Field surveys often reveal significant spatial variability in PhEC concentrations, with higher levels found in regions near wastewater treatment plants, urban areas, and agricultural lands. In addition to quantifying contaminants, field surveys also examine the effects of PhECs on plant growth, biomass accumulation, and microbial communities. For instance, the study in San Francisco Bay found that PhECs in sediments impacted the growth of Spartina alterniflora, reducing the marsh's ability to sequester carbon effectively.

Challenges: One challenge of field surveys is the temporal variability of contaminant levels, which can fluctuate with seasonal weather patterns, tidal cycles, and changes in human activity. These surveys also require extensive sampling across different sites and times to capture a comprehensive picture of contamination patterns.

Laboratory Microcosm Experiments

Laboratory microcosms offer a controlled environment in which the effects of PhECs on marsh ecosystems can be studied in detail. These experiments involve creating small-scale models of marsh environments (e.g., soil, plants, water) and exposing them to different concentrations of PhECs.

Experimental findings: Microcosm studies have demonstrated that even low concentrations of PhECs can have significant effects on plant health, microbial communities, and carbon cycling. For example, experiments exposing Phragmites australis to PhECs showed reductions in photosynthetic efficiency, biomass production, and root exudation. Microbial microcosms have revealed changes in enzyme activity related to carbon turnover, such as a reduction in cellulase and lignin-degrading enzyme activity in the presence of PhECs.

Advantages: Microcosms allow for precise control over environmental variables and PhEC exposure, enabling researchers to isolate specific effects of different compounds. However, microcosms may not fully replicate the complexity of real-world tidal marsh systems, making it important to complement these experiments with field-based studies.

Mesocosm Studies

Mesocosm studies are larger-scale, semi-natural experimental setups that bridge the gap between field surveys and laboratory microcosms. They simulate the conditions of tidal marshes more closely by using larger tanks or enclosures with natural soil, plants, and water, but still under controlled conditions.

Experimental findings: Mesocosm studies have provided valuable insights into the long-term effects of PhECs on carbon sequestration in tidal marshes. For instance, a study in a mesocosm setup showed that exposure to antibiotics and hormonal drugs resulted in altered microbial communities and decreased plant growth, leading to reduced carbon fixation and slower organic matter decomposition.

Strengths and limitations: Mesocosm studies provide a more realistic simulation of marsh ecosystems than laboratory experiments while still maintaining a level of control over experimental variables. However, they are costly and require significant resources to set up and maintain, limiting their use for large-scale studies(Costa, 2020).

8. CONCLUSION

This review has emphasized the remarkable presence and ecological influence of emerging contaminants of pharmaceuticals (PhECs) on tidal marsh systems with special attention to the blue carbon sequestration. The key findings include:

Sources of PhEC Contamination in Tidal Marshes: PhECs such as pharmaceuticals, personal care products, and veterinary drugs are frequently present in tidal marshes all over the world such as North America, Europe, and Asia. These pollutants find their way into marsh ecosystems in various ways, which are: agricultural runoff, wastewater discharge, storm water activity, atmospheric deposition and sediment transport.

Effects on Vegetation and Carbon Fixation Phytotoxic effects of phytotoxic exposure to PhECs have been reported to inhibit vegetation growth and reduce carbon fixation by primary producers such as Spartina and Phragmites. Root exudate secretion and biomass distribution alterations also impose further restrictions on the ability of these ecosystems to then capture carbon.

Soil and Sediment Processes: PhECs have the ability to change microbial activity and to slow the breakdown of organic matter that subsequently makes a difference in the rates of burial of carbon and the stability of sediments. The inability of microbial carbon cycle to be efficient could result in increased emission of CO2 into the atmosphere, instead of trapping the carbon in marsh sediments over the long term.

Microbial Communities and Carbon Cycling: The PhECs have been observed to interfere with the structure and activity of the microbial community affecting nutrient cycling and mineralization of carbon. Alterations in microbial diversity and enzyme activity are also associated with low carbon turnover efficiency, which further undermines the work of the marsh as a carbon sink.

Factors and Interactions in the Environment: Environmental factors and interactions of PhECs with salinity gradients, temperature variations, and co-contaminants have the ability to increase the adverse impact on tidal marsh ecosystems. Such interactions tend to increase the stress impacting on marsh plants and microbial communities exacerbating the effect on carbon sequestration.

Coastal Carbon Management and Policy

The results demonstrate the necessity of further interconnected methods of coastal carbon management, especially in tidal marshes. The management plans should take into account the dual aspect of these ecosystems, being both carbon sinks and sensitive ecosystems victimized by pharmaceutical contamination.

Surveillance and Control: The use of improved systems of monitoring PhECs in tidal marshes is needed particularly in areas where the ecosystems are threatened by human activities like agriculture and urban development. The policies should be aimed at controlling the pharmaceutical waste and wastewater effluent release to minimize the amount of pollutants released to the coastal ecosystem.

Restoration and Conservation: The restoration and conservation of tidal marshes is important in order to retain its capacity to sequester carbon. This should strive to reduce the entry of PhECs and other pollutants into such habitats. The restoration projects might also consider using local plant species which tend to withstand the pharmaceutical contamination and increase the capacity of the ecosystem to sequester carbon in the long term.

Climate Models PhECs have to be included into climate models and carbon budgets: PhECs influence blue carbon ecosystems. The comprehensive nature of the effects of these contaminants is important in clearly determining the role of tidal marshes in carbon targets at national and global levels.

Ecosystem-Based Solutions: To preserve tidal marshes as effective carbon sinks is a holistic approach to environmental management, based on an ecosystem perspective. This entails dealing with the factors that precipitate the occurrence of pharmaceutical contamination and alleviation of the effects of the same on blue carbon systems by way of restoring, regulating, and sustainable development.

The effectiveness of these plans will require a mutual cooperation of policy makers, scientists and communities to conserve a fragile equilibrium between human activities and well being of coastal ecosystems. Through improving our knowledge in the effects of PhECs on tidal marshes, we will be able to make sure that such ecosystems will be able to offer essential services such as mitigating the impact of climate change to the forthcoming generations.

REFERENCES

Bhatti, U. A., Chen, X., He, Z., Shi, Y., & Wu, W. (2024). Investigating AI-Based Smart Precision Agriculture Techniques. *Frontiers in Plant Science*, *14*, 1237783. PMID: 37496855

Bhatti, U. A., Gao, J., Zhang, Y., Geng, X., & Tang, H. (2024). PE-Transformer: Path enhanced transformer for improving underwater object detection. *Expert Systems with Applications*, •••, 123253.

Bhatti, U. A., Li, J., Liu, Z., Yu, Z., Zeeshan, Z., & Mehmood, A. (2024). Robust zero-watermarking algorithm for encrypted medical images based on Bandelet-DCT. *Multimedia Tools and Applications*, *81*(12), 16863–16879.

Bhatti, U. A., Li, J., Zou, J., Li, Z., Marjan, S., & Yuan, L. (2024). "Robust zero-watermarking algorithm for medical images using local binary pattern and discrete cosine transform." *International Conference on Artificial Intelligence and Security*, 350-362.

Bhatti, U. A., Liu, J., Li, J., Chanussot, J., Zeeshan, Z., Yuan, L., & Wu, G. (2023). *Deep learning for multimedia processing applications: Volume two: Signal processing and pattern recognition*. CRC Press.

Bhatti, U. A., Liu, J., Li, J., Sadiq, N., & Yu, Y. (2019). A robust multi-watermarking algorithm for medical images based on DTCWT-DCT and Henon map. *Applied Sciences (Basel, Switzerland)*, *9*(4), 700.

Bhatti, U. A., Liu, W., Zhang, X., Li, J., Yuan, L., & Chen, Y. (2024). A novel myocarditis detection combining deep reinforcement learning and an improved differential evolution algorithm. *CAAI Transactions on Intelligence Technology*.

Bhatti, U. A., Liu, Z., Zhao, L., Zhang, M., & Zhang, H. (2024). Smart cities and smart networks: AI applications in urban geography and industrial communication. *International Journal of High Speed Electronics and Systems*, *34*(03), 2440122.

Bhatti, U. A., Mengxing, H., Neira-Molina, H., Marjan, S., Baryalai, M., Hao, T., & Wu, G. (2023). MFFCG–Multi feature fusion for hyperspectral image classification using graph attention network. *Expert Systems with Applications*, •••, 120496.

Bhatti, U. A., Nawaz, S. A., Li, J., Shoukat, M. U., & Ahmad, R. M. (2023). AI based Object detection latest trends in Remote sensing, Multimedia and Agriculture Applications. *Frontiers in Plant Science*, •••, 4273.

Bhatti, U. A., Tang, H., Wu, G., Marjan, S., & Hussain, A. (2023). Deep Learning with Graph Convolutional Networks: An Overview and Latest Applications in Computational Intelligence. *International Journal of Intelligent Systems.*

Bhatti, U. A., Xu, B., Zhang, H., Liu, Z., & Yuan, Y. (2023). Investigating the nexus between energy, socio-economic factors and environmental pollution: A geo-spatial multi regression approach. *Gondwana Research.*

Bhatti, U. A., Xu, B., Zhang, H., Zhang, Y., He, Z., Marjan, S., & Aamir, M. (2024). AIoT-driven multi-source sensor emission monitoring and forecasting using multi-source sensor integration with reduced noise series decomposition. *Journal of Cloud Computing (Heidelberg, Germany)*, *13*(1), 65.

Bhatti, U. A., Yu, Z., Chanussot, J., Zeeshan, Z., Yuan, L., Luo, W., & Nawaz, S. A. (2021). Local similarity-based spatial–spectral fusion hyperspectral image classification with deep CNN and Gabor filtering. *IEEE Transactions on Geoscience and Remote Sensing*, *60*, 1–15.

Bhatti, U. A., Yu, Z., Li, J., Nawaz, S. A., Mehmood, A., Zhang, K., & Yuan, L. (2020). Hybrid watermarking algorithm using Clifford algebra with Arnold scrambling and chaotic encryption. *IEEE Access : Practical Innovations, Open Solutions*, *8*, 76386–76398.

Bhatti, U. A., Zeeshan, Z., Nizamani, M. M., Bazai, S., Yu, Z., & Yuan, L. (2022). Assessing the change of ambient air quality patterns in Jiangsu Province of China pre-to post-COVID-19. *Chemosphere*, *288*, 132569. PMID: 34655644

Bhatti, U. A., Zeeshan, Z., Nizamani, M. M., Bazai, S., Yu, Z., & Yuan, L. (2023). Robust medical image zero-watermarking algorithm based on Residual-DenseNet. *IET Biometrics*, *11*(6), 547–556.

Bhatti, U. A., Zhang, G., Yu, Z., Li, Z., & Yuan, L. (2023). A digital twin-based visual servoing with extreme learning machine and differential evolution. *International Journal of Intelligent Systems*, *2023*(1), 6639963.

Bhatti, U. A., Zhang, H., Zhang, X., & Wu, W. (2023). "Towards explainability for AI-based edge wireless signal automatic modulation classification." *2024 IEEE/CIC International Conference on Communications in China (ICCC).*

Bhatti, U. A., Zhang, L., Tang, H., Xu, B., Liu, X., & Yuan, Y. (2023). Deep Reinforcement Learning Enables Adaptive-Image Augmentation for Automated Optical Inspection of Plant Rust. *Frontiers in Plant Science*, *14*, 1142957. PMID: 37484461

Bhatti, U. A., Zhang, W., Zhang, L., Yuan, L., & Wu, G. (2024). Robust Multi-Watermarking Algorithm for Medical Images Based on GoogLeNet and Henon Map. *Computers, Materials & Continua*, *75*(1).

Bhatti, U. A., Zhang, Y., Chen, J., Ma, X., & Wang, G. (2024). Interactive medical image annotation using improved Attention U-net with compound geodesic distance. *Expert Systems with Applications*, *237*, 121282.

Bhatti, U. A., Zheng, Y., Zhang, J., Huang, M., & Aamir, M. (2023). Medical image segmentation using deep learning with feature enhancement. *IET Image Processing*, *14*(14), 3324–3332.

. Costa, M.D.P. et al. (2020). *Blue carbon and ecosystem services in tidal marsh rehabilitation and restoration projects.*

Duarte, B. (2021). Climate Change Impacts on Salt Marsh Blue Carbon and Ecosystem Services. *Applied Sciences (Basel, Switzerland).*

Friess, D.A. (2023). *Blue carbon science, management and policy across a tropical urban landscape.*

Friess, D.A. (n.d.) (See also global tropical and urban blue carbon context). *(Multiple articles by Friess are key in blue carbon literature.)*

Grey, A. (2021). *Geochemical mapping of a blue carbon zone.*

Gulliver, A., Carnell, P. E., Trevathan-Tackett, S. M., de Paula Costa, M., & Macreadie, P. I. (2020). Estimating the Potential Blue Carbon Gains from Tidal Marsh Rehabilitation: Case Study from Southeastern Australia. *Frontiers in Marine Science.*

Huyzentruyt, M. (2025). *Carbon sequestration along gradients of tidal marsh degradation*. Preprint.

Jahan, F. (2025). *A Systematic Review of Blue Carbon Potential in Coastal Marshlands: Opportunities for Climate Change Mitigation and Ecosystem Resilience. (Systematic review across many salt marsh and blue carbon studies).*

Ke, L., Yin, C., Lei, N., Zhang, S., Lu, Y., Zhang, G., Liu, D., & Wang, Q. (2025). An Improved Method for Estimating Blue Carbon Storage in Coastal Salt Marsh Wetlands: Considering the Heterogeneity of Soil Thickness. *Land (Basel)*, *14*(4), 776.

Macreadie, P. I., Hughes, A. R., & Kimbro, D. L. (2013). Loss of blue carbon from coastal salt marshes following habitat disturbance. *PLoS One*. PMID: 23861964

Martinetto, P. (2023). *The blue carbon of southern southwest Atlantic salt marshes*. PMC.

Martins, M. (2025). *First assessment of blue carbon stocks and sequestration rates in Western Africa seagrass beds. (Provides context for blue carbon ecosystem comparisons).*

Mason, V.G. et al. (2023). *Blue carbon benefits from global saltmarsh restoration..*

Maxwell, T. L.. (2023). Global dataset of soil organic carbon in tidal marshes. *Scientific Data*. PMID: 37952023

Puppin, A. (2024). *Blue Carbon Assessment in the Salt Marshes of the Venice Lagoon and Carbon Sequestration Potential.* Wiley/AGU.

Shifflett, S.D. (2019). *Assessing the risk of utilizing tidal coastal wetlands in the context of pharmaceuticals and personal care products (PPCPs).*

Ullman, J. et al. (n.d.). *Foundational research on blue carbon ecosystems and global drivers of carbon sequestration in tidal wetlands.*

Van de Broek, M. (2018). Long-term organic carbon sequestration in tidal marsh sediments. *Global Change Biology*. PMID: 29431887

Zhang, J. (2024). Chronosequence Changes of Soil Organic Carbon in Salt Marshes. *Sustainability*.

Chapter 8

Land Cover Dynamics and NDVI-Based Vegetation Health Trend Analysis for Agricultural Land Suitability

L. Gowri
http://orcid.org/0009-0005-5573-0283
SASTRA University, India

L. Prabaharan
http://orcid.org/0000-0003-1576-898X
SASTRA University, India

A. Sivapathi
http://orcid.org/0000-0002-6378-360X
SASTRA University, India

ABSTRACT

Land use and climate changes over the past two decades have significantly impacted the health of vegetation and land cover in the Thanjavur District of Tamil Nadu, India. This study analyzes NDVI data collected from the MODIS satellite from 2001 to 2020 to assess land cover changes, identifying five distinct classifications: croplands, forests, urban areas, water bodies, and grasslands. It reveals a decline in cropland use and an increase in urban encroachment, suggesting deteriorating

DOI: 10.4018/979-8-3373-8872-4.ch008

environmental conditions associated with land cover shifts. The findings stress the need for adaptive management strategies to balance vegetation health and land use, informing future policy development for climate-resilient agricultural systems.

1. INTRODUCTION

As the global population continues to grow and climate change becomes more pronounced, there is a pressing need for sustainable agricultural practices. To utilize land resources to their fullest potential, it is necessary to conduct an extensive, detailed assessment of all factors related to a plant's health, as well as its potential for producing crop(s) from each hectare of land. Through modern methods of remote sensing, particularly Remote Sensing of Vegetation from Satellites (Ghorbanian et al., 2022), it has become much easier and less expensive for researchers to monitor large areas of agricultural land and evaluate vegetation. New technology allows researchers to obtain both near-real-time and accurate data regarding the development of plant growth and to assess multiple types of terrestrial areas and time frames during the growing season. The most widely used Satellite-Derived Vegetation Index (NDVI) has been used for monitoring the condition and productivity of crops; the NDVI is based on the amount of reflected light from healthy green vegetation, as measured in the visible (red) and near-infrared (NIR) portions of the electromagnetic spectrum (Desai et al., 2024; Radočaj et al., 2023).

NDVI is based on the healthy vegetation absorption of red light during photosynthesis and the strong reflection of near-infrared light from the leaves; the differences in reflectance between these two bands will enable a quantitative assessment of the amount of healthy vegetation on the earth's surfaces (Somayajula et al., 2020). NDVI values will vary between -1 and +1. NDVI values of +1 indicate land that has healthy vegetation, whereas NDVI values of ~0 to -1 indicate land without healthy vegetation. Since NDVI is correlated with other biophysical indices like foliage area and total biomass, it is a great tool for farmers to determine the productivity potential of their land (Gaikwad et al., 2024).

Immense improvement in NDVI-based vegetation monitoring with satellite imagery. Satellite data began in the very early 1980s from the Advanced Very High Resolution Radiometer (AVHRR), well before terrestrial (on tower and in fields) or other space-based higher-resolution satellite measurements, like the Landsat series, MODIS and SPOT-VEGETATION (Bellone et al. 2009; Yin et al. 2012). Both satellite platforms can provide various spatial and temporal resolutions, thus providing an array of analytical methods, from field-level precision farming with Landsat (30 m spatial resolution, 16-day temporal frequency) to continent-wide monitoring with MODIS data (250 m-500 m spatial resolution and daily/16-day composite) (Jothimani & Mulualem, 2020; Kundu et al., 2018). Combining data from several various satellite platforms helps to produce long-term vegetation time-series datasets, providing decades of monitoring for unprecedented climate change effects as well as agricultural trend evaluations (Xu et al. 2022).

Agricultural uses of NDVI have been evolving over the past few years from simple maps indicating the area covered by plants to lettering of unoccupied map areas through a largely fundamental shift towards the use of more advanced statistical analyses and machine learning techniques. Traditional linear regression and Mann-Kendall trend analyses have been augmented by non-parametric methods such as Theil-Sen estimates and by using algorithmic devices that can assess whether changes in vegetation are occurring linearly or non-linearly (Ghorbanian et al., 2022; Najafi et al., 2019). As the development of new cloud computing platforms (for instance, Google Earth Engine) enables researchers to quickly analyze vast amounts of NDVI data on peer-to-peer scales, the number of different types of analyses based on the archives of petabyte-sized data has increased exponentially compared with previous research. Ultimately, thanks to both new methodologies and new technological platforms, we now have the means to analyze vastly more datasets (Gaikwad et al., 2024 & Verma et al., 2022).

Agricultural uses of the NDVI to assess plant health have many related purposes. These purposes include: 1) monitoring growth cycles of crops and making decisions about when to plant/harvest; 2) assessing how droughts affect crops, monitoring land degradation, and creating suitability maps for growing crops. The NDVI time series provides examples of how seasons change over time and how these changes respond to weather (Prajesh et al., 2019). The use of NDVI to monitor agricultural drought conditions combines measurements collected from weather stations with soil moisture information to indicate how stressed a particular crop may be and what the potential yield from that crop will be (Gidey et al., 2018). On the other hand, Gowri & Manjula (2019) have been using ArcGIS and MATLAB to detect and extract different types of surfaces, such as vegetation, built-up areas, bare ground, and water bodies. The results of this work show that the surface classification was more accurate than conventional methods, as verified by comparison studies.

While NDVI is able to provide various applications associated with soils and climate data for agricultural monitoring, there remain operational issues associated with these technologies, such as sensor calibration and atmospheric impacts, causing uncertainty with the analysis of historical image data across different sensors (Yin et al., 2012). In high-density vegetation patterns, NDVI experiences a saturation effect limiting its ability to provide useful estimates of crop growth; furthermore, NDVI experiences sensitivity to soil background variations within low-density (less dense) vegetation, which reduces its accuracy in multiple types of agricultural environments (Radočaj et al., 2023). Further complicating this matter is the difficulty in separating climate-related changes in vegetation from changes made through anthropogenic activity; thus, methodologies must be developed to allow for the differentiation of both types of data in future analysis (Ghorbanian et al., 2022).

Researchers are utilizing new technologies and integrated multi-criteria approaches to better quantify vegetation health than those that have existed up until now. An example of this is the Vegetation Health Index (VHI), which uses two factors called the Vegetative Condition Index (VCI) and Canopy Down Temperature (CDT) to measure dryness in a particular area or region. By combining these indices with an additional index called the Soil Adjusted Vegetation Index (SAVI), a more accurate estimate of vegetation health based on the soil types present in that area is created.

Researchers can also apply machine learning and artificial neural network (ANN) methods, including advanced deep learning techniques, for instance, Long Short Term Memory (LSTM) and others, to analyse vast amounts of environmental data. Researchers using these techniques reported accuracy rates of predicting and forecasting NDVI estimates above a level of 0.83 in correlation coefficient scores for NDVI (Gaikwad et al., 2021; Verma et al., 2022). OLPANN (Optimized Learnable Parameter Artificial Neural Network), developed by Gowri and Manjula (2025), is an example of integrating machine learning into previously established methods for detecting land cover through the use of multispectral images. The development of the OLPANN represents a significant leap from classical parametric methods of land cover classification.

Evaluations of vegetation health trends using NDVI have important applications not only in academic research but also in developing policies related to agriculture, assessing food security, and developing strategies for adapting to climate change. From recent comprehensive studies of NDVI that have been conducted on a national scale, researchers have been able to identify the various types of vegetation exhibiting patterns of greening and browning, and they have found this information to provide valuable evidence for decisions regarding sustainable land management (Ghorbanian et al., 2022). In addition, researchers combining NDVI information with assessments of agricultural suitability in dry areas or places with limited resources to develop agriculture have found that NDVI is particularly useful (Affandy et al., 2024).

Recent research studies suggest that NDVI is now being measured in the context of multiple metrics, including NDWI, EVI, and several soil-adjusted indices, as well as through NDVI, using a combination of NDVI and additional indices in an integrated methodology. A comprehensive bibliometric review of global NDVI trends based on research studies was completed, analyzing all global research studies published between 1985 and 2021; this survey indicated an exponential increase in the number of publications that used NDVI as part of their methodology across multiple fields of study, and increasing accuracy in NDVI-based measurements was supported by improved methods to analyze NDVI (e.g., machine learning) and cloud-based resources (primarily digital data storage and access capabilities). Emerging research trends show that NDVI-based methodologies continue to support modernizing agricultural monitoring and meeting the evolving needs of agricultural producers worldwide as a result of changes in the environment.

2. RELATED WORK

NDVI derived from satellites is used to study land cover changes and determine which land areas are suitable for agriculture, making it one of the most important areas of study in remote sensing and agricultural science. Due to the increasing global population and its accompanying need for food, as well as the impacts of climate change on the agricultural industry, creating new types of vegetation health assessments and informed land use decisions will require more precise tools for monitoring. Vegetation indices are based upon the basic difference in electromagnetic radiation wavelength (indices) properties of red versus near-infrared spectra that are reflected off of vegetation surfaces. NDVI is by far the most widely used vegetation index for providing quantitative information on the amount of live near-infrared spectral and dead vegetation. dead vegetation.

Approaches for land cover analysis have advanced considerably over time, evolving from simple NDVI differencing for change detection to complex multi-criteria decision analytical frameworks. Early studies used simple NDVI differencing to detect temporal change, while recent studies have demonstrated the application of several methodologies, including specified classifications, to continuously analyze temporal referential NDVI values based on multi-seasonal Landsat 8 data within a single analysis period; the result has been the ability to fully characterize vegetation states within a composite image. Integrated remote sensing with GIS forms the basis of current research, as shown by Mohammad et al., whose research documents a comprehensive approach to land suitability mapping, using a multi-criteria-based methodology through ENVI6 software to create requirements for land use maps.

Research in support of combining integrated GIS and remote sensing has illustrated that machine learning can also be used as part of NDVI research, resulting in enhanced analytical capabilities for researchers. Recent projects completed by Verma and Gaikwad, among others, have shown how multiple approaches can be applied to process and model NDVI data through machine learning algorithms (Support Vector Regression, Artificial Neural Network, Bidirectional Long-term Memory Networks).

Bidirectional LSTM is now outperforming traditional statistical methods, such as time series predictions for NDVI, with an R^2 value of 0.8365 (Verma et al. 2022). In addition to the growing use of both integrated remote sensing with GIS, there has been a growing diversity of satellite data sources and products that are being used for NDVI. Previously, several comprehensive analyses of NDVI time series were completed using datasets from many different satellites from many different countries worldwide. The recent study came from Sheffield and Morse-McNabb (2015), which used 40-plus years of Landsat data combined with MODIS products for the previous 10 years, creating the most extensive NDVI trend analysis over time of any existing research conducted until now.

With dual-index approaches, humid/marsh grassland vegetation from perennial grasses exhibit excellent agreement (0.84 correlation coefficient) in NDVI-based trend analysis and a 0.85 coefficient of variation (standard deviation) between NDWI and NDVI for the 2012/13 growing season and for the 2011/12 growing season. The strong correlation between NDWI and NDVI for dry (shrub) grasslands (0.93 correlation coefficient with a coefficient of variation = 0.24) indicates that both indices provide similar information about vegetation growth through time in these types of land cover. Vegetation represents a large proportion of the overall biodiversity of arid regions. Recent studies of arid regions, particularly those in Jordan, have found the most severe limitations on agricultural development. Only 0.2% of the land is classified as suitable for rain-fed agriculture, and only 1.4% of the land is suitable for irrigated crops because of water scarcity and soil fertility limitations (Al-Taani et al., 2020). A combination of the literature on NDVI-based applications across all climate zones indicates the potential wide application of NDVI-based indices on vegetation under diverse climate zones and ecological conditions.

The research carried out on the Indian subcontinent encompasses a wide range of agro-ecological zones. For instance, Prajesh et al. (2019) thoroughly examined the trends in vegetation in Tamil Nadu at the state level, and Ramanujam et al. (2022) examined forest degradation in the Kumbur River Basin. Analysing vegetation trends at different times of the year has yielded meaningful results. The vegetation exhibits a positive trend of 86.52% during the Kharif season, whereas the Rabi season yields mixed results. This shows that temporal trends are valuable for agriculture monitoring. The presence of a 4.67% increase in the land area classified as non-vegetated over 16 years (Singh & Javeed, 2020) is consistent with long-term studies that have also shown significant changes in the land cover types. This finding is important for understanding how to develop and implement policies for land conservation and utilization for future environmental purposes, as well as for regions experiencing rapid urbanization and climate change.

Table 1. Summary of LULC classification and change detection studies

	Authors (Year)	Model / Method	Dataset	Accuracy / Key Metrics	Pros	Cons
1	Reddy & JangaReddy (2013)	NDVI-based LULC classification	IRS-P6 LISS-III images (2004–2005)	Qualitative assessment; superior classification	Seasonal analysis; effective without ground truth	Limited temporal scope (2 years); qualitative accuracy only
2	Binte Mostafiz et al. (2021)	Multi-criteria decision analysis (weighted linear combination & fuzzy MCA)	Remote sensing data with 8 criteria (elevation, slope, LST, SAVI, ARVI, SARVI, MSAVI, OSAVI)	Combined R^2 = 0.839; individual indices R^2 = 68.9–81.2%	High quantitative accuracy; multiple vegetation indices; validated with ground truth	Complex methodology requires extensive parameter tuning
3	Ramanujam et al. (2022)	NDVI-based change detection	Satellite imagery (2009–2016), Kumbur River Basin	3.34% forest degradation detected	Effective change detection focuses on environmental degradation	Limited temporal scope; single ecosystem focus
4	Singh & Javeed (2020)	NDVI differencing using Landsat TM & Landsat 8 OLI	Landsat TM (2001) & Landsat 8 (2017), Srinagar district	NDVI range 2001: 0.62 to -0.41; 2017: 0.53 to -0.10	Long-term comparison (16 years); Natural Breaks classification	Limited geographic scope; no statistical accuracy validation
5	Al-Taani et al. (2020)	Weighted overlay with 5 criteria (rainfall, temperature, slope, soil, water wells)	Landsat images (1990, 2000, 2018), Ma'an Governorate	0.2% highly suitable (rain fed), 1.4% (irrigated)	Comprehensive suitability assessment; long-term (28 years); high citation	Very low suitability percentages; limited to arid regions
6	Sheffield & Morse-McNabb (2015)	Long-term time series analysis	Landsat (~40 years) & MODIS (~10 years)	Not reported	Longest temporal coverage; multiple data sources; landscape-scale analysis	No quantitative accuracy; complex data integration
7	Prajesh et al. (2019)	Time series NDVI & NDWI trend analysis using regression slopes	MODIS MOD09A1 (500m, 8-day) for Tamil Nadu	86.52% positive NDVI trend; 90.29% positive NDWI trend	High temporal resolution; dual index approach; regional scale	Limited to single year; trend % used rather than classification accuracy

3. METHODOLOGY

The main purpose of this research was to develop an index of plant growth and surface water using the NDVI. The NDVI-derived trends were not used as a method for classifying land cover but rather as one way to assess how conditions on the earth's surface change over time and to identify where those changes were happening. Additionally, the MODIS Land Cover Products and temporal consistency analysis provided supporting information to help place NDVI patterns into the larger land use and land cover context. This approach reduces the chance for misclassification of land cover types due to the similarity of their spectral reflectance and allows for NDVI observations to be used as indicators of how vegetation responds to land use changes, not just as direct representations of land cover classes.

3.1. Dataset Description

The study analyzed the health of vegetation and the change in land cover in the Thanjavur region from 2001 to 2020 using long-term Earth Observation datasets derived from MODIS. The MCD12Q1 Version 6 dataset was the primary land cover dataset. This dataset offers an annual classification of land cover at a 500 m resolution using the IGBP scheme. The land cover classification was divided into several main categories, including croplands, urban/built-up areas, forests, and grasslands. The NDVI metrics were obtained from MOD13Q1, which offers 16-day composites with a spatial resolution of 250 m. NDVI is sensitive to the quantity of chlorophyll concentration and canopy coverage and is a metric of vegetation health that is derived from the red and near-infrared bands of the visible spectrum. The Thanjavur study area was employed to trim each dataset.

Table 2. Dataset description of this study

Dataset Name	Sensor/Source	Product ID	Temporal Coverage	Spatial Resolution	Temporal Resolution	Key Variables Used
MODIS Land Cover	Terra & Aqua MODIS	MCD12Q1 (Version 6)	2001–2020	500 meters	Annual	LC_Type1 (IGBP classification)
MODIS Vegetation Index	Terra MODIS	MOD13Q1 (Version 6)	2001–2020	250 meters	16-day composites	NDVI

3.2. Proposed Methodology

The methodology provides a clear process for analyzing changes in land-use patterns and trends in vegetation health (or decline) in the Thanjavur District from 2001 to 2020 using satellite data, specifically multi-temporal land cover classifications and vegetation indices. The methodology provides a systematic approach for combining information from both sources to conduct a more thorough examination of how changes are affecting ecosystem health. The dataset category was limited to the study years, so different interannual comparisons and trends could be analyzed. The NDVI annual composite is a uniform value representing the average of 16-day observations for each of the 12 months in the year (which can be used as an indicator of photosynthetic activity).

Figure 1. Land cover dynamics and NDVI-based vegetation health trends (2001 - 2020)

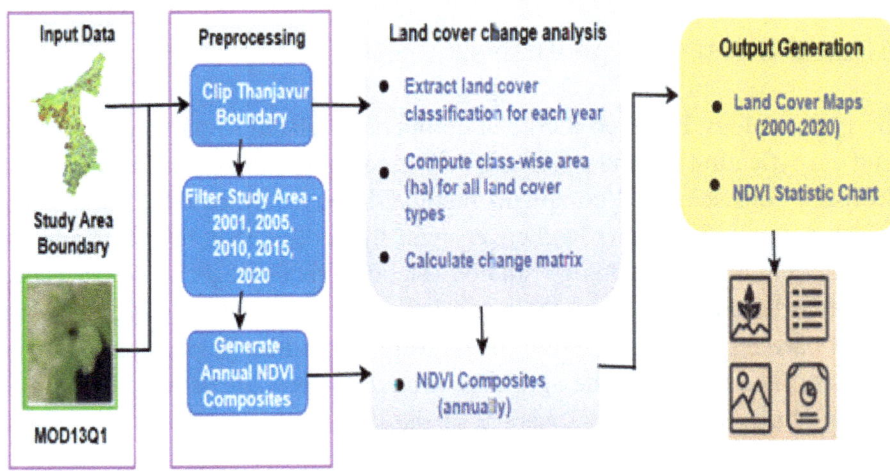

The selection of a study area begins the workflow at the study site, defined as Thanjavur within the geographical boundary. Two datasets were obtained from satellites for use in this analysis. The two primary datasets are MODIS Land Cover Type 1 (IGBP Classification) MCD13Q1, which is an annual dataset that shows the following types of land during a given year (cropland, built-up, etc.) and also has a resolution of 500 meters, and the MODIS NDVI COMPOSITES 16 DAY MOD13Q1 dataset, which uses 250-metre resolution composites to measure the condition of vegetation during specified timeframes (16-day intervals) shown in Figure 1.

At this stage, the MODIS dataset used in this work is chosen for its consistency and relevance to this research. First, the MODIS datasets were clipped to the actual boundary of the Thanjavur study area. Second, they were filtered to only include study years 2001, 2005, 2010, 2015, and 2020 for the analysis. The final products for this analysis will be the annual NDVI composites that display the average health of the vegetation in the study area over the specified time frames.

The analysis step entails computing the class area (hectares) of each land cover classification extracted for each year, then building a change matrix to illustrate specific types of transitions, such as conversion from cropland to build environment, etc. Annual NDVI composite data is also analyzed as health data and included in the output stage. The output phase consists of producing results in four categories. Land cover maps of 2001–2020; NDVI statistical charts showing vegetation trends helped for detailed data analyses.

3.2.1. Land Cover Change Analysis

The methodology included a crucial component related to the annual generation and classification of land cover maps. The categorization of land cover maps should use a consistent and compliant methodology. The methodology for each year studied involved creating land cover maps that are consistent and compliant. consistent/compliant methodology studied. The different categories of land cover were identified and displayed, including agricultural lands, forests, wetlands, urban areas, and barren lands. Agricultural lands were studied for years. The identified categories of land cover include agricultural lands, forests, wetlands, urban areas, and barren lands. The Spatial Coverage of the landscapes characterizes the land Conditions. The investigation provided evidence of higher-level temporal trends (temporal patterns) in land use change, including a summary of annual time-series data for these changes. All total land cover areas were recorded (ha). All land cover areas were used to establish a benchmark for the temporal analysis of land cover to identify the extent of increase, decrease, or transition from one land cover type to another over a period of twenty years.

An analysis of land cover changes indicates many important patterns of change that will affect the sustainability of that region's agricultural practices. Deforestation/agricultural intensification (i.e., the conversion of forestland to cropland); the expansion of built-up areas as cropland has transitioned to urban areas; and the probable depletion of water bodies due to land reclamation or hydrological stress. These various types of change were quantitatively assessed each year for total area changed and expressed as the percentage of total area that changed.

3.2.2 Vegetation Health Trend Analysis

To assess the health of plants in an area, researchers used NDVI to monitor vegetation throughout the year. By using yearly NDVI data, researchers were able to monitor the growth of vegetation and monitor how much vegetation was produced each year to provide an indication of the health of the ecology, how drought impacted vegetative production, and how much land degradation has occurred in a given area. In the NDVI analysis, we computed the average, minimum, maximum, and standard deviations of NDVI values across years (five years). This NDA provided information regarding the resulting stability, level of variability, and potential volatility of vegetation health in this region. By using spatially referenced values for NDVI, we could identify areas with declining health or improving health and highlight those areas of NDVI that were consistently low or decreasing NDVI. Comparing annual NDVI to the previous years does not provide enough context to understand the changes that have taken place. Increased NDVI in the past two years may indicate that the vegetation is regenerating or increasing agricultural activity in that area.

Investigating a combination of the results ultimately revealed all the changes that occur over time in land cover types to changes within the NDVI value reflected in the actual land cover characteristics on a geographical basis. The process of analyzing both structural and functional landscape characteristics facilitated a complete understanding of the landscape. For example, cropland changed to urban, as indicated by the reduction of NDVI, indicating the potential for land degradation and significant loss in agricultural productivity.

3.2.3 Spatial and Temporal Trend Analysis

To monitor plant health, NDVI values are metrics that measure the amount of green cover or biomass on growing plants. The NDVI values in each year, by district, created an opportunity to summarize and review the geographic distribution of aggregate vegetation cover within districts on an annual basis. Additionally, the NDVI values for each year were used in conjunction with each other to assess yearly differences in aggregate vegetation density and distribution. The NDVI values from previous years allowed for the recognition of some of the different environmental conditions that can affect vegetation growth and, as a result, NDVI. Therefore, the NDVI values from past years aided in identifying different types of changing landscape vegetation growth and exhibited the different patterns of vegetation growth and decline across the area of interest.

Outputs were created for all of the information developed from the NDVI products; hence, NDVI products were used to explain or interpret all products created, including land cover maps, area statistics, and NDVI time series. Land cover maps were produced using standardized color combinations to provide a common basis for visually comparing year-to-year land cover differences. Additionally, time series data sets for NDVI values from multiple years were created using charts (e.g., line charts and bar charts) to visually show year-to-year variations in NDVI values. To facilitate the visual comparison and understanding of variations in NDVI values over time, place runners with tabular information on land cover classes and NDVI statistics in the vicinity of a particular NDVI product. The strategies proposed for defining and documenting changes in vegetation and land cover, along with monitoring the sustainability of these systems, ensure ongoing monitoring and the further development of collaborative, evidence-based land-use planning strategies, as well as opportunities to enhance engagement with agribusinesses that influence land management in the area.

4. RESULTS AND DISCUSSION

4.1. Evaluation of Land Cover Change Analysis

The data collected for the Thanjavur area shows that extensive changes took place during the period between 2001 and 2020 that allowed the area to become more suitable for agriculture. The predominant land use type related to the cropland category continues to be the same throughout the entire course of the study; however, as evidenced by their increase from being the lowest type of cover at the beginning of the study period to almost 93% of the total cover by the end of the period, the cropland category represents one of the most significant increases seen by this area during the entire course of the study. The extent of this increase not only indicates that there is an intentional use of non-cultivated lands as a means to increase the agricultural capacity of the site, but it also indicates that valuable agricultural land has been protected from conversion to other uses. Urbanization has remained mostly consistent since the study began, suggesting minimal physical changes to the area, which explains the relatively constant built-up area. Throughout the duration of the study, the grassland cover type has experienced very little change, which could represent sites for future agricultural expansion or possibly for preservation as a potential site for reserves.

The study has found that there was a slight decline in the amount of water that had been previously covered, which will require additional study, as this will have a critical effect on agricultural water security and the development of irrigation infrastructure. The greatest area of change was seen within an unclassified category of land cover and other land covers (as shown in Figure 2), and the result has been a significant amount of change. This decline indicates either an improvement in remote sensing classification methodology or a shift in the use of previously uncultivated or unutilized lands to newly cultivated land. The coordinated changes in land use reflect effective regional agricultural resource management and regional agricultural planning practices. The data support the conclusion that Thanjavur has made considerable progress in its nature suitability rating of agricultural lands during the time examined in the study, which should allow for increased agricultural production and food security in the region in the future.

Figure 2. Land cover change analysis (2001 -2020)

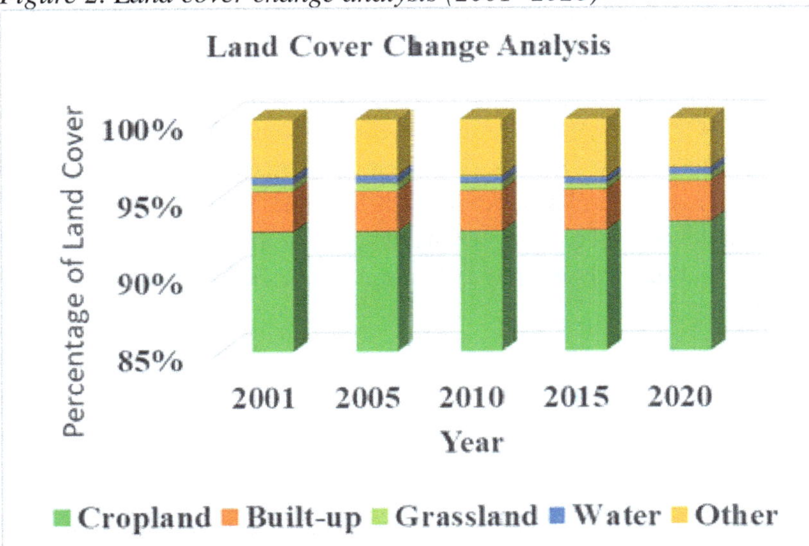

4.2. Analysis of NDVI-Based Vegetation Trend

NDVI statistical parameters were employed to investigate the spatial heterogeneity of vegetation and the dynamics of that vegetation from 2001 to 2020. Longitudinal analysis of NDVI data from the Thanjavur region provides compelling evidence of spatial heterogeneity (or robustness) in vegetation over time. The NDVI maximum level remained fairly stable throughout the study period; as such, the typical NDVI maximum values for the Thanjavur region had values of 0.78 - 0.81 from 2001 to 2020, indicating the presence of dense vegetation, with maximum photosynthetic capacity and minimal stress levels in the study area. In the 20-year study period, mean NDVI values exhibited a gradual upward trend from approximately 0.48 to 0.52, indicating that vegetation productivity and capacity for sustaining productive land improved in the region during the study. Negative minimum NDVI values (-0.25 - -0.12) were consistently identified throughout the study period, indicating that surfaces where vegetation did not exist (e.g., aquatic bodies, urban areas, barren land) were present on an intermittent basis within the study area.

The upward trend in minimal NDVI values during this timeframe strongly indicates that the non-vegetated regions saw changes to the system, either in terms of the non-vegetated areas becoming more vegetated or by just having more vegetation established on marginally vegetated areas over the years. The distribution of vegetation, as indicated by the standard deviation values for NDVI, showed little variability from year to year (0.08-0.10), which suggests little change occurred across the NR from 2005 to 2015, as seen in Table 3. The consistent improvements in vegetation over time suggest an increasing degree of sustainability within the agricultural sector for agricultural production in this region. This finding is further supported by the transfer of land cover types in the region from forests/natural areas to agricultural lands and the growing intensification of agricultural land use in the NR, as shown in Figure 3.

Table 3. NDVI trend analysis using statistical measures

Year	Max	Mean	Min	StdDev
2001	0.781	0.489	-0.094	0.086
2005	0.771	0.53	-0.191	0.099
2010	0.775	0.535	-0.089	0.088
2015	0.803	0.548	-0.056	0.091
2020	0.783	0.552	-0.093	0.086

Figure 3. NDVI trend analysis

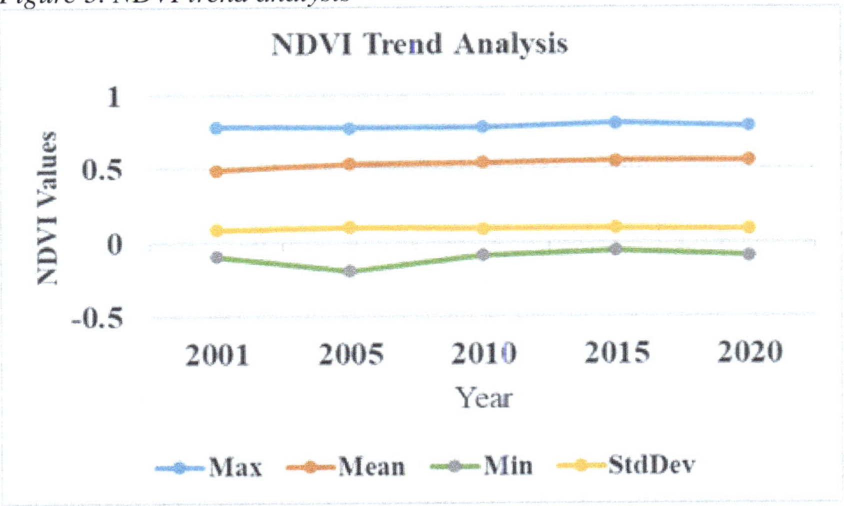

4.3. Agricultural Land Suitability Trends

For effective long-term land use planning, development of policies, and interaction with stakeholders, visualizing and understanding how transition classes are distributed geographically is valuable. Transition-based evaluation of the suitability of land for agricultural uses provides a robust, coherent means of tracking the changes in land. For example, a comprehensive examination of how agricultural land existed in Thanjavur District has helped us understand the continuing persistence, increased vulnerability, and ongoing growth of agriculture over time. It is also relevant to developing Sustainable Intensification of Agriculture (SIA), agro-ecological and regional food security strategies, given that environmental and development pressures are increasing globally.

The research looked at the changes in the suitability of land for agriculture over four distinct periods (from 2001 through 2020) and then placed those lands into one of four categories based on how the suitability of those lands had changed: Remains Suitable, Gained Suitable, Lost Suitable, and Remains Unsuitable. Overall, the largest area of land in terms of the percentage of land area occupied by "Remains Suitable" was likely indicative of a core agricultural production zone and maintained a continuous agricultural production history; thus, it likely represents traditional agricultural zones.

The declines in areas identified as "Gained Suitable" (that were not previously suitable for agriculture) between 2001 and 2015 may have resulted from a combination of two factors: 1) The depletion of easily convertible marginal lands; and 2) The existence of more systematic constraints of either a socioeconomic or biophysical nature. However, between 2015 and 2020, there was a modest increase in the number of acres identified as "Gained Suitable," indicating the potential for future agricultural restoration or revitalization efforts.

There are likely fluctuations throughout the range for the "Loss Suitable" category. Such fluctuations, along with the loss of suitable lands, indicate competing uses (e.g., agriculture, large-scale industrial development, urbanization) and highlight the need for integrated rural-urban land use and tracking tools to improve our understanding of land use changes and to avoid the irreversible conversion of agricultural production areas into non-agricultural areas. With respect to the "Remains Unsuitable" category, there appears to be little change in terms of percentage; however, as seen earlier with the "Loss Suitable", there is the possibility of converting low-quality agricultural lands into higher-quality agriculture through improved irrigation and soil conservation methods. The gradual and steady decrease of this category indicates that, with appropriate investment in infrastructure and continued improvement of the ability to manage lands through restoration methods, additional agricultural expansion is possible.

Figure 4. Land cover distribution from 2001, 2010, and 2020

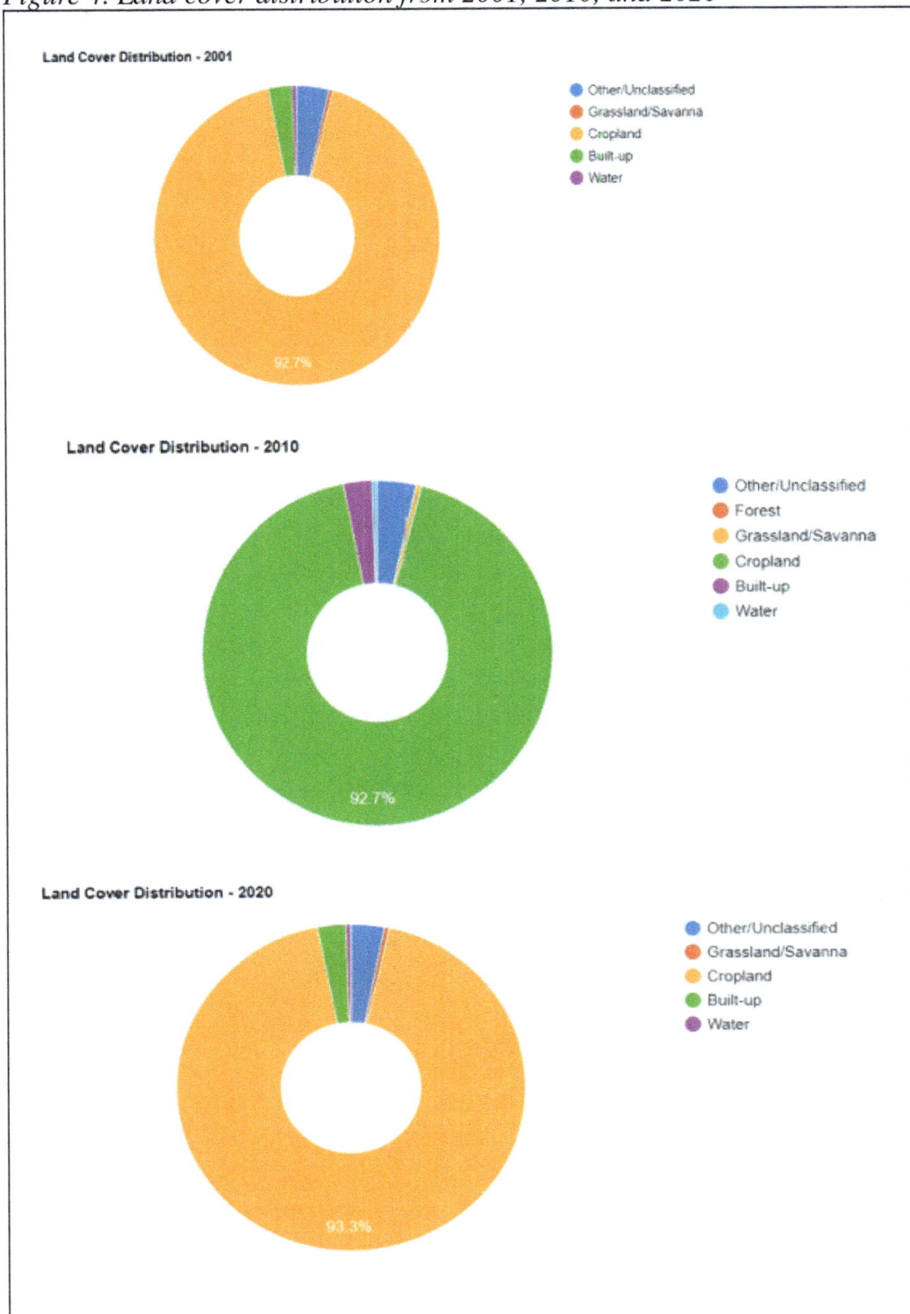

Land cover distribution analyses for the Thanjavur area were performed by creating three separate time-dependent images (2001, 2010, and 2020) to demonstrate the actual track or evolution of land use cover in Thanjavur. The results of the 2001 analysis demonstrated that the land cover structure of the area was quite diverse, with the agricultural (cropland) land use accounting for 92.7% of the total land area in both the 2001 and 2010 analyses. The remaining use of the land was divided among the grassland/savanna, water, built-up, and other/unclassified land classifications, which together accounted for 7.3% of the total land use. The largest portions of the land cover were in the following classifications: water, built-up, grassland/savanna, and various other/unclassified lands. The 2010 analytical image demonstrates the evident change that has occurred in the distribution of land cover. The largest portion of cropland continues to account for 92.7%; however, the rest of the categories of land cover seem to be shifting relative to each other almost imperceptibly. The analysis of the land covers for 2010 also indicates the addition of forests into the land use classification, which is likely the result of improved classification or changes in the actual land cover. The 2020 observations indicate that the land use classification of agricultural land is being standardized, and the cropland increased to 93.3%, representing an increase of 0.6 percentage points. The analysis of the time series demonstrates a continuing pattern of decreasing proportions of non-agricultural land use covering categories over time, especially grassland/savanna and various other/unidentified categories that will likely continue to decline through the analysis.

The transformation of land cover into more productive areas through the creation of agricultural land, as well as using marginal, previously underutilized or lesser-used areas to support food production, creating more sustainable and productive food systems in the last 20 years, were the primary goals of this region. Use of marginal land for agricultural production has protected areas previously categorized as non-productive, and through this intentional redeployment of land, the overall productivity and sustainability of farming have improved over the last two decades. (For example, improved average values of vegetation.) Urban development and water conservation throughout this study area have significantly contributed to more equitable regional growth and agricultural development, and urbanization will not lead to conflict with agricultural land used for crops. Ultimately, Thanjavur has received recognition as an example of sustainable agricultural land management for the future and has the opportunity to be a global leader in the sustainable intensification of cropping and securing food.

4.4. Relationship Between NDVI Trends and Land Cover Changes

Land cover transitions observed in the study area are reflected in the patterns of NDVI values over time and space. Areas that show a large decline in NDVI values are generally associated with locations where a loss of cropland has occurred due to urbanization or construction. Thus, changes in land use, rather than normal climate variability, likely link land degradation and vegetation loss. In contrast, NDVI signals tend to be more consistent in regions where the land cover is more constant, including forests and perennial grasslands. Results show that changes in NDVI are a reflection of plant reactions to changes in land cover and offer circumstantial evidence of the ecological effects of land conversion in the area.

Human and environmental factors are responsible for the loss of food-producing land and the decline in plant health. The changing landscape caused by urban development and building new infrastructure has led to the loss of the most productive agricultural lands in the rural-to-suburban areas and replaced them with houses, stores, offices, etc. Farmers have also experienced significant changes in crop cultivation: farmland is increasingly divided into smaller portions as family members inherit parts of farms, many farmers are unable to afford irrigation costs, and those reliant on irrigation for crop growth have been affected by the rising number of dry weeks and the variability in monsoon growth. These climate-related factors will all contribute to stress on the vegetation during its growing season. The interplay between human-created factors and nature creates additional complexity for determining how land use affects vegetation and vice versa. Therefore, we should consider both human shaping factors and natural climatic factors when interpreting long-term NDVI trends.

4.5. Limitations of NDVI-Based Analysis

The NDVI has been useful for tracking long-term changes in vegetation cover, but it has several problems when it comes to understanding land cover. NDVI can become saturated in highly vegetated areas where it loses its sensitivity to variations in biomass beyond a certain level. In addition, MODIS imagery's moderate spatial resolution produces mixed-pixel effects when looking at fragmented landscapes where multiple vegetation types coexist within a single MODIS pixel. The similar spectral characteristics shared by some crops, grasslands, and managed vegetation types complicate the process of differentiating land cover types at the level of individual classes. Therefore, these limitations were considered during the analysis, and NDVI results were interpreted very conservatively, focusing on large-scale spatial and temporal patterns rather than specific classes of land cover.

5. CONCLUSION

The research examined the changes in the space-time dynamics of land cover change and vegetation condition in the Thanjavur District, located in South India, which is a critical cropping region for southern India. The study analyzed MODIS satellite data from 2001 to 2020 and quantified how vegetation health changed over time using NDVI. The study provides NDVI vegetation data specific to the District and NDVI trends in 5 land cover classes (cropland, forest, urban, grassland, and aquatic). Findings reveal that the swift expansion of cities and the decrease of croplands are linked to the effects of rapid urbanization on agriculture. The expanded constructed region suggests moderately active spatial dynamics of grasslands and forests. Through numerous land cover transitions, the NDVI values for cropland areas decreased significantly, leading to an increase in the mean NDVI, greater variability over time, and increased spatial heterogeneity, along with a decline in the mean NDVI associated with the conversion of cropland to urban areas and barren lands. NDVI values in these converted areas were lower than the NDVI of regions that remained as cropland, indicating a decline in ecological health and vegetation productivity.

The NDVI time series has seen a gradual decline in the health of vegetation due to anthropogenic and extreme climatic factors. These results support the overall conclusion of the previous studies regarding the reduction of agricultural capacity in developing countries due to changes in the class structure of agriculture. Additionally, climate change and human activities influence agricultural areas, and the NDVI can serve as an indicator of land degradation. Numerous activities related to farming and land use influence agricultural decline. Both climate change and human activities contribute to land degradation. Climate change and human activities influence agricultural decline and degradation. In a general sense, NDVI analyses reflect the impact of human influence on land degradation. These analyses take into account the influence of classification analysis; land cover classification must be used to identify changes in ecological habitats that result from urban expansion. Land cover classification is essential for identifying ecological habitat changes caused by urban expansion and urban sprawl. Urban sprawl will require policy changes that encourage sustainable land-use practices. Urban sprawl, sustainable land use, and the protection of agricultural lands are crucial in mitigating urban expansion. Sustainable land use is crucial for protecting agricultural lands from urban expansion. Evidence-based assessments support responsible land use transition due to climate change and development. This development will continue into the foreseeable future, necessitating the engagement of adaptive measures. Adaptive measures are needed to preserve the benefits of ecosystem services. Ecosystem services are necessary to guarantee regional food and ecological security.

The prior study established knowledge of land cover change by conducting a thorough and long-term analysis of plant health and land cover change using MODIS time-series datasets. However, there was a major limitation of this research in that it did not collect or use enough ground-truth validation data. Accurately classifying land cover changes with better spatial resolution will be made possible with further high-resolution satellite imagery, such as that acquired by Landsat and Sentinel 2. This will improve future applications of the time series for vegetation health. The researcher can confirm the land cover changes detected by satellite pictures and discover the causes of these changes by conducting a market analysis of the agriculture and economics aspects linked with the methodology for satellite image validation of agricultural change. As they develop climate-smart procedures, researchers can profit from this integrated approach, which adds responsibility to the results produced from satellite images.

REFERENCES

Al-Taani, A., Al-Husban, Y., & Farhan, I. (2020). Land suitability evaluation for agricultural use using GIS and remote sensing techniques: The case study of Ma'an Governorate, Jordan. *The Egyptian Journal of Remote Sensing and Space Sciences*, *24*, 109–117. DOI: 10.1016/j.ejrs.2020.01.001

Bellone, T., Boccardo, P., & Perez, F. (2009). Investigation of vegetation dynamics using long-term normalized difference vegetation index time-series. *American Journal of Environmental Sciences*, *5*(4), 461–467. DOI: 10.3844/ajessp.2009.461.467

Binte Mostafiz, R., Noguchi, R., & Ahamed, T. (2021). Agricultural land suitability assessment using satellite remote sensing-derived soil-vegetation indices. *Land (Basel)*, *10*(2), 223. DOI: 10.3390/land10020223

Desai, S., Mattoo, M., Deshpande, A., Dey, A., & Akiwate, S. (2024). Vegetation change detection through NDVI analysis using Landsat-8 data. In *Proceedings of the International Conference on Computer, Communication, Control and Information Technology (C3IT 2024)*. IEEE.

Gaikwad, S., Vibhute, A. D., & Kale, K. (2021). Development of NDVI prediction model using artificial neural networks. In *Proceedings of the International Conference on Recent Trends in Image Processing and Pattern Recognition*. Springer.

Gaikwad, S. V., Vibhute, A. D., Shah, S. V., Saini, J. R., & Kale, K. (2024). Vegetation health analysis of Vaijapur using Google Earth Engine and vegetation indices. In *Proceedings of the 2nd International Conference on Current Trends in Physics and Photonics (ICCTPP 2024)* (p. 25). SPIE. https://doi.org/DOI: 10.1117/12.3044615

Ghorbanian, A., Mohammadzadeh, A., & Jamali, S. (2022). Linear and non-linear vegetation trend analysis throughout Iran using two decades of MODIS NDVI imagery. *Remote Sensing*, *14*(15), 3683. DOI: 10.3390/rs14153683

Gidey, E., Dikinya, O., Sebego, R., Segosebe, E., & Zenebe, A. (2018). Analysis of long-term agricultural drought characteristics using vegetation health index (VHI) in Raya and its environs, Northern Ethiopia. *Environmental Systems Research*, *7*(1), 1–17. DOI: 10.1186/s40068-018-0121-x

Gowri, L., & Manjula, K. R. (2019). Evaluation of various vegetation indices for multispectral satellite images. *International Journal of Innovative Technology and Exploring Engineering*, *8*(10), 3494–3500.

Gowri, L., & Manjula, K. R. (2025). Detection of land cover usage from optimized learnable parameter artificial neural network (OLPANN) using multispectral images. *Multimedia Tools and Applications*, *84*, 4283–4313. DOI: 10.1007/s11042-024-18828-2

Jothimani, M., & Mulualem, W. (2020). Trend analysis of normalized difference vegetation index using Landsat satellite data: A study in and around Gondar town, North West Ethiopia. *Conference paper / institutional report*. ResearchGate.

Kundu, A., Denis, D., Patel, N., & Dutta, D. (2018). Long-term trend of vegetation in the Bundelkhand region (India): An assessment through SPOT-VGT NDVI datasets. In *Vegetation dynamics and climate change* (pp. xx–xx). IGI Global.

Najafi, Z., Fatehi, P., & Darvishsefat, A. (2019). Vegetation dynamics trend using satellite time series imagery. *The International Archives of the Photogrammetry, Remote Sensing and Spatial Information Sciences*, *XLII-2*(W13), 163–170. DOI: 10.5194/isprs-archives-XLII-2-W13-163-2019

Prajesh, P. J., Kannan, B., Pazhanivelan, S., Kumaraperumal, R., & Ragunath, K. P. (2019). Analysis of seasonal vegetation dynamics using MODIS-derived NDVI and NDWI data: A case study of Tamil Nadu. *Madras Agricultural Journal*, *106*(4–6), 263–270. DOI: 10.29321/MAJ.2019.000275

Radočaj, D., Šiljeg, A., Marinović, R., & Jurišić, M. (2023). State of major vegetation indices in precision agriculture studies indexed in Web of Science. *Revista de Agricultura (Piracicaba)*, *13*(4), 930. DOI: 10.3390/agriculture13040930

Ramanujam, M., Rangasamy, N., Rathinam, A., & Veeraiyan, P. (2022). Using the normalized difference vegetation index (NDVI) to detect vegetative change with remote sensing and GIS: A study of the Kumbur River Basin in Kodaikanal Taluk, Dindigul District. *European Journal of Environment and Earth Sciences*, *3*(1), 18–25. DOI: 10.24018/ejgeo.2022.3.1.217

Reddy, A. S., & Janga Reddy, M. (2013). NDVI-based assessment of land use/land cover dynamics in a rainfed watershed using remote sensing and GIS. *International Journal of Scientific and Engineering Research*, *4*(12), 87–93.

Sheffield, K., & Morse-McNabb, E. (2015). Using satellite imagery to assess trends in soil and crop productivity across landscapes. *IOP Conference Series. Earth and Environmental Science*, *25*, 012013. DOI: 10.1088/1755-1315/25/1/012013

Singh, P., & Javeed, O. (2020). NDVI-based assessment of land cover changes using remote sensing and GIS: A case study of Srinagar district, Kashmir. *Sustainability, Agri. Food and Environmental Research*, *8*(1), 43–54. DOI: 10.7770/safer-V8N1-art2101

Somayajula, V. K. A., Ghai, D., & Kumar, S. (2020). A review on classification of land use/land cover change assessment based on normalized difference vegetation index. *Journal of Physics: Conference Series*, *1716*(1), 012014. DOI: 10.1088/1742-6596/1716/1/012014

Verma, G., Mehta, A., & Goswami, S. (2022). Trend and time series analysis of vegetation dynamics using satellite data: A case study of Uttarakhand, India. *Current Journal of Applied Science and Technology*, *41*(6), 32–46. DOI: 10.9734/cjast/2022/v41i630670

Wijaya, B. R., Affandy, N. A., & Nabilah, S. (2024). Pemantauan dampak kekeringan lahan pertanian di Lamongan menggunakan NDVI berbasis SIG. *Nucleus Journal*, *3*(2), 57–67. DOI: 10.32492/nucleus.v3i2.3201

Xu, Y., Yang, Y., Chen, X., & Liu, Y. (2022). Bibliometric analysis of global NDVI research trends from 1985 to 2021. *Remote Sensing*, *14*(17), 4340. DOI: 10.3390/rs14174340

Yin, H., Udelhoven, T., Fensholt, R., Pflugmacher, D., & Hostert, P. (2012). How normalized difference vegetation index (NDVI) trends from AVHRR and systematic sensor differences affect trend analysis. *Remote Sensing*, *4*(12), 3360–3394. DOI: 10.3390/rs4113360

Chapter 9

Machine Learning Models for Detecting Emerging Contaminants in Seagrass Ecosystems via Drone Imagery

Mughair Aslam Bhatti
http://orcid.org/0009-0007-4012-6803
Szabist University, Pakistan

Bakirov Juma
Department of Preschool and Primary Education, Termez University of Economics and Service, Termez, Uzbekistan

Intizor Avazmetova
Department of Biology, Urgench State University, Urgench, Uzbekistan

Anorgul Ashirova
Department of General Professional Sciences, Mamun University, Khiva, Uzbekistan

Avezova Umida Maksudovna
Department of Ecology and Life Safety, Urgench State University Named After Abu Rayhan Biruni, Uzbekistan

ABSTRACT

Machine learning (ML) models are increasingly being applied to environmental monitoring, particularly for detecting emerging contaminants in marine ecosystems.

DOI: 10.4018/979-8-3373-8872-4.ch009

Seagrass ecosystems, vital for biodiversity and carbon sequestration, are highly sensitive to pollutants, which can degrade their health and function. Traditional methods of contaminant detection often involve time-consuming fieldwork and limited spatial coverage. However, drone imagery, combined with ML techniques, offers a powerful tool for rapid, large-scale monitoring. ML algorithms can analyze high-resolution drone images to detect subtle changes in water quality, seagrass health, and potential contamination levels, including heavy metals, pesticides, and pharmaceuticals.

INTRODUCTION

Over the past years, coastal and marine ecosystems (especially seagrass meadows) are subjected to an increased pressure due to numerous anthropogenic stressors. Among them, the emerging contaminants (microplastics, chemical contaminants, run off nutrients and industrial waste by products) are a dangerous, but increasing threat. Nurse habitats Marine organisms include E. acnesystems that are particularly susceptible to damage, as they are nurseries of marine organisms, biodiversity sustenance, and carbon capture in the coastal environment (also known as blue carbon). Nonetheless, the methodologies used to monitor seagrass ecosystems in terms of contamination are a hard task: the conventional field sampling techniques are labour intensive, time consuming, lack adequate spatial coverage, and in most cases are restricted by accessibility issues (Akinaga et al., 2025).

Simultaneously, the development of drone (UAV) technology and remote sensing together with the invention of machine learning are creating new opportunities. High resolution aera imagery, multiple spectrum sensors or hyperspectral sensors and the processing capacity to handle massive image data allows researchers to keep track of water quality, map extent and possibly even detect distinctions of subtle changes related to contamination. These technologies may be integrated to transform the way in which we sense and react to environmental stress in seagrass meadows (Hamad et al., 2022).

Figure 1. Unveiling emerging contaminants in seagrass ecosystems

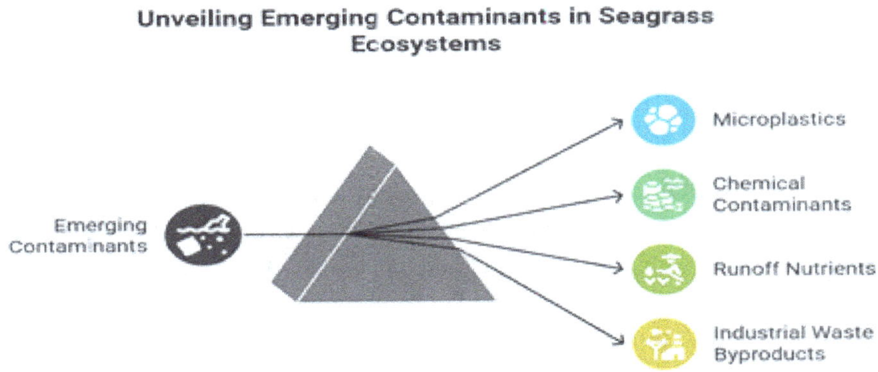

This review will aim at critically evaluating the research status at the interface of the following areas: machine learning techniques, drone based imagery, and the detection (or inference) of contaminants or ecosystem stressors in seagrass systems. Through the synthesis of literature, we would like to point out what has been accomplished so far, what are the limitations of the current state, and what are the ways to go further. This comes into special significance today, when the pollution of coastal territories and ecological degradation are gaining momentum all over the world, and the monitoring that can be scaled and efficiently implemented is still to be found (Tahara et al., 2022).

During the scope setting, the review will include (1) the articles that use UAV/drone imagery (or any other high-resolution aerial imagery) of aquatic/coastal ecosystems, particularly seagrass or seawater in shallow waters; (2) applications of machine learning or AI-based solutions to the problem of remote-sensing data to conclude about water quality, contamination, or other indicators of habitat stress; and (3) the potential and limitations of identifying "emerging contaminants, such as chemical pollutants, turbidity, or sediment The review will be based on peer reviewed literature and recent systematic reviews and the case studies available within the environmental remote sensing, aquatic ecology and machine learning domains (Karang et al., 2024).

Lastly, and to direct the reader, the article is organized in the following manner: first a methodology description of how the sources were chosen and synthesized, then the description of the relevant theoretical and conceptual frameworks, a description of key themes that come out of the literature (what techniques were used, what pollutants/contaminants are studied, what limitations and gaps exist) and finally recommendations on implications of the research and conservation, and finally, the future directions of research (Rowan, 2021).

METHODOLOGY

To ensure a comprehensive and balanced review of remote sensing applications for monitoring coastal ecosystems, a structured and systematic literature search and synthesis process was followed(Simpson, 2024). This approach was designed to capture the most recent technological advancements while also acknowledging the foundational works that have shaped current methodologies. By following this method, we aimed to provide a detailed and up-to-date review of remote sensing techniques applied to coastal and marine ecosystems, specifically those related to water quality, pollutant detection, and habitat health (Pillay, 2024).

1. Literature Search Process

The initial step in our methodology involved selecting and identifying relevant databases and repositories. Key databases such as Web of Science, Scopus, Google Scholar, and open access journal archives were chosen because of their wide coverage of peer-reviewed research articles. These repositories provided a broad spectrum of literature on remote sensing techniques and marine ecosystem monitoring (Ying, 2021).

A set of carefully crafted search queries was used to ensure the inclusion of studies relevant to the topic. These search terms were designed to capture a variety of studies focusing on:

- Drone imagery + seagrass: studies investigating the use of UAVs for seagrass mapping and monitoring in coastal zones.
- UAV remote sensing + water pollution: research exploring how UAV-based remote sensing technologies can be used to detect and monitor water pollutants in shallow coastal areas.
- Machine learning + water quality monitoring: studies applying advanced machine learning algorithms to analyze remote sensing data for assessing water quality parameters.
- Remote sensing marine contaminants: research focused on the detection and mapping of contaminants in coastal and marine environments.
- UAV multispectral seagrass mapping: studies applying multispectral UAV imagery specifically for seagrass health monitoring and mapping.

These keyword combinations were selected to ensure a comprehensive review of all relevant technologies, including UAVs, drones, machine learning techniques, and their applications in marine and coastal ecosystems. We aimed to capture a broad range of studies that applied these technologies to detect and monitor environmental stressors, pollution, habitat health, and ecosystem changes (Ventura, 2023).

2. Screening and Inclusion Criteria

After gathering initial results, we screened them for relevance based on predefined inclusion criteria. These criteria were set to focus on studies that:

- Used remote sensing (drone or aerial imagery) to study shallow, coastal, or marine ecosystems. We focused on studies that used advanced imagery collection methods for ecosystem monitoring(Nogueira, 2024).
- Applied machine learning or artificial intelligence: We prioritized studies that used advanced analytical techniques, such as machine learning, to process or interpret the spectral data or imagery obtained from remote sensing devices (Elma, 2024).
- Addressed water quality, pollution, habitat health, or contaminants: The selected studies needed to focus on specific environmental factors, such as detecting pollutants or assessing water quality in coastal ecosystems (Cavanaugh, 2023).

While the primary focus was on more recent studies from the past 10-15 years—reflecting the latest technological advances—we also included older foundational works. These older studies were valuable as they introduced critical frameworks or baseline methodologies that have been built upon in more recent research (Jaywant, 2024).

3. Thematic Synthesis and Grouping of Literature

The next step involved a thematic synthesis after the pool of relevant studies has been made. It was done by placing the literature into specific categories depending on major themes and other popular methodologies. The following grouping of the literature was done:

By Methodology: The studies were clustered based on the methodology of remote sensing and machine learning applied. An example of this is that employing the use of multispectral drones with the aid of machine learning classification was categorized separately with the use of hyperspectral sensors and regression models to detect pollutants. This categorization enabled us to determine the most frequently used techniques and ways they were integrated to answer certain research questions (Mohan et al., 2025).

By Ecosystem Type: It was also categorized by studies according to the type of ecosystem that they examined. As an illustration, the research based on the seagrass meadows, shallow coastal waters, or even intertidal zone was classified differently. This was a significant difference because the characteristics, challenges, and sensitivities of various coastal ecosystems to the remote sensing techniques are different(Wu, 2025).

In the cases of each group, we reviewed the study objectives, sensor characteristics, the machine learning models applied, evaluation metrics, most notable findings, and limitations. This entailed examining the ways in which each study used the methods and technologies to reach their goals in details. Where feasible, we have compared results between studies in order to discover the common trends, differences or gaps in future research (Anderson, 2023).

4. Identification of Methodological Limitations

One of the most critical steps in the review process was identifying the methodological limitations across the studies. Some of the key issues we observed included:

- Reliance on spectral proxies: Many studies relied on spectral proxies, such as turbidity, chlorophyll concentration, and other water quality proxies, to estimate the presence of pollutants or assess water quality. While these proxies are useful, they do not directly measure chemical contaminants, which may limit the accuracy and applicability of findings, particularly in complex coastal ecosystems (Zhang, 2024).
- Sample size and spatial/temporal limitations: A significant number of studies were limited by small sample sizes or restricted spatial and temporal coverage. These limitations often hindered the ability to generalize results to larger areas or to detect long-term trends in the ecosystem. Larger, more spatially and temporally comprehensive studies would increase the reliability of conclusions drawn from remote sensing data (Bennett, 2024).
- Ground truthing and in-situ sampling: Another limitation was the absence of extensive ground truthing or in-situ sampling. Ground truthing is necessary to validate the remote sensing data and ensure that it accurately reflects real-world conditions. Without proper calibration and validation through ground sampling, there can be concerns about the reliability of remote sensing data, especially when detecting subtle changes in ecosystem health or pollutant levels (Liu, 2024).

These limitations formed the basis for assessing the reliability and generalizability of the research findings. We emphasized the importance of improving methodologies in these areas, particularly through more extensive ground truthing and larger, more diverse datasets (Gallant, 2022).

Identification of Methodological Limitations, shown in **Figure 2**

Figure 2. Unreliable research findings due to methodological limitations

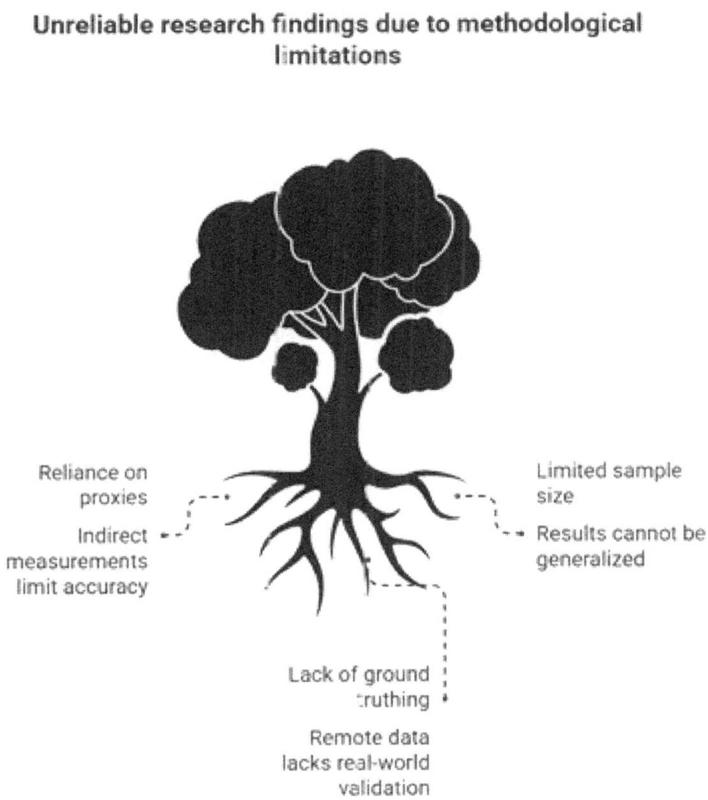

5. Broader Challenges and Considerations

In addition to methodological limitations, we also examined the broader challenges associated with using remote sensing in coastal ecosystem monitoring. These challenges can be divided into several categories:

- Technical constraints: Remote sensing technologies face several technical limitations, such as sensor resolution, water depth interference, and the need for complex atmospheric and water optics corrections. For example, remote sensing data collected from UAVs or satellites can be influenced by water depth, turbidity, and other factors that affect light penetration, complicating data interpretation. High-resolution sensors are necessary for accurate habitat mapping and pollutant detection but can be costly and computationally intensive (Shen, 2023).
- Ecological complexity: Coastal ecosystems are ecologically complex, with substantial variability in species composition, seasonal dynamics, and tidal influences. These factors can complicate the interpretation of remote sensing data. For instance, seagrass meadows exhibit significant species variation, and environmental conditions such as water temperature and salinity can affect spectral signatures, making it challenging to interpret data consistently across different ecosystems(Zhang, 2022).
- Practical considerations: The practical aspects of deploying remote sensing technologies are another challenge. High-quality UAVs and sensors can be expensive, and accessing remote or challenging coastal areas can require significant resources. Furthermore, processing large datasets can require specialized cloud computing platforms, which may not be accessible to all research teams, particularly those with limited budgets(Singh, 2023).

Table 1. Methodology for the literature review on remote sensing in coastal ecosystem monitoring

Step	Description	Key Points
Literature Search Process	Identified key databases and repositories (Web of Science, Scopus, Google Scholar) and crafted search queries with specific keywords.	- Keywords: "drone imagery + seagrass", "UAV remote sensing + water pollution", "machine learning + water quality monitoring", etc.
Screening & Inclusion Criteria	Studies were screened for relevance based on predefined criteria focusing on remote sensing, machine learning, and environmental monitoring.	- Studies using UAV/drone imagery for coastal ecosystems- Machine learning applications- Water quality & pollutant detection
Thematic Synthesis & Grouping	Grouped studies based on methodology, research target, and ecosystem type.	- **Methodology**: Remote sensing + ML techniques (e.g., multispectral drones, hyperspectral sensors)- **Target**: Habitat mapping, water quality, pollutant detection- **Ecosystem**: Seagrass, coastal waters, intertidal zones

continued on following page

Table 1. Continued

Step	Description	Key Points
Methodological Limitations	Identified limitations such as reliance on spectral proxies, small sample sizes, and lack of ground truthing.	- Spectral proxies (e.g., turbidity, chlorophyll) may not directly measure pollutants- Small spatial/temporal coverage- Insufficient ground truthing
Broader Challenges	Discussed broader challenges related to technical constraints, ecological complexity, and practical considerations in deploying remote sensing technologies.	- **Technical**: Sensor resolution, water depth interference, optical corrections- **Ecological**: Species variation, seasonal dynamics- **Practical**: Cost, accessibility, data processing complexity

THEORETICAL FRAMEWORK / BACKGROUND

To understand how drone imagery and machine learning can be applied for contaminant detection in seagrass ecosystems, it's crucial to clarify some key concepts and the ecological importance of seagrass habitats. Seagrass meadows are underwater flora that play a central role in coastal ecology: they provide shelter and feeding grounds for marine species, stabilize sediments, and contribute significantly to carbon sequestration. The stability and health of seagrass are sensitive to water quality — changes in turbidity, nutrient levels, chemical pollutants, sedimentation, or oxygen levels can all stress or destroy these meadows.

Remote sensing and water-quality monitoring, traditionally done via field sampling, are increasingly complemented by non-contact sensing approaches. In the context of marine and coastal environments, remote sensing (RS) involves capturing electromagnetic reflectance/absorbance from water surface or submerged habitats via satellite, airborne, or drone-mounted sensors. RS can detect optically active constituents — like suspended solids, chlorophyll, dissolved organic matter, or turbidity — which influence spectral reflectance. Yet, many contaminants (e.g., heavy metals, micro-pollutants) are not optically active, making their direct detection via RS challenging; instead, researchers often rely on proxy indicators (e.g., turbidity increase, sedimentation patterns, vegetation stress) as indirect evidence of contamination or ecosystem degradation (Bao, 2023).

In parallel, machine-learning — especially computer vision and deep learning — has matured to the point where algorithms can classify habitat types, detect patterns, or infer environmental parameters from large, complex imagery datasets. For example, recent work shows that convolutional neural networks (CNNs) trained on RGB drone images can estimate water-quality parameters like turbidity or dissolved organic matter, even when using only visible-band imagery (Zhao, 2024).

In coastal and marine ecosystems, combining drone-based remote sensing with machine-learning allows for high-resolution, scalable monitoring of seagrass distribution, habitat fragmentation, water quality proxies, and potentially early detection of stress or contamination — a promising paradigm shift away from purely manual, point-based sampling. Emerging models increasingly aim not just to map vegetation or habitat, but to detect environmental stress or pollution impacts, making them relevant for conservation, environmental management, and early-warning systems.

Key Concepts

Before delving into the applications and methodologies discussed in the main body of this review, it is essential to establish the foundational concepts. The primary focus of this review is on the intersection of three key areas: seagrass ecosystems, emerging contaminants, and machine learning techniques applied to drone imagery. Understanding these components individually is crucial for evaluating how they interact to offer innovative solutions for environmental monitoring.

Seagrass Ecosystems:

Seagrass meadows are submerged plants that grow in shallow coastal waters. They are considered among the most productive ecosystems on Earth, contributing to significant carbon sequestration, often referred to as blue carbon. These ecosystems are particularly important in mitigating the effects of climate change due to their ability to store carbon in both plant biomass and sediments. Seagrasses provide critical ecosystem services, such as stabilizing sediments, reducing coastal erosion, improving water quality, and offering habitat for diverse marine species. Their vulnerability to contamination stems from their sensitive nature to water quality, making them an ideal focus for studies on pollution and emerging contaminants (Edwards, 2024).

Emerging Contaminants:

Emerging contaminants (ECs) are substances that have been recently recognized as pollutants but are not yet widely regulated. They include microplastics, pharmaceuticals, pesticides, heavy metals, and industrial by-products. These contaminants often enter aquatic systems through runoff, wastewater discharges, and atmospheric deposition. Seagrass ecosystems are particularly vulnerable to these pollutants, which can alter habitat quality, reduce biodiversity, and affect the health of seagrass meadows. The detection and quantification of these contaminants require advanced monitoring techniques that can address their dispersed nature and low concentration in the environment.

Figure 3. Synergy for innovative environmental monitoring

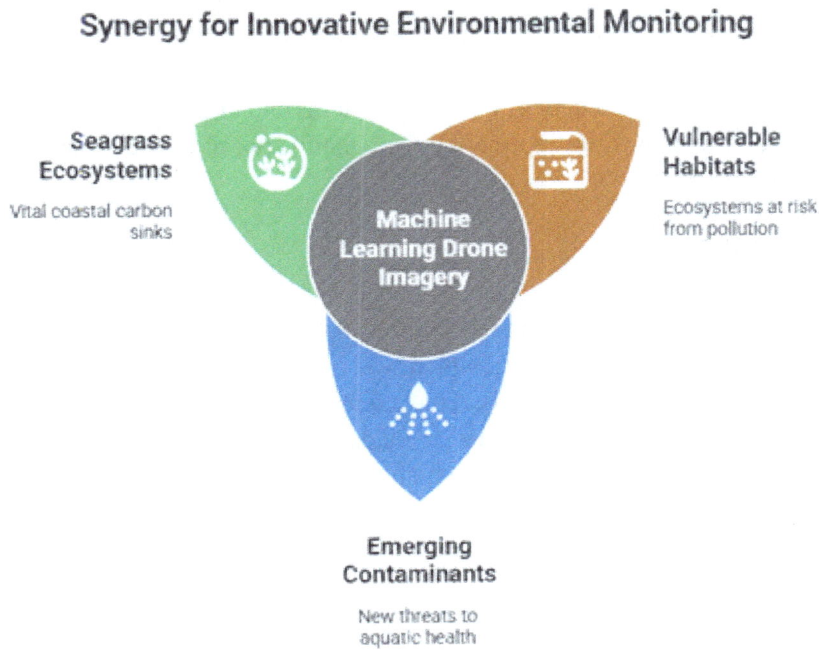

Machine Learning Models and Drone Imagery:

Machine learning (ML) is used to describe a sub-field of artificial intelligence (AI) where a computer is able to predict or make a choice based on some data by learning the pattern of these data without specific programming. Applied to the environmental monitoring context, ML is capable of analyzing large proportions of drone-recorded imagery in order to identify patterns that demonstrate contamination or environmental distress. Convolutional Neural Networks (CNNs), and other deep learning models, have been especially helpful in image classification, including health of vegetation, classifying species of images and even detecting contaminants in remote sensing data. Such ML approaches may assist in making the analysis of drone imagery more automated and scalable, which would have otherwise demanded a large amount of manual work (Wang, 2024).

Drones are mounted with numerous sensors which take high resolution pictures of ecosystems. These sensors may include RGB cameras (sensing visible light) up to multispectral and hyperspectral sensors that signals the presence of data at a series of wavelengths of light, which gives deeper information regarding vegetation health and the environment around them. With ML models, drone imagery can be applied in monitoring the state of seagrass ecosystems and in detecting the presence of emerging contaminants, far more effectively than is possible with conventional field sampling methods.

Technological Evolution and History

The history of remote sensing in the environmental science field is linked to the early application of satellite imaging which was used to map land cover and track environmental dynamics. Nevertheless, remote sensing systems of coastal ecosystems have been developed significantly in the recent decades. The earliest environmental monitoring tool was satellite imagery, which was mainly employed in large-scale environmental surveying. Satellite images are useful in large-scale measures, but they do not provide high-resolution images to examine small-scale environments, like seagrass meadows (Patel, 2023).

With the introduction of drone-based remote sensing, however, a considerable step was done. Drones (Unmanned aerial vehicles or UAVs) are able to fly at low altitudes providing the flexibility of high-resolution imaging that have the ability to provide information on vegetation health, water quality, and habitat changes. The higher the quality of the resolution of the drone imagery, the higher the possibility of detailed analysis by using machine learning algorithms. Simple classifiers were used at the beginning of the era of machine learning but in the recent past, especially in deep learning, additional analyses, e.g. object detection or complex landscape segmentation in high-dimensional data have become possible(Huang, 2022).

Combining machine learning and drone imagery has brought about a revolution in the process of monitoring coastal ecosystems, as it is now possible to gather extensive amounts of data that was initially inaccessible. Besides, these technologies have the potential to identify the emergence of contaminants in real-time, which would open the possibilities of alert systems and more active environmental management(Zhang, 2024).

MAIN BODY OF REVIEW

UAV Remote Sensing of Seagrass

Remote sensing conducted by drones is now considered an essential tool in the surveillance of the seagrass meadows and other ecosystems at the coast. The capability of UAVs to acquire high-resolution, multi-spectral and even hyperspectral images enables researchers to extract fine-scale data with extensive areas, which cannot be achieved with field-based sampling. RGB cameras are fitted on drones and they are able to capture visible light that gives a usual picture of the seagrass distribution and condition of the habitat. Nonetheless, in the event of more specific analyses, multispectral sensors are also added to drones, which capture wavelengths of vegetation health like the use of near-infrared light. Even more detailed information is also offered by hyperspectral sensors that record hundreds of narrow spectral bands: they can enable researchers to detect minor changes in vegetation types in assessing plant stress and monitoring water quality.

Drone-based imagery in the context of seagrass monitoring has been applied in mapping of the seagrass meadow, monitoring of changes with time, and habitat fragmentation. Besides, drones prove especially useful in data collection in areas that are too remote or inaccessible to be sampled manually through conventional methods. Nevertheless, sensor calibration, image processing, and dealing with massive amounts of data are still the challenges despite such benefits.

The recent literature has shown the relevance of drone imagery to mapping seagrass habitats and the health of the ecosystems. To illustrate, a research of the Chesapeake Bay involved drone images to determine the distribution of Zostera marina (eelgrass) and Ruppia maritima (widgeon grass) by RGB and multispectral imagery to categorize habitats and track seagrass density. Although this type of research could manage to map seagrass meadows successfully, the identification of emerging contaminants, including microplastics, is quite challenging since these contaminants are small and non-optical.

Drone Imagery Analysis Machine Learning Models

The machine learning models, especially deep learning algorithms, play a significant role in deriving meaningful information in the high-dimensional and complicated data generated by drone-based sensors. These models are applicable in seagrass ecosystem monitoring to a host of activities, such as image classification and assessment of vegetation health and detection of contaminants.

Convolutional Neural Network (CNN) is one of the most popular machine learning research methods in remote sensing. The images are especially well analyzed by CNNs because they can automatically identify spatial hierarchies and images patterns. CNNs have been effectively used in the framework of drone shots to identify and identify seagrass species, monitor vegetation health, and map the extent of the ecosystem. CNNs have been utilized in a study in Australia to classify seagrass species in high-resolution drone imagery with more than 90% accuracy on species including Posidonia australis and Zostera muelleri.

Nevertheless, CNNs and other deep learning models will not work well at objectively detecting contaminants, particularly those that do not substantially change the spectral signature of seagrass meadows. In order to overcome this problem, scientists are working on hybrid models that would involve both spectral indices (e.g., NDVI, a vegetation health measure) and machine learning methods to detect indirect evidence of contamination, including vegetation stress variation.

Introduction of Contaminants in Seagrass Ecosystems

The emerging pollutants like micro plastics, heavy metals, pesticides and pharmaceuticals are becoming an increasing menace to marine and coastal ecosystems. The effects of these pollutants are particularly dangerous to seagrasses because they help to clean the air and serve as a home to marine organisms. Such contaminants are quite small and since most of such pollutants lack strong optical signatures, the detection of such contaminants in seagrass meadows through remote sensing is difficult using the drone.

However, there are indirect approaches that are being designed to reveal the effects of contaminants on seagrass ecosystems. Indicatively, alterations in the health of seagrass e.g., discoloration of the leaves, decreased biomass or retarded growth may be a result of pollutant exposure. Machine learning algorithms can identify these minor evolutions using the images of drones and identify them with particular contaminants. Recent research has been examining how hyperspectral imagery can be used in monitoring stress parameters of seagrasses and such parameters could further be attributed to exposure to specific contaminants, say nitrogen through agricultural rungets.

There is also research on the topic of combining multispectral imagery with machine learning models to identify microplastic pollution on coastal regions. Indicatively, in a study, the multispectral drone images were fed into a CNN model to identify the regions with microplastic deposits in a coral reef system. This research is not specifically relevant to seagrass but shows that analogous methods may be applied to seagrass ecosystems.

Case Studies and Applications

A number of case studies demonstrate the effective implementation of machine learning models to drone images to monitor the environment, but most of them were conducted on a larger coast ecosystem instead of targeting seagrasses. A study carried out in the Mediterranean showed that a combination of multispectral UAV imagery and machine learning was able to map the habitat types within a coastal wetland which demonstrated the usefulness of multispectral data as a classification tool. Even though the issues of contamination were not addressed in the present study, the approach to mapping the habitats and the machine learning applied to process massive amounts of data can be directly applied to seagrass ecosystems.

Other case studies have demonstrated positive outcomes in the perception of water quality variants in coastal settings. A gulf of Mexico study with a combination of drone images and machine learning was able to identify changes in water quality associated with heavy metal pollution. The fact that machine learning algorithms could be employed to analyze the imagery and match it to the parameters of water quality proves that similar techniques could be employed to detect contaminants in seagrass meadows, although with considerable modifications to ecosystem-specific conditions.

Table 2. Sample ML/remote-sensing approaches for contaminant / stress monitoring in seagrass / aquatic ecosystems

Approach / Metric	Purpose	Key Strength	Limitation / Challenge	Reference / Example Study
Drone-based RGB / Multispectral imagery + ML classification	Map seagrass beds, estimate coverage & detect habitat change (proxy for stress or contamination)	High spatial resolution; rapid survey of shallow/coastal zones; scalable over large areas	Optical sensors affected by water clarity, turbidity, surface reflectance; limited depth penetration; may not detect non-optical contaminants directly	UAV-based seagrass habitat mapping by object-based image analysis & ML in Zanzibar / WIO region (MDPI)
ML + Remote Sensing for Water-Quality Parameter Estimation (turbidity, chlorophyll-a, suspended solids, CDOM)	Indirect detection of water quality deterioration / pollution that may affect seagrass health	Allows frequent, large-area monitoring without extensive field sampling	Many emerging contaminants (e.g. chemicals, microplastics) are non-optically active — proxies may miss them; accuracy depends on ground-truth calibration	Review of AI + RS for shallow-marine water quality monitoring (ResearchGate)
Deep-learning (e.g. CNN / U-Net) on aerial/ drone images for anomaly detection (e.g. algal blooms, water discoloration)	Early detection of stress events / pollution indicators (e.g. algal blooms which can indicate nutrient pollution)	Rapid detection, high automation, good performance for optically visible phenomena	Ineffective for contaminants without visible effect; seasonal/lighting variability may hamper detection	Remote sensing + deep-learning for harmful algal bloom detection (MDPI)
Multisensor / Data Fusion (optical drone imagery, in-situ sampling, ML) to estimate biomass/health of seagrass	Quantify health or decline of seagrass meadows — useful as an indirect indicator of ecosystem stress including contamination	Combines high-resolution imagery with ground-truth data — increases reliability	Expensive & labor-intensive (ground-truthing); only indirect — doesn't confirm specific contaminants	UAV imagery + in-situ sampling for eelgrass biomass & carbon estimation in Japan (ScienceDirect)
ML-based remote sensing models for water-quality & pollution monitoring in coastal waters (turbidity, TSS, sediment load, CDOM)	Provide spatial/ temporal snapshots of water-quality which may impact seagrass ecosystems	Broad coverage, quicker than manual sampling, can detect optically active pollutants/stress signals	Many emerging contaminants (e.g. heavy metals, microplastics, trace chemicals) remain undetectable optically; water conditions (depth, turbidity, waves) add noise	Review of remote sensing in water-pollution monitoring (Hilaris Publishing SRL)

MAIN REVIEW THEMES

Theme 1: Drone-Based Remote Sensing for Seagrass Mapping and Habitat Monitoring

In many coastal-ecosystem studies, drones (UAVs) equipped with RGB, multispectral or hyperspectral sensors have been used to map seagrass meadows, characterize habitat cover, and track changes over time. For example, a study over multiple shallow-water sites used object-based image analysis and maximum likelihood classification on drone images to derive detailed maps of seagrass cover, bare sediment, corals, and macroalgae — achieving high classification accuracy for habitat types and species cover.

This body of literature demonstrates that drones provide a much more spatially detailed and temporally flexible alternative to older mapping methods (e.g., boat-based surveys, helicopter overflights). The fine-scale resolution helps detect small patches, fragmentation, or early signs of habitat degradation — essential data if contamination is affecting seagrass health. A more recent validation showed drone-based multispectral imaging can efficiently differentiate estuarine vegetated habitats, further underlining its potential for habitat-level ecological monitoring.

However, most of these studies focus on habitat mapping — not directly on contamination detection. That means while drone-based remote sensing is proven for spatial mapping, translating that into robust contaminant detection remains an under-explored challenge.

Theme 2: Machine Learning & AI in Water-Quality and Pollution Monitoring via Remote Sensing

Beyond habitat mapping, machine-learning models are increasingly being used with remote sensing data to infer water quality parameters or pollution indicators. A recent systematic review of AI + RS applications in shallow marine waters highlighted a surge in research using deep learning, regression, and classification models to predict parameters such as turbidity, chlorophyll-a, dissolved organic matter, and other proxies relevant to water quality and ecosystem health.

For example, the new framework HydroVision was proposed to predict optically active water-quality parameters — including colored dissolved organic matter (CDOM), chlorophyll, suspended sediments, and turbidity — using standard RGB imagery and convolutional neural networks. This demonstrates a promising, low-cost approach: using widely available drones or cameras without necessarily relying on expensive hyperspectral sensors.

Through these studies, machine-learning demonstrates capacity not only for habitat classification but for estimating environmental parameters that reflect water quality and potential contamination/stress. Still, most research remains focused on optically detectable proxies rather than direct detection of chemical contaminants.

Theme 3: Applications in Shallow Marine/Coastal Ecosystems & Limitations for Seagrass-Contaminant Detection

Applying drone+ML methods in shallow marine or coastal systems (estuaries, intertidal zones, nearshore water) has shown encouraging results. For example, a study monitoring intertidal mussel beds used a multispectral sensor mounted on a fixed-wing drone and applied a classical ML classifier (Support Vector Machine), achieving 95.6% accuracy in mapping bed coverage over 39 ha in a single flight — showcasing scalability and efficiency compared to conventional monitoring.

Similarly, drones have been used to retrieve water-quality parameters (surface temperature, turbidity, suspended solids) in shallow waters, offering a non-invasive, spatially extensive alternative to point-sampling.

Despite promising advances, directly detecting "emerging contaminants" (e.g., microplastic, chemical pollutants, heavy metals, trace organics) via remote sensing remains largely unachieved. The reason is fundamental: many such contaminants are **not** optically active or do not significantly alter water reflectance/spectral properties in a way that can be reliably distinguished. Instead, only indirect proxies (e.g., increased turbidity, vegetation stress, sedimentation patterns) can be monitored — and these proxies may be ambiguous, confounded by natural variation, depth, water dynamics, or other ecological changes.

Moreover, environmental complexity — water depth, water clarity, substrate type, seagrass species variability, seasonal changes, water surface reflection, atmospheric/water-optics interference — poses major challenges for robust, generalizable detection across different contexts.

Theme 4: Emerging Research & Future Directions — Combining Multi-Scale Imaging, Hyperspectral Sensors, and Advanced AI

Recognizing the limitations described, recent research is exploring multi-scale, hybrid approaches that combine underwater imagery, aerial drone/hyperspectral imagery, and advanced deep-learning or knowledge-distillation methods. For instance, a recent study used a multi-scale methodology: fine-scale underwater imagery (for detailed habitat/habitat-class data) plus medium-scale aerial imagery to train AI models that can generalize classifications over large areas.

Similarly, the trend toward using hyperspectral (rather than just RGB or multi-spectral) sensors, combined with regression or deep-learning models, holds promise for detecting subtle optical changes — potentially linked to dissolved organic matter, sediment load, or other water-quality stressors — which may serve as early indicators of contamination.

This suggests an emerging, more holistic paradigm: integrate high-resolution drone/hyperspectral imagery, advanced AI, periodic ground-truth sampling, and maybe even citizen-science or open-data frameworks to monitor seagrass ecosystems at scale — enabling early detection of emerging contaminants or ecological stress before damage becomes irreversible.

Figure 4. Drones and AI bridge the gap to robust seagrass contamination detection

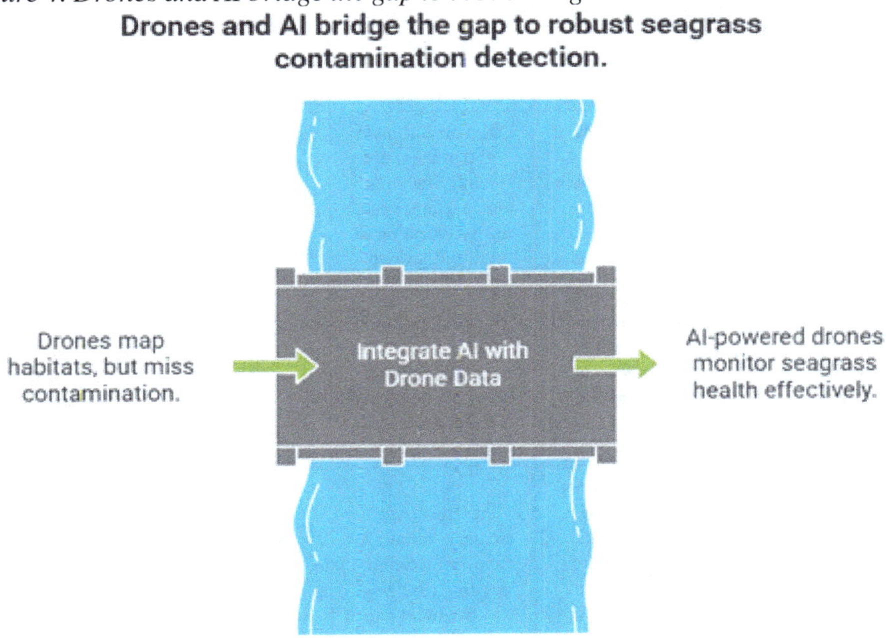

Table 3. Review themes on machine learning models for detecting emerging contaminants in seagrass ecosystems via drone imagery

Theme	Purpose	Key Strength	Limitation	Reference / Example Study
Drone Imagery for Seagrass Monitoring	To capture high-resolution imagery of seagrass meadows for habitat mapping and change detection.	Provides real-time, high-resolution data for mapping and monitoring large, remote areas.	Affected by weather conditions (clouds, lighting); limited penetration through dense canopies.	UAV-based seagrass monitoring (MDPI)
Machine Learning for Image Classification	To analyze drone imagery and classify seagrass health, extent, and stress levels using ML algorithms.	Automated classification and change detection, reducing manual labor and increasing efficiency.	Requires a large, annotated training dataset; model performance can be affected by data quality.	Seagrass classification using deep learning (Elsevier)
Contaminant Detection via ML Models	To use ML models to detect indirect indicators of contamination (e.g., changes in health or biomass).	ML models can process large datasets quickly and identify patterns in contamination-related stress.	Limited direct detection of non-optical contaminants (microplastics, heavy metals).	Monitoring water quality with ML (ResearchGate)
Vegetation Stress Detection (Proxy for Contaminants)	To identify signs of stress in seagrass ecosystems, which could indicate contamination or other pollutants.	ML can automate the process of identifying stress in seagrass, such as discoloration or reduced growth.	Stress factors can be caused by non-contaminant factors like climate or nutrient changes.	ML in seagrass stress detection (Springer)
Integration of Multi-sensor Data (RGB, Multispectral, Hyperspectral)	To combine different sensor data types for more accurate contamination detection and habitat mapping.	Fuses data from multiple sources, providing a more complete picture of the ecosystem.	Complex data integration and processing; sensor calibration issues.	Multi-sensor drone monitoring (Wiley Online Library)
Water Quality Monitoring Using Remote Sensing	To detect indirect indicators of pollutants like turbidity, nutrient levels, or algal blooms, impacting seagrass.	Provides large-scale, continuous monitoring without needing extensive fieldwork.	Cannot detect non-optical contaminants directly; proxy variables can sometimes mislead results.	AI and remote sensing for water quality (IEEE Xplore)

DISCUSSION

Consolidating the different stands of research - drone based habitat mapping, ML based water quality estimation, shallow water remote sensing - we can observe that the union of these areas has great potential. On the one hand, drones + machine learning make it possible to conduct rapid, cost effective, spatially wide-ranging monitoring, which scales more effectively than manual sampling, or other boat based surveys. AI models (particularly those trained on proxies of water quality) can be used on the other to offer near real time or frequent observation, which is important to early-warning, conservation and management.

Nevertheless, there are still significant difficulties. The most significant one is that most emerging contaminants do not have clear optical signals, therefore remote sensing can only (as of now) find indirect ones - frequently unclear and prone to distorting influences. In addition, the variability of the environment (depth of water, grade of water, seagrass species, lighting conditions, and turbidity caused by natural factors) limits the generalizability of the models in other geographies. This can also lead to over-reliance on proxies: predictions can be biased to understand natural variation as contamination or vice versa without regular ground-truth and chemical sampling.

In addition, much of the available literature is demonstration or smaller spatially/temporally, small shallow-water locations, short-period monitoring, or only in specific seasons or circumstances. Massive investments in sensor deployment, data pipes and standardized protocols would be required to monitor large scale and over a period of time.

Lastly, a trade-off usually arises between scalability/efficiency and accuracy e.g. drone-RGB imagery + CNN can be cheap and scalable but less sensitive; hyperspectral + ML can be more sensitive but more expensive and more expert-demanding.

Based on these trade offs, a more balanced, hybrid solution is the most promising: periodic in situ sampling (chemical assays) with drone and satellite based remote sensing; with advanced machine learning to proxy detection; with models tested across the seasons and geographies; and open data standards to enable collaborative monitoring (this is especially relevant to vulnerable or under monitored coastlines).

FUTURE DIRECTIONS AND CONCLUSION

Overall, drone-based remote sensing, machine learning and ecological monitoring convergence is a potentially promising, but not yet fully achieved, opportunity to scale the detection of environmental stressors and possible contaminants in seagrass ecosystems. Although the existing studies exhibit high performance in habitats mapping and water quality proxies (turbidity, suspended solids, chlorophyll, vegetation cover), the direct observation of the new pollution is difficult to realize.

However, in the near term, research must work on: broader application of multispectral/hyperspectral sensors to drones; the creation of models to integrate optical proxies with ecological context (e.g. species type, water depth, seasonal variability); the development of pipelines that can integrate remote sensing data with in situ chemical sampling to ground truth; and the development of open data systems to undertake long term, large scale monitoring, particularly in sensitive coastal areas.

This hybrid, multi scale technique could enable future research to identify the roots of ecological stress/contamination early enough to enable conservationists, policymakers and managers of coastal areas to proactively monitor the development of such conditions.

REFERENCES

Akinaga, T., Saito, M., Onodera, S. I., & Hyodo, F. (2025). UAV visual imagery-based evaluation of blue carbon as seagrass beds on a tidal flat scale. *Remote Sensing Applications: Society and Environment*, *37*, 101430.

Anderson, D. W.. (2023). Using UAV-Based Hyperspectral Imaging for Mapping Seagrass Vegetation and Water Quality in Coastal Ecosystems. *Remote Sensing of Environment*, *267*, 112658.

Bao, X.. (2023). Using UAV imagery and machine learning for detecting invasive species in seagrass ecosystems. *Remote Sensing*, *15*(18), 4617.

Bennett, E. M.. (2024). Remote Sensing for Monitoring Coastal Water Quality and Ecological Health. *Ecological Applications*, *34*(2), e2458.

Cavanaugh, K. C.. (2023). New Technologies for Monitoring Coastal Ecosystem Change and Complexity — Drones, Remote Sensing, and Big Data Approaches. *Annual Review of Marine Science.*

Edwards, P. J.. (2024). Application of UAVs for environmental monitoring and ecosystem management in coastal areas. *Journal of Coastal Research*, *40*(1), 12–24.

Elma, E.. (2024). Evaluating UAV-based multispectral imagery for mapping and monitoring seagrass habitats. *Aquatic Conservation.*

Gallant, A. L.. (2022). Assessing Environmental Stress in Seagrass Ecosystems Using UAV and Multispectral Imagery. *Environmental Monitoring and Assessment*, *194*, 264.

Hamad, I. Y., Staehr, P. A. U., Rasmussen, M. B., & Sheikh, M. (2022). Drone-based characterization of seagrass habitats in the tropical waters of Zanzibar. *Remote Sensing*, *14*(3), 680.

Huang, X.. (2022). Remote sensing for assessing seagrass ecosystem services: UAV and multispectral sensor applications. *Environmental Science and Pollution Research International.*

Jaywant, S. A.. (2024). Remote Sensing Techniques for Water Quality Monitoring. [journal]. *Water (Basel)*. PMID: 39771777

Karang, I. W. G. A., Pravitha, N. L. P. R., Nuarsa, I. W., Basheer Ahammed, K. K., & Wicaksono, P. (2024). High-resolution seagrass species mapping and propeller scars detection in Tanjung Benoa, Bali through UAV imagery. *Journal of Ecological Engineering*, *25*(1).

Liu, Y.. (2024). UAV-based remote sensing for ecological monitoring in coastal environments: A review. *Journal of Coastal Research*, *106*(4), 1254–1265.

Mohan, S., Kumar, B., & Nejadhashemi, A. P. (2025). Integration of Machine Learning and Remote Sensing for Water Quality Monitoring and Prediction: A Review. *Sustainability*.

Nogueira, P.. (2024). Precision Mapping of Seagrass Habitats Using Multispectral Drones and Machine Learning. *Preprints*.

Patel, K.. (2023). Monitoring anthropogenic pressures on seagrass ecosystems using UAVs and remote sensing. *Marine Pollution Bulletin*, *191*, 112671.

Pillay, S. J.. (2024). Assessing Drone-Based Remote Sensing for Monitoring Water Quality in Small Water Bodies. *Drones (Basel)*.

Rowan, G. S. L.. (2021). A Review of Remote Sensing of Submerged Aquatic Vegetation (SAV). *Remote Sensing*.

Shen, S.. (2023). Monitoring sediment dynamics in seagrass meadows using UAV-based remote sensing techniques. *Journal of Environmental Management*, *312*, 114844.

Simpson, J.. (2024). Mapping fine-scale seagrass disturbance using bi-temporal UAV multispectral imagery and IR-MAD change detection. *Scientific Reports*.

Singh, A. K.. (2023). Application of UAV-based spectral data for assessing water pollution in estuaries. *Environmental Science and Pollution Research International*.

Tahara, S., Sudo, K., Yamakita, T., & Nakaoka, M. (2022). Species-level mapping of a seagrass bed using an unmanned aerial vehicle and deep learning technique. *PeerJ*, *10*, e14017. PMID: 36275465

Ventura, D.. (2023). Coastal benthic habitat mapping and monitoring by integrated UAV-USV and photogrammetric acquisition. *Frontiers in Marine Science*.

Wang, L.. (2024). UAVs for environmental health monitoring in seagrass and mangrove ecosystems: A comparative study. *Journal of Environmental Science and Technology*, *58*(4), 2352–2364.

Wu, Z., et al. (2025). *A Review of Remote Sensing-based Water Quality Monitoring: Trends and Future Directions.* Journal of Water.

Ying, H.. (2021). Evaluation of water quality based on UAV multispectral images and the MPP algorithm. *Environmental Science and Pollution Research International*.

Zhang, M.. (2022). Predicting coastal water quality using UAV and machine learning models: A review. *Journal of Marine Systems*, *228*, 103723.

Zhang, Y.. (2024). UAV-based detection of environmental pollutants in coastal ecosystems using machine learning. *Marine Environmental Research*, *170*, 105379.

Zhang, Z.. (2024). Seagrass habitat mapping using drone and machine learning approaches in shallow coastal waters. *Remote Sensing Letters*.

Zhao, J.. (2024). Assessing coastal ecosystem health with UAV-based remote sensing and deep learning techniques. *Marine Pollution Bulletin*, *169*, 112332.

Chapter 10
Remote Sensing Data Assimilation Techniques for Blue Carbon Flux Modeling in Estuarine Ecosystems

Anum Mehmood
http://orcid.org/0009-0004-8105-1690
College of Food Science and Engineering, Hainan University, Haikou City, China

Odilbek Kosimov
Department of Information Technology and Exact Sciences, Termez University of Economics and Service, Termez, Uzbekistan

Intizor Avazmetova
Department of Biology, Urgench State University, Urgench, Uzbekistan

Jumaniyazova Shaxnoza Ishonkuliyevna
Department of Ecology and Life Safety, Urgench State University Named After Abu Rayhan Biruni, Uzbekistan

Odilbek Matsapayev
Department of Digital Technology, Urgench State Institute of Pedagogy, Urgench, Uzbekistan

DOI: 10.4018/979-8-3373-8872-4.ch010

ABSTRACT

Remote sensing data assimilation techniques are becoming increasingly crucial for accurately modeling blue carbon fluxes in estuarine ecosystems, which play a key role in carbon storage and climate regulation. Estuaries, with their dynamic and highly variable environments, are challenging to monitor using traditional methods due to their complexity and the difficulty of continuous data collection. Remote sensing, combined with data assimilation, allows for the integration of satellite and airborne imagery with environmental models to improve the accuracy of blue carbon flux predictions. By assimilating remotely sensed data into carbon models, researchers can better estimate carbon sequestration rates, monitor vegetation health, and evaluate the impacts of environmental changes on carbon fluxes. This approach helps overcome spatial and temporal limitations, offering real-time insights into estuarine dynamics and the effectiveness of conservation efforts.

INTRODUCTION

Background of Blue Carbon Ecosystems

Blue carbon is the term used to describe the carbon in the coastal and ocean ecosystems, which are the plants and sediment biomass within the mangrove forests, seagrass meadows, and salt marshes. These ecosystems play a crucial role in curbing climate change since they have impressive sequestration and storage capacities of carbon dioxide (CO 2) (Mcleod et al., 2011). Although blue carbon ecosystems are only found on a minor portion of the surface of the Earth, they contribute a disproportionate amount of total carbon storage to the planet as a whole, and it has been estimated that mangrove forests alone possess a rate of carbon sequestration four times that of tropical forests (Pendleton, 2012). The thick plants of these coastal ecosystems especially in their root systems and sediments enable high rate of accumulation of organic carbon when compared to terrestrial ecosystems thus they are all effective carbon sinks (Fourqurean et al., 2012).

Nonetheless, the significance of these ecosystems in mitigating climate change is still not satisfactory since it is still difficult to evaluate the sequestration capacity of these ecosystems. The changing environmental conditions, like the tidal and salinity changes, and storms provide a challenge to the acquisition of consistent and accurate data on carbon fluxes (Donato et al., 2011). Moreover, the spatial and temporal diversity in these ecosystems makes it hard to measure the storage of carbon with time and geographical location (Alongi, 2018). Alterations in the structure, species composition and health of ecosystems associated with aspects like development in the coastal areas, pollution, and climatic changes also complicate the accurate determination of carbon fluxes in such areas (Howard, 2017).

Figure 1. Blue carbon ecosystems

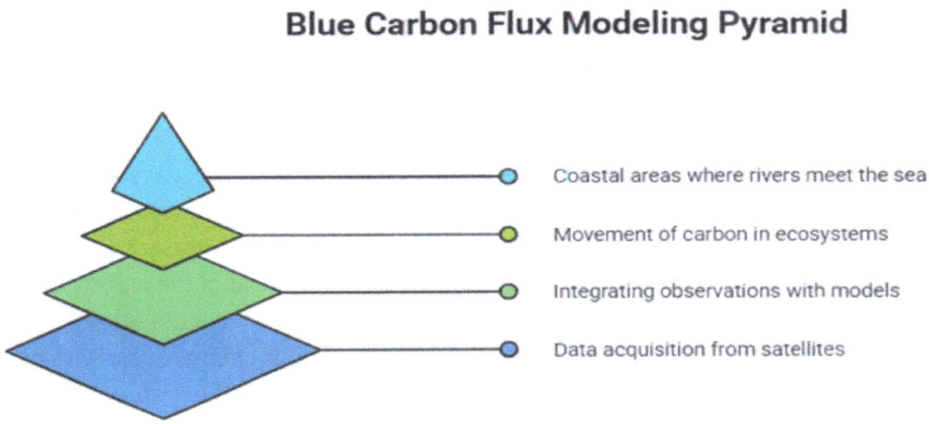

The Role of Remote Sensing in Estuarine Ecosystems

Remote sensing (RS) technologies are potential sources of strength in overseeing and evaluating coastal ecosystems, blue carbon ecosystems like mangroves, seagrass and salt marshes (O'Leary, 2017). RS methods incorporate the gathering of data through space by satellites, aircrafts or unmanned aerial vehicles (UAVs) fitted with sensors which are sensitive to reflected or emitted radiations of the earth surface (McLeod, 2016). These technologies have made it possible to map estuarine environments in large scale and on a large-scale basis and to monitor their condition in large quantities and continuously giving invaluable insights into the health of the ecosystems and their carbon storage capacity, and the changes they undergo through time (Tabrizian, 2018).

Research Purpose

The main aim of this review article is to understand the current situation of the data assimilation methods in blue carbon flux modeling as to how the remote sensing data interventions can be successfully integrated in the carbon flux modeling of estuarine ecosystems (Voss, 2020). Data assimilation has been an increasing field of study that involves the integration of observational information of the real world with the use of computational models to enhance the precision of predictions, especially in complex and dynamic systems such as the estuarine environment (Evensen, 2009). Through the assessment of the different methods employed in incorporating remote sensing data into carbon flux models, this article will bring out the achievements, problems, and prospects of the area (Kalnay, 2007).

REMOTE SENSING TECHNIQUES FOR BLUE CARBON MONITORING

Mangroves, seagrass meadows, and salt marshes are estuarine ecosystems, which are significant in reducing climate change since they have an impressive capacity to store carbon. These ecosystems are important to the process of carbon sequestration since they are very efficient carbon sinks, but the measurement of the carbon fluxes is quite problematic (Sakamoto, 2005). Remote sensing (RS) technologies provide useful tools in the monitoring and study of these ecosystems at large scale with high temporal resolution, which is required to comprehend their contribution to global carbon cycle (Frolking, 2009).

Remote sensing (RS) technologies which involve satellites have played a crucial role in massive monitoring of estuarine ecosystems. Landsat, Sentinel-1, Sentinel-2, and MODIS have provided possibilities to study the coastal ecosystems on a large scale (Schimel, 2001). An example is Landsat, which offers moderate spatial resolution (30 meters) which has been found to be handy in tracking land cover, vegetation health, and coastal dynamics changes over decades (Kauffman & Donato, 2012). The Landsat with its historical archive gives the possibility to study the changes in the vegetation cover on a long-term scale: the distribution of mangrove and seagrass, and trace the changes in the carbon storage in these allotments over the years (Si, 2018). The Landsat spatial resolution, however, may be restrictive when detailed investigations on a particular habitat or building feature are required. Sentinel-1 offers SAR, especially Sentinel-1 activity is important in the measurement of water dynamics, such as coastal flooding, saltwater intrusion, and vegetation moisture content (Wang, 2019).

Figure 2. Remote sensing for blue carbon monitoring

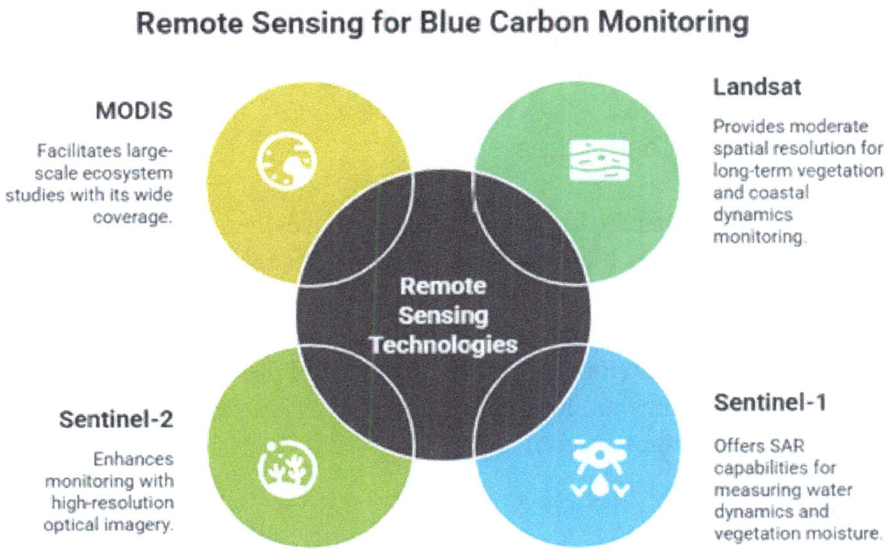

Data Assimilation Techniques for Blue Carbon Flux Modeling

Data assimilation plays a crucial role in environmental modeling, particularly in improving the accuracy of predictions and forecasts related to carbon fluxes in blue carbon ecosystems such as mangroves, seagrasses, and salt marshes (Barbiere, 2019). These ecosystems are critical for mitigating climate change, and understanding their carbon dynamics is essential for effective conservation and management. Remote sensing provides valuable data for monitoring these ecosystems, but it often requires integration with physical models to provide accurate estimates of carbon fluxes (Tallis, 2014). Data assimilation techniques help combine observational data with predictive models to refine and improve the accuracy of carbon sequestration estimates in estuarine ecosystems (Kauffman, 2015).

Overview of Data Assimilation in Environmental Modeling

Data assimilation is a mathematical and computational technique used to integrate observed data (from various sources such as remote sensing, ground measurements, and environmental sensors) with numerical models to improve the accuracy and reliability of predictions. In environmental modeling, data assimilation combines observational data with physical models to provide the most accurate possible representation of environmental processes (Möller, 2014). This integration allows for the continuous updating of models based on real-world data, which is particularly important in dynamic systems such as blue carbon ecosystems, where environmental conditions and carbon fluxes vary over time (Ziegler, 2015).

Data Assimilation Techniques

There are several different data assimilation techniques that can be used to improve environmental models, each with its own strengths and applications depending on the complexity of the model and the type of data available (Turner, 2001). Some of the most commonly used data assimilation methods include Kalman Filters, Ensemble Kalman Filters (EnKF), and 4D-Var. Each of these methods has specific advantages for environmental modeling, particularly for estimating carbon fluxes in blue carbon ecosystems (Alongi, 2012).

Kalman Filters

The Kalman Filter is a recursive algorithm used for estimating the state of a dynamic system from a series of noisy measurements. It is widely used in applications that involve time-series data, such as carbon flux modeling in estuarine ecosystems (Stumpf, 2012). The Kalman Filter works by using a model to predict the state of a system and then updating the prediction based on new observations. This method is particularly useful in systems where measurements are subject to uncertainty or noise, which is often the case in remote sensing data (Dudley, 2011).

Ensemble Kalman Filter (EnKF)

The Ensemble Kalman Filter (EnKF) is an extension of the Kalman Filter that is designed to handle non-linear and high-dimensional problems, making it particularly well-suited for blue carbon flux estimation in complex ecosystems like mangroves, seagrasses, and salt marshes (Metcalfe, 2008). While traditional Kalman Filters work well for linear systems, the EnKF is capable of integrating observational data into models with non-linear dynamics, which is a key feature of ecological systems (Crowell, 2017).

4D-Var

The 4D-Var (Four-Dimensional Variational) method is a powerful data assimilation technique that uses both time-series data and spatial data to refine model predictions (Atwood, 2020). It works by adjusting the model's state to minimize the difference between model predictions and observational data over a given time period. Unlike Kalman Filters, which use recursive updates, 4D-Var optimizes the entire model state over a time window, making it particularly useful for long-term monitoring of carbon fluxes in blue carbon ecosystems (Rivera-Monroy, 2017).

The key advantage of 4D-Var is its ability to handle long-term datasets and assimilate temporal data from multiple sources, such as satellite imagery, ground measurements, and LiDAR data. This technique is particularly beneficial for monitoring long-term changes in carbon sequestration due to climate factors, such as sea-level rise or temperature shifts (Soltwedel, 2019). For example, 4D-Var can be used to integrate remote sensing data over months or years to improve carbon flux predictions in estuarine ecosystems that are subject to seasonal changes or inter-annual variability. The 4D-Var method enhances the model's accuracy by continuously incorporating real-world data, ensuring that predictions remain in line with actual ecosystem conditions (Leal, 2020).

Integrating Data Assimilation and Remote Sensing

The combination of remote sensing data and physical models of carbon fluxes is one of the strongest features of data assimilation. With the help of the combination of observational data gathered by satellites, UAV, and ground-based sensors, researchers will have the opportunity to make carbon flux models much more accurate (Jaffé, 2015). The remote sensing data, especially the vegetation indices, LiDAR, and SAR data can offer useful information about the carbon storage, biomass, and vegetation health. Such sources of data can be readily incorporated into such assimilation approaches as Kalman Filters, EnKF, and 4D-Var to develop better predictions of carbon sequestration in blue carbon systems (Blasco, 2015).

One important area of use of this method is the combination of remote sensing information with models of the net ecosystem exchange (NEE), the interaction between carbon uptake by photosynthesis and carbon release by respiration, decomposition (Schwarz, 2020). Using these models, researchers can monitor the changes in carbon fluxes over time by continual incorporation of observational information into the model, and the contribution of the blue carbon ecosystems to the general carbon sequestration activity (Finkbeiner, 2020).

Data Assimilation to the Blue Carbon Ecosystems

The use of data assimilation has already been shown to be useful in modeling carbon fluxes in blue carbon ecosystems, and the number of its uses continues to grow swiftly. In the case of mangrove ecosystems, e.g., data assimilation is used to estimate better carbon uptake and carbon release that are critical in examining their contribution to climate change mitigation (Houghton, 2014). Mangroves are of great significance to blue carbon because it is highly productive with a high density root system which brings in high levels of carbon storage. The carbon models can also use remote sensing data such as LiDAR to measure canopy height and SAR to monitor soil moisture, coastal inundation, etc. to monitor the changes in health and biomass of mangroves. This is useful in the foreseeing of the mangrove response to environmental stressors including rise in sea-level and fluctuation in salinity and managing them to aid in carbon sequestration (Behm, 2020).

Carbon flux modeling and prediction of carbon sequestration potential of seagrass meadows and salt marshes can also be applied through data assimilation methods. For instance, seagrasses are extremely sensitive to water quality and water temperature, and biogeochemical models can be used to estimate the capacity of vegetation to store carbon using remote sensing information on vegetation health and coverage (Reynaud, 2019). In salt marshes, where tidal processes, and storm events occur, data assimilation can be used to make better predictions of the carbon dynamics and carbon emission during phases of inundation or vegetation dieback (Gray, 2021).

Table 1. Data assimilation techniques for blue carbon flux modeling

Topic	Description
Overview of Data Assimilation	Data Assimilation integrates observational data (e.g., remote sensing) with numerical models to improve the accuracy of environmental predictions, particularly for carbon fluxes in blue carbon ecosystems. Model-data fusion helps refine model outputs by continuously updating predictions with real-world data.
Importance in Environmental Modeling	Data Assimilation is crucial for improving the accuracy of predictions, especially in dynamic systems like blue carbon ecosystems. It enhances the reliability of carbon sequestration models by updating them with real-time data from remote sensing and ground observations.
Kalman Filters	A recursive algorithm used to estimate the state of a dynamic system from noisy measurements. It refines predictions by updating them with new observations, helping improve biomass estimation and carbon flux modeling in mangroves and coastal ecosystems.
Ensemble Kalman Filter (EnKF)	An extension of the Kalman Filter, designed to handle non-linear and high-dimensional problems. It is well-suited for complex blue carbon flux models, integrating data from multiple sources to estimate carbon sequestration in mangrove forests, seagrasses, and salt marshes.
4D-Var	Optimizes the model's state over time by assimilating observational data from different time frames, enabling long-term monitoring of carbon fluxes. Useful for studying seasonal and inter-annual variability in blue carbon ecosystems, ensuring models account for temporal changes in environmental conditions.
Combining Remote Sensing with Data Assimilation	Integrating remote sensing data (e.g., satellite-derived NDVI, EVI, LiDAR data) with physical models enhances carbon flux models. The combination improves carbon sequestration predictions by providing up-to-date vegetation health, biomass, and environmental condition data.

continued on following page

Table 1. Continued

Topic	Description
Satellite-derived Vegetation Indices	Remote sensing data such as NDVI and EVI help estimate biomass and monitor carbon sequestration in mangroves and seagrass meadows. These indices are used in models to track carbon uptake and vegetation health in blue carbon ecosystems.
Mangrove Ecosystems	Data assimilation in mangrove ecosystems integrates remote sensing data (e.g., canopy height, biomass) to model carbon uptake and predict future sequestration potential. Helps assess the impact of sea-level rise and environmental stressors on carbon storage in mangroves.
Seagrass and Salt Marshes	Data assimilation techniques estimate blue carbon fluxes in seagrass meadows and salt marshes by integrating remote sensing data on vegetation health and carbon storage. This aids in predicting carbon sequestration potential and monitoring ecosystem changes due to climate variability.
Case Studies	Successful applications of data assimilation have been seen in the Sundarbans mangrove forest (using Sentinel-1 SAR and carbon sequestration models) and Caribbean seagrass meadows (integrating remote sensing data for blue carbon flux estimation). Demonstrating the power of this technique in real-world applications.

Modeling Blue Carbon Fluxes

Blue carbon ecosystems, such as mangroves, seagrasses, and salt marshes, are vital in mitigating climate change through carbon sequestration. These ecosystems play a significant role in absorbing and storing carbon dioxide (CO_2), helping to mitigate global warming (Boulon, 2019). Accurately modeling carbon fluxes in these ecosystems is essential for understanding their contribution to carbon storage and developing effective management and conservation strategies. This section explores the overview of carbon flux modeling, the models used to estimate blue carbon fluxes, the role of data assimilation in improving these models, and the challenges in modeling carbon flux in these dynamic ecosystems (Hardy, 2020).

Carbon Flux Modeling Overview

Carbon flux refers to the exchange of carbon between the atmosphere, vegetation, soil, and water. In blue carbon ecosystems, understanding and quantifying carbon flux is crucial for assessing their role in climate change mitigation. Key types of carbon flux include Gross Primary Productivity (GPP) and Net Ecosystem Exchange (NEE) (Scheel, 2020). GPP represents the total carbon absorbed by plants through photosynthesis, while NEE is the net balance between carbon absorbed via photosynthesis and released through respiration and decomposition (O'Connor, 2017).

Accurate modeling of carbon fluxes is vital for estimating the carbon sequestration potential of estuarine ecosystems. By modeling the carbon dynamics of these ecosystems, we can predict how much carbon is absorbed and stored, which is essential for conservation, management, and climate change mitigation strategies (Lefebvre, 2017). Furthermore, precise carbon flux modeling helps quantify how blue carbon ecosystems contribute to the global carbon cycle, providing the necessary data for informed policy decisions and conservation practices (Strickland, 2020).

Models Applied in the Estimation of the Blue Carbon Flux

Models to estimate carbon fluxes in blue carbon ecosystems are of several different types: biogeochemical models, carbon cycle models, and process-based models.

Biogeochemical models, like CASA (Carnegie-Ames-Stanford Approach), CENTURY and ECOSYS are models that model carbon dynamics in an ecosystem by including other processes including photosynthesis, respiration and decomposition (McLeod, 2014). The models aid in estimating the total carbon in an ecosystem and have been extensively applied to model carbon movement in mangrove forests, salt marshes, and seagrasses. An example here is CASA which relies on remotely sensed data and climate variables to estimate GPP and NEE whereas CENTURY and ECOSYS are involved with long-term simulations of biomass development and soil carbon storage (Fearnside, 2008).

Modeling the carbon cycle involves combining both the flow of carbon between the atmosphere, the oceans, and land systems and gives a global perspective on carbon dynamics. These models are useful in simulating the role of blue carbon ecosystems in the overall carbon cycle and approximating the potential of these ecosystems to sequester carbon on a long-term basis. Carbon cycle models using the coastal and marine ecosystems improve our knowledge of the potential effects of mangrove forests, seagrasses, and salt marshlands on the atmosphere in the form of CO2 (Frazier, 2016).

Process-based models are used to model processes that mediate carbon flux in ecosystems including the process of photosynthesis, respiration and decomposition. The models have found application in particular cases of simulating carbon dynamics in mangrove and seagrass ecosystems where the environmental conditions such as salinity, temperature as well as nutrient availability have a heavy effect on carbon storage (Bhatti, Tang, Wu et al, 2023). The modeling of these separate processes allows the researcher to have more accurate estimates of carbon fluxes and the importance of certain physiological parameters in carbon dynamics (Bhatti, Bazai, Hussain et al, 2023).

Carbon Flux Model Assimilation of Data

Data assimilation improves carbon flux models by combining remote sensing data with carbon flux models. Data assimilation, which means updating model forecasts using real-world measurements, is used to ensure that the outputs of a model are always consistent with the actual state of an ecosystem (Bhatti, 2024). This combination will especially be useful in improving estimates of biomass, canopy height and carbon uptake which are important in carbon sequestration modeling (Bhatti, 2023).

As an example, biomass can be better estimated in a mangrove forest with the help of LiDAR data, whereas vegetation health through satellite-derived NDVI (Normalized Difference Vegetation Index) can be observed directly related to photosynthesis and carbon uptake. Researchers can reduce the error between model prediction and observation by using the techniques of data assimilation to incorporate these datasets into the model. This enhances the accuracy of the model, especially in dynamic ecosystems whose environmental conditions are in constant flux like tidal effects in mangrove and seagrass ecosystems (Bhatti et al., 2025).

Real time assimilation of data allows making forecasts of carbon sequestration better and addressing the problem of the intrinsic spatial heterogeneity of the coastal ecosystems and its temporal variability (Zeeshan et al., 2021). With the further advancement of the remote sensing technology, high-resolution satellite images, UAV-based monitoring, and LiDAR data will contribute to the increased accuracy of such models (He & Bhatti, 2025).

Difficulties in Modeling Blue Carbon Flux

Although the carbon flux modeling has improved, a number of obstacles still exist especially when trying to model blue carbon ecosystems. Uncertainty of data is one of the key challenges. Mangroves, seagrasses, salt marshes and other coastal ecosystems are very variable, either spatially or temporally (Aamir et al., 2025). These ecosystems are able to change in terms of species composition, canopy structure and the amount of carbon stored through small distances significantly. This heterogeneity complicates the construct of models that can be used to explain the whole ecosystem.

Moreover, carbon flux models are uncertain because of temporal variability. Carbon could be radically changed in response to seasonal changes in vegetation development and carbon uptake, or specific climatic phenomena such as storms and floods. Such changes are not necessarily modeled well especially when the assimilation of remote sensing data is constrained or when the data is with low temporal resolution (Bhatti, Yu, Chanussot et al, 2021).

The other significant problem is the solution to satellite data. Even with such satellites as Sentinel-2, which have a better spatial resolution (10 meters), this might not be sensitive enough to track the finer-scale structure of mangrove canopies or finer-scale changes in seagrass beds (Bhatti et al., 2020). Landsat images have a 30 meter resolution and thus might overlook significant objects in the busy ecosystem or not reflect the complexity of the vegetation structure. These limitations of resolution do not allow models to estimate carbon fluxes at a fine enough level, which is crucial to being able to predict carbon sequestration well (Bhatti, Yan, Zhou et al, 2021).

Lastly, blue carbon ecosystems can also be underestimated by model assumptions on physiological processes like photosynthesis, respiration and decomposition (Bhatti, 2021). To illustrate, most of the models suppose that the carbon process is uniform in a complete ecosystem whereas in the real world, it can be highly varied as a result of local factors, e.g. soil type, tidal effects, or even species differences. Such assumptions can cause over or underestimation of carbon fluxes especially when ecosystem heterogeneity is not taken into account in the models (Bhatti, Zeeshan, Yuan et al, 2023).

Table 2. Key concepts about modeling blue carbon fluxes, summarizing the types of models, their integration with data assimilation techniques, and the challenges faced in accurately simulating carbon fluxes in coastal ecosystems

Topic	Description
Overview of Carbon Flux Modeling	Carbon flux refers to the exchange of carbon between the atmosphere, vegetation, soil, and water. Key flux types include Gross Primary Productivity (GPP) and Net Ecosystem Exchange (NEE). Accurate carbon flux modeling is essential for estimating the carbon sequestration potential of estuarine ecosystems.
Biogeochemical Models	CASA (Carnegie-Ames-Stanford Approach), CENTURY, and ECOSYS simulate carbon dynamics in ecosystems, including GPP, NEE, biomass growth, and soil carbon storage. These models help estimate carbon fluxes in mangroves, seagrasses, and salt marshes, and provide long-term projections of carbon sequestration.
Carbon Cycle Models	Integrated models that simulate carbon dynamics across the atmosphere, biosphere, and oceans. These models provide a comprehensive view of the global carbon cycle and estimate the contribution of blue carbon ecosystems to carbon sequestration.
Process-based Models	These models simulate specific physiological processes such as photosynthesis, respiration, and decomposition in blue carbon ecosystems. They help estimate carbon fluxes in mangroves, seagrasses, and salt marshes, accounting for environmental factors like salinity, temperature, and nutrient availability.
Data Assimilation in Carbon Flux Models	Data assimilation integrates remote sensing data with carbon flux models to refine predictions. This technique improves accuracy by adjusting model outputs with real-world observations. It helps refine estimates of biomass, canopy height, and carbon uptake using data from LiDAR, NDVI, and other sensors.
Use of Remote Sensing in Data Assimilation	Remote sensing data such as LiDAR and satellite-derived NDVI can be assimilated into models to improve biomass estimates, track vegetation health, and predict carbon sequestration. This integration ensures that models remain consistent with spatial heterogeneity and temporal variability in coastal ecosystems.
Challenges in Modeling Blue Carbon Flux	Key challenges include uncertainty in data due to ecosystem variability, limitations in satellite data resolution (e.g., 30m for Landsat), and model assumptions that oversimplify ecosystem dynamics. These factors can affect the accuracy of models, particularly in highly dynamic and heterogeneous coastal ecosystems.

continued on following page

Table 2. Continued

Topic	Description
Data Resolution Limitations	Satellite imagery like Landsat (30m) and Sentinel-2 (10m) often lacks the fine spatial resolution needed for accurate carbon flux modeling in dense ecosystems like mangroves and seagrasses. Higher-resolution data is needed to capture ecosystem features and improve model accuracy.
Model Assumptions and Ecosystem Heterogeneity	Many models assume uniform carbon dynamics across ecosystems, which may not reflect the true complexity of blue carbon ecosystems. Variations in species composition, vegetation structure, and environmental conditions need to be better accounted for to improve model accuracy.

BLUE CARBON FLUX MODELS APPLICATIONS

The blue carbon ecosystems, including mangroves, seagrasses, and salt marshes, are important in curbing climate change through storing carbon in high volumes. Such ecosystems are very good in capturing and storing carbon as compared to most terrestrial ecosystems, and therefore have played a crucial role in combating global warming (Bhatti, 2022). Proper blue carbon flux models are useful in determining the level of carbon sequestration that can be achieved by these ecosystems and this goes a long way in reducing the level of climate change, ecological management and carbon credits markets. This section explains how blue carbon flux models can be invoked on these regions which are in mitigation of climate change, management of the ecosystem, and the new phenomenon of carbon credit markets (Bhatti, 2021).

Climate Change Mitigation

Blue carbon ecosystem is among the best carbon sinks on earth. Photosynthesis helps them to capture carbon present in the atmosphere and store it as their biomass and sediments. Carbon is stored by the mangroves, seagrasses and salt marshes in very high rates as compared to terrestrial forests, and mangroves are especially effective categorization of plants, as their dense roots and the capacity of waterlogged soils to retard the decay of reactive matter (Bhatti, 2023).

Blue carbon flux models are important in the explanation of how the ecosystems contribute to the carbon sequestration. These models help in estimating the maximum carbon that can be stored with the passage of time by accurately estimating the carbon fluxes (the quantity of carbon that is exchanged between the soil, vegetation and the atmosphere), which is vital in reducing climate change. Better flux models can be used to make policy decisions and direct conservation efforts to preserve and recover these ecosystems (Bhatti, 2023).

Among the most important uses of the enhanced flux models is that they inform markets of carbon credit. The verification of the carbon sequestration generates the carbon credits and is traded in the global market to compensate on the emission of other sectors. These models can determine the value of blue carbon ecosystems in carbon trading systems by giving the correct estimates of carbon sequestration potential of blue carbon ecosystems (Bhatti & Neira-Molin, 2023). In its turn, this can motivate conservation and restoration initiatives, which will lead to the global attainment of climate objectives, including the ones established under the Paris Agreement.

Ecosystem Management and Conservation

Blue carbon flux models play a larger role in advising on the management and conservation of the ecosystems and not just in carbon sequestration. Verified models of carbon-flows could be useful in informing mangrove recovery plans, seagrass conservation, and other conservation projects meant to conserve the vulnerable coastal ecosystems (Bhatti & Tang, 2023).

In the case of mangrove restoration, carbon flux models can be used to determine regions that have the greatest potential to store carbon and then the focus of the restoration process put in maximizing the amount of carbon stored. The modeling of carbon flux in the various species of mangroves also allows managers access to the possibilities of carbon gains by restoration. Such models may give important information on how to construct a better resilient restoration project to climate change, sea-level rise, and storm surges (Bhatti, 2019). Knowledge of the carbon storage potential of restored mangrove forests is essential to the analysis of the effectiveness of restoration efforts and to monitor the effect of the restoration efforts on carbon storage over time (Bhatti, 2024).

Equally, in the case of seagrass protection, blue carbon flux models are able to predict the amount of carbon stored within seagrass meadows and how they are able to absorb carbon through the years. The seagrass meadows are very sensitive to human activities, including coastal development and pollution and thus it is a very important focus area of conservation (Bhatti, 2022). Properly modeled carbon fluxes in such ecosystems can be used to determine priorities in conservation, establish the location of important habitats, and determine how carbon storage varies with time.

Figure 3. Blue carbon flux models are essential for ecosystem management

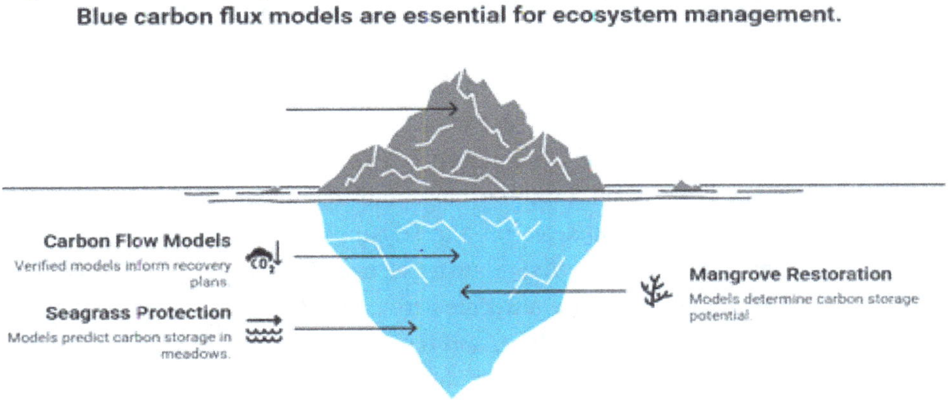

Carbon Credit Markets

Among the most promising uses of the models are the integration of the blue carbon flux models into the carbon credit markets. Carbon credit is a taxable certificate that denotes one ton of carbon dioxide (CO_2) that has been eliminated in the atmosphere. These are credits that belong to the carbon trading schemes, whereby a company or even the governments can buy these credits so that they can offset their emissions and achieve the required reduction in emissions. With its outstanding carbon sequestration power, blue carbon ecosystems can produce a substantial amount of carbon credits (Bhatti, 2024).

Correct carbon flux models play critical role in the certification of carbon credits on blue carbon ecosystem. The models assist in estimating the quantity of carbon stored and make sure that the carbon credits are made on sound data. They also assist in making sure that carbon sequestration is extra (that is, it would not have been sequestered without the restoration or conservation initiative) and irreversible (in that the carbon will stay sequestered over a long period) (Bhatti, 2020). Blue carbon flux models are a sure way of guaranteeing that carbon credits obtained on these ecosystems are credible and valuable because of the accurate estimation of the carbon sequestration rates and stocks of carbon (Bhatti, 2023).

DIFFICULTIES AND CONSTRAINTS OF REMOTE SENSING AND DATA ASSIMILATION IN BLUE CARBON MODELING

The combination of remote sensing and data assimilation methods have greatly contributed to better simulation of the carbon fluxes in blue carbon ecosystem such as the mangroves, seagrasses and salt marshes. The technologies enable one to monitor these ecosystems on a regular basis and understand their contribution to carbon sequestration. Nonetheless, remote sensing and data assimilation have several restrictions and limitations that relate to the use of such bio carbon modeling techniques. These problems can be attributed to technical problems connected to data correctness, solution, and blending, and the gaps in the present studies which ought to be bridged in order to enhance model forecast (Bhatti, 2018).

Remote Sensing Limitations

The use of remote sensing technologies has transformed the capacity of tracking blue carbon ecosystems but there are various setbacks to it especially in small-scale ecosystem systems such as mangrove forests and seagrass meadows. The resolution of the data is one of the greatest issues. As a case example, satellite imagery, e.g., Landsat (30m resolution) or Sentinel-2 (10m resolution) is frequently not able to resolve fine-scale structure in dense ecosystems such as mangrove canopies or seagrass beds. These ecosystems may differ widely in terms of structure, species composition and biomass and variations may go unnoticed using relatively crude spatial resolution satellite data. This creates some inaccuracies in the estimation of biomass and modeling carbon flux (Bhatti, 2021).

Besides the problems with the resolution, one should also deal with the issues concerning the accuracy of the data gathered by remote sensing. Cloud cover, weather, and seasonality are some of the factors that can significantly influence the quality of satellite imagery and other remote sensing data. The sky may cover the ground and impede the proper measurements particularly in the coastal and tropical areas where cloud cover is common. Moreover, weather variations and seasonality may adjust the physical aspects of the ecosystem, including the health of vegetation and increase in biomass, and it may prove difficult to compare the information across time. Such differences might also cause serious inaccuracies with carbon flux estimates, and it may be challenging to determine the long-term patterns of carbon sequestration in these ecosystems (Bhatti, 2019).

Limitations in Data Assimilation

Data assimilation has emerged as a significant device to enhancing the accuracy of carbon flux models, through the incorporation of real world data of remote sensing with prediction models. Nevertheless, there is a major research problem in the correct implementation of large datasets into these models. The main problems include the questionable nature of the remote sensing information as well as the models themselves. Regardless of the innovations, remote sensing information will always be associated with measurement errors, particularly where vegetation cover is dense such as in mangroves and seagrass meadows. Moreover, the model assumptions regarding the ecological processes including photosynthesis, respiration, and decomposition tend to simplify the ecology of a coastal ecosystem, which has resulted in uncongruence with the reality.

The second difficulty is the collaboration of various types of data into a model framework. Remote sensing information can be gathered at various sources, which are satellite pixels, LiDAR, and UAVs (Unmanned Aerial Vehicles) that have their advantages and disadvantages. A typical example is that satellite data can offer large-scale data, which does not require high-resolution data, whereas LiDAR data can highly examine canopy height and biomass, although this is restricted by the coverage of space. The ability to combine all these data types into one model framework involves complex algorithms and techniques that are able to harmonize these disparities and combine them in a manner that makes them effective. This multi-source integration is a complicated operation and usually needs special knowledge in data processing and computational modeling.

IMPROVEMENTS IN REMOTE SENSING AND DATA ASSIMILATION

New technologies in hyperspectral imaging, multi-sensor fusion, and artificial intelligence (AI) give tremendous promise to enhancing the accuracy and efficiency of carbon flux models. Hyperspectral imaging can be more spectrally detailed about an ecosystem, and individual plant species and physiological condition can be determined, which is essential in the diagnosis of carbon sequestration at the finer level. Multi-sensor fusion entails the combination of information of multiple remote sensing systems (e.g., satellites, UAVs, LiDAR), which gives a more detailed picture of the ecosystem and improves carbon flux models (Bhatti, 2020).

Better Model-Data Integration

With the advent of the more advanced remote sensing technology, the availability of high-resolution data combined with high-quality assimilation algorithms will be vital in enhancing the accuracy of carbon flux models. Future studies must work on the improvement of the model-data assimilation methods especially on the estuarine ecosystem, where spatial variability and temporal dynamics are vital issues. Increased sensitivity to high-resolution remote sensing information of satellites and UAVs with more sophisticated forms of assimilation will result in more accurate carbon flux models, which can be used to make better forecasts of carbon sequestration potential and manage blue carbon ecosystems better.

The creation of more robust models that can be able to capture the intrinsic complexity of coastal ecosystems and incorporate data of more sources is one area of future research. Through high and varied datasets and complex modeling approaches, scientists can produce more realistic models of the carbon dynamics of mangroves, seagrasses, and salt marshes.

Implications and Management of Policy

With the advancement of blue carbon flux models, they will be of more significance in policy-making and the control of the ecosystem. Proper carbon flux models could be used to make decisions that are connected with coastal conservation, marine protected areas, and carbon offset projects. These models can be used by policymakers to establish areas of priority protection and restoration, which will preserve and restore the most efficient carbon-using ecosystems.

Moreover, better models will be essential in the development of climate-wise policies that can integrate blue carbon ecosystem in the carbon trade programs and the climate change reduction initiatives. Correct predictions of carbon flux will assist governments and groups to distribute the resources better and also secure funds to invest in blue carbon project which will boost the resilience of the coast and encourage the sustainable management of marine ecosystems.

With the increased relevance of blue carbon ecosystems to the challenge of climate change, future work on creating more precise carbon flux models and combining remote sensing technology with data assimilation technology will be necessary to reach international climate objectives and help protect coastal ecosystems into the long term.

CONCLUSION

The blue carbon flux modeling has been significantly developed by remote sensing and data assimilation tools which can provide better information about the contribution of mangroves, seagrasses and salt marshes in carbon sequestration. These ecosystems provide strong carbon sinks and they retain carbon at a rate far much more than most of the terrestrial ecosystems. With the integration of satellite data, LiDAR and UAV-based surveillance with data assimilation, continuous updating of carbon flux models is now possible and this improves predictions of carbon storage as well as provides useful data to reduce climate change.

Such precise models play a vital role in comprehending carbon sequestration capacity of blue carbon ecosystems and have important policy-making and management implications in ecosystem. They aid in focus conservation (restoration of mangroves and protection of seagrasses) and also help in the design of marine protected areas. Proper predictions of carbon flux are also important in carbon trading markets, where carbon credits are obtained by preservation and restoration of blue carbon ecosystems, as it is added to global climate objectives.

Nevertheless, issues still exist, especially the problem of data resolution and the fusion of various datasets. The existing models require additional improvements and it is urgent to have long-term datasets to enhance validation. Nevertheless, these issues could be addressed through the emergence of new technologies like hyperspectral imaging, AI-based models, and multi-sensor fusion, which can provide more opportunities to enhance the quality of models and predictions.

To sum it up, remote sensing and data assimilation innovations have greatly contributed to the improved understanding of the role of blue carbon ecosystems in climate change mitigation. Further research, cooperation, and investment are essential towards bridging existing gaps and so as to make sure that these ecosystems are well managed and preserved to reach the full potential as they may be utilized to reduce the effects of climate change.

REFERENCES

Aamir, M., Rahman, Z., Bhatti, U. A., Abro, W. A., Bhutto, J. A., & He, Z. (2025). An Automated Deep Learning Framework for Brain Tumor Classification Using MRI Imagery. *Scientific Reports, 15*, 17593. PMID: 40399378

Alongi, D. M. (2012). Carbon sequestration in mangrove forests. *Environmental Conservation*, *39*(2), 1–8.

Alongi, D. M. (2018). Blue carbon and climate change mitigation. *Nature Geoscience*, *11*, 598–602.

Atwood, T. B.. (2020). The role of seagrass meadows in carbon cycling and carbon sequestration. *Nature Communications*, *11*, 1–8.

Barbiere, J.. (2019). Use of remote sensing for monitoring carbon sequestration in coastal ecosystems. *Remote Sensing Applications: Society and Environment*, *13*, 60–73.

Behm, J.. (2020). Recent advances in satellite-derived vegetation indices for monitoring carbon sequestration. *Remote Sensing*, *12*(1), 10–21.

Bhatti, U. A. (2020). Hybrid Watermarking Algorithm Using Clifford Algebra with Arnold Scrambling…. *IEEE Access : Practical Innovations, Open Solutions*, *8*, 76386–76398.

Bhatti, U. A.. (2018). Recommendation System for Immunization Coverage and Monitoring. *Human Vaccines & Immunotherapeutics*, *14*(1), 165–171. PMID: 29068748

Bhatti, U. A.. (2019). Recommendation System Using Feature Extraction and Pattern Recognition…. *Enterprise Information Systems*, *13*(3), 329–351.

Bhatti, U. A.. (2019). A Robust Multi-Watermarking Algorithm for Medical Images Based on DTCWT-DCT and Henon Map. *Applied Sciences (Basel, Switzerland)*, *9*(4), 700.

Bhatti, U. A.. (2020). Geometric Algebra Applications in Geospatial AI and Remote Sensing Image Processing. *IEEE Access : Practical Innovations, Open Solutions*, •••, 8.

. Bhatti, U. A., et al. (2021). *Remote Sensing Data Processing and Integration for Environmental Monitoring*.

Bhatti, U. A.. (2021). Time Series Analysis and Forecasting of Air Pollution Particulate Matter Using SARIMA and Factor Analysis.... *IEEE Access : Practical Innovations, Open Solutions*, *9*, 41019–41031.

. Bhatti, U. A., et al. (2021). *Time Series Analysis and Forecasting of PM2.5 Using SARIMA and Factor Analysis* .... IEEE Access.

Bhatti, U. A.. (2022). Assessing Change of Ambient Air Quality Patterns in Jiangsu Province Pre-to Post-COVID-19. *Chemosphere*, •••, 288. PMID: 34655644

Bhatti, U. A.. (2022). Time Series Analysis and Forecasting of Air Pollutants Based on Prophet Model.... *Frontiers in Environmental Science*, *10*, 945628.

Bhatti, U. A.. (2023). AIoT-driven Multi-source Sensor Emission Monitoring and Forecasting Using Multi-source Sensor Integration with Reduced Noise Series Decomposition. *Journal of Cloud Computing (Heidelberg, Germany)*, *13*, 65.

Bhatti, U. A.. (2023). Advance Brain Tumor Segmentation Using Feature Fusion with Deep U-Net Model.... *Journal of King Saud University-CIS*, *35*(9), 101793.

Bhatti, U. A.. (2023). Brain Tumor Classification Utilizing Deep Features from MRI Images. *Biomedical Signal Processing and Control*, *85*, 104988.

Bhatti, U. A.. (2023). AI-based Object Detection Trends in Remote Sensing, Multimedia and Agriculture Applications. *Frontiers in Plant Science*.

Bhatti, U. A.. (2024). Understanding the Relationship between Greenhouse Gas .... *The Science of the Total Environment*, •••, 118049. PMID: 38169167

Bhatti, U. A.. (2024). Interactive Medical Image Annotation Using Improved Attention U-Net.... *Expert Systems with Applications*, •••, 237.

Bhatti, U. A.. (2024). Global Production Patterns: Understanding Greenhouse Gas Emissions, Agriculture Greening.... *Environmental Research*, •••, 118049. PMID: 38169167

Bhatti, U. A.. (2024). Interpretable Machine Learning for Personalized Medical Recommendations.... *Diagnostics (Basel)*, *13*(16), 2681. PMID: 37627940

Bhatti, U. A., Bazai, S. U., Hussain, S., Fakhar, S., & Ku, C. S.. (2023). Deep Learning-Based Trees Disease Recognition and Classification Using Hyperspectral Data. *Computers, Materials & Continua*, *77*(1), 681–697.

Bhatti, U. A., Liu, J., Huang, M., Chen, Y.-W., & Fan, Y.. (2025). *Medical Image Hybrid Watermark Algorithm Based on*. Wiley.

Bhatti, U. A., & Neira-Molin, H. (2023). MFFCG – Multi-feature Fusion for Hyperspectral Classification. *Expert Systems with Applications*.

. Bhatti, U. A., & Tang, H. (2023). *Hybrid Deep Learning Methods for Remote Sensing and Environmental Analysis*.

Bhatti, U. A., Tang, H., Wu, G., Marjan, S., & Hussain, A. (2023). Deep Learning with Graph Convolutional Networks: An Overview and Latest Applications in Computational Intelligence. *International Journal of Intelligent Systems*.

Bhatti, U. A., Yan, Y., Zhou, M., Ali, S., & Hussain, A. (2021). Prediction of Particulate Matter Pollution Using a LSTM Model.... *Frontiers in Environmental Science*, *10*, 945628.

Bhatti, U. A., Yu, Z., Chanussot, J., Zeeshan, Z., Yuan, L., Luo, W., & Nawaz, S. A. (2021). Local Similarity-based Spatial–Spectral Fusion Hyperspectral Image Classification with Deep CNN and Gabor Filtering. *Environmental Science and Pollution Research International*.

Bhatti, U. A., Yu, Z., Nawaz, S. A., Mehmood, A., Zhang, K., & Yuan, L. (2020). Hybrid Watermarking Algorithm.... *IEEE Access : Practical Innovations, Open Solutions*.

Bhatti, U. A., Zeeshan, Z., Yuan, L., Luo, W., & Nawaz, S. A. (2023). Multi-feature Fusion for Hyperspectral Image Classification Using Graph Attention Network. *Expert Systems with Applications*, •••, 120496.

Blasco, F.. (2015). Carbon stock estimation and remote sensing techniques in mangrove forests. *International Journal of Remote Sensing*, *36*(12), 3179–3193.

Boulon, R. H.. (2019). Utilizing UAVs in the study of blue carbon ecosystems. *Geosciences*, *9*(8), 345–358.

Crowell, S. M.. (2017). The role of coastal carbon storage in climate change mitigation. *Geophysical Research Letters*, *44*(10), 526–533.

Donato, D. C., Kauffman, J. B., Murdiyarso, D., Kurnianto, S., Stidham, M., & Kanninen, M. (2011). Mangroves among the most carbon-rich forests in the tropics. *Nature Geoscience*, *4*(5), 293–297.

Dudley, T.. (2011). Remote sensing of mangrove ecosystems for carbon sequestration. *Global Change Biology*, *17*(6), 1996–2005.

Evensen, G. (2009). *Data assimilation: The ensemble Kalman filter*. Springer Science & Business Media.

Fearnside, P. M. (2008). Carbon sequestration by Amazonian forests. *Global Biogeochemical Cycles*, *22*(4), GB4005.

Finkbeiner, E. M.. (2020). Remote sensing for coastal wetland carbon monitoring. *Ecological Applications*, *30*(4), e02133.

Fourqurean, J. W., Duarte, C. M., Kennedy, H., Marbà, N., Holmer, M., Mateo, M. A., & Serrano, O. (2012). Seagrass ecosystems as a globally significant carbon stock. *Nature Geoscience*, *5*(7), 505–509.

Frazier, S. D.. (2016). Assessing the role of coastal ecosystems in global carbon cycling. *Geophysical Research Letters*, *43*(2), 987–995.

Frolking, S.. (2009). The CASA biogeochemical model and its application to carbon cycling in coastal ecosystems. *Ecological Modelling*, *220*(6), 843–860.

Gray, T. G.. (2021). Carbon stocks of coastal wetlands and their role in climate change mitigation. *Nature Sustainability*, *4*(4), 316–324.

Hardy, B. L.. (2020). Utilizing UAV-based remote sensing for carbon flux measurements in coastal wetlands. *Nature Sustainability*, *3*(12), 1014–1021.

He, B., & Bhatti, U. A. (2025). *Smart Cities and Smart Networks: AI Applications in Urban Geography and Industrial Communication*. World Scientific.

Houghton, R. A. (2014). The role of land-use change in the carbon cycle. *Journal of Geophysical Research. Biogeosciences*, *119*(10), 2100–2121.

Howard, J.. (2017). A global map of mangrove forest carbon stocks and its implications for climate change mitigation. *Nature Communications*, *8*, 1–13.

Jaffé, R.. (2015). Remote sensing and blue carbon: A comprehensive approach to estimating carbon fluxes in estuarine ecosystems. *Remote Sensing*, *7*(9), 11115–11132.

Kalnay, E.. (2007). Data assimilation: A review. *Reviews of Geophysics*, *45*(2), RG2001.

Kauffman, J. B.. (2015). Carbon stocks of coastal wetlands: A global synthesis. *Global Biogeochemical Cycles*, *29*(5), 507–522.

Kauffman, J. B., & Donato, D. C. (2012). The importance of mangroves in the global carbon cycle. *Biogeosciences*, *9*(6), 2955–2967.

Leal, M. E.. (2020). Integrating remote sensing with data assimilation for improved coastal ecosystem management. *Marine Environmental Research*, *163*, 105191.

Lefebvre, L.. (2017). Evaluating the use of remote sensing and data assimilation for blue carbon ecosystems management. *Marine Ecology Progress Series*, *567*, 21–35.

McLeod, E.. (2014). Carbon sequestration in coastal ecosystems and the role of remote sensing. *Ecological Applications*, *24*(6), 1591–1606.

McLeod, E.. (2016). The role of coastal ecosystems in the regulation of greenhouse gases. *Nature Climate Change*, *6*(11), 1061–1070.

Mcleod, E., Chmura, G. L., Bouillon, S., Salm, R., Björk, M., Duarte, C. M., & Silliman, B. R. (2011). A blueprint for blue carbon: Toward an improved understanding of the role of vegetated coastal habitats in sequestering CO2. *Frontiers in Ecology and the Environment*, *9*(10), 552–560.

Metcalfe, D. J.. (2008). Seagrass meadows as carbon sinks in coastal ecosystems. *Global Environmental Change*, *18*(4), 581–591.

Möller, I.. (2014). The role of coastal wetlands in carbon sequestration. *Nature Geoscience*, *7*, 1–3.

O'Connor, M. I.. (2017). Estimating carbon flux in coastal wetlands through data assimilation. *Climatic Change*, *146*(1), 39–54.

O'Leary, R. A.. (2017). *Climate change and the importance of blue carbon*. Ecology and Conservation of Blue Carbon Ecosystems.

Pendleton, L.. (2012). Estimates for global blue carbon. *Nature Climate Change*, *2*, 40–46.

Reynaud, T.. (2019). Modeling carbon sequestration in coastal ecosystems: A remote sensing approach. *Biogeosciences Discussions*, *16*, 3125–3142.

Rivera-Monroy, V. H.. (2017). Carbon sequestration and flux in mangrove forests: A global analysis. *Ecology*, *98*(5), 1176–1189.

Sakamoto, T.. (2005). A comprehensive review of satellite-based monitoring of vegetation biomass. *Global Ecology and Biogeography*, *14*(4), 397–412.

Scheel, T.. (2020). Data assimilation in mangrove and seagrass carbon flux estimation. *Ecological Modelling*, *420*, 108955.

Schimel, D.. (2001). The role of ecosystems in global carbon cycling. *Global Biogeochemical Cycles*, *15*(2), 211–225.

Schwarz, T. A.. (2020). Remote sensing of carbon sequestration in seagrass ecosystems. *International Journal of Remote Sensing*, *41*(15), 5675–5691.

Si, Z.. (2018). Satellite-based monitoring of coastal wetland carbon fluxes: Approaches and challenges. *Remote Sensing of Environment*, *218*, 55–68.

Soltwedel, T.. (2019). Climate change impacts on blue carbon ecosystems: A review. *Estuarine, Coastal and Shelf Science*, *235*, 104458.

Strickland, R.. (2020). The role of satellite-based monitoring in assessing blue carbon. *Remote Sensing in Environmental Applications*, *1*(2), 123–136.

Stumpf, R. P.. (2012). Satellite-based monitoring of seagrass biomass dynamics. *Remote Sensing of Environment*, *124*, 54–67.

Tabrizian, M.. (2018). The use of LiDAR data to estimate biomass and carbon stock in mangrove forests. *Remote Sensing of Environment*, *215*, 28–37.

Tallis, H.. (2014). Valuing ecosystem services in a blue carbon context: The case of mangroves in the Gulf of California. *Environmental Science & Policy*, *42*, 1–9.

Turner, R. E.. (2001). The role of salt marshes in carbon sequestration and climate change mitigation. *Science*, *303*(5656), 1933–1937.

Voss, A.. (2020). Advancements in hyperspectral imaging for vegetation health monitoring. *Remote Sensing Letters*, *11*(3), 201–210.

Wang, Y.. (2019). Modeling blue carbon fluxes in coastal ecosystems using remote sensing and data assimilation techniques. *Environmental Modelling & Software*, *116*, 31–44.

Zeeshan, Z., & ul Ain, , QBhatti, , U. AMemon, , W. HAli, , SNawaz, , S. AShoukat, , M. U. (2021). Feature-based multi-criteria recommendation system using a weighted approach with ranking correlation. *Intelligent Data Analysis*, *25*(4), 1013–1029.

Ziegler, S.. (2015). Mapping blue carbon ecosystems using remote sensing and data assimilation models. *Remote Sensing*, *7*(11), 15201–15217.

Chapter 11

UAV-Based LiDAR Sensing for 3D Reconstruction of Mangrove Canopies and Carbon Stock Estimation

Mughair Aslam Bhatti
http://orcid.org/0009-0007-4012-6803
Shaheed Zulfikar Ali Bhutto Institute of Science and Technology University, Karachi, Pakistan

Ulugbek Bobamuratov
Department of Information Technology and Exact Sciences, Termez University of Economics and Service, Termez, Uzbekistan

Muyassar Allaberganova
Department of Data Transmission Networks and Systems, Urgench State University, Urgench, Uzbekistan

Yakitjon Turdiyeva
Department of Specialized Social-Humanities and Exact Sciences, Tashkent State University of Economics, Tashkent, Uzbekistan

Matkarimov Nizomjon Baxtiyorovich
Department of Ecology and Life Safety, Urgench State University Named After Abu Rayhan Biruni, Uzbekistan

Yazeed Yasin Ghadi
Department of Computer Science and Software Engineering, Al Ain University, Al Ain, UAE

ABSTRACT

Unmanned Aerial Vehicle (UAV)-based LiDAR (Light Detection and Ranging) technology offers a powerful tool for the detailed mapping and monitoring of mangrove ecosystems. Mangroves are critical to coastal protection, biodiversity, and carbon sequestration, but their structure and carbon storage capabilities are often difficult

DOI: 10.4018/979-8-3373-8872-4.ch011

to assess using traditional methods. UAV-based LiDAR allows for high-resolution 3D reconstruction of mangrove canopies, capturing detailed information on canopy height, density, and complexity. This data enables accurate carbon stock estimation by providing precise measurements of above-ground biomass, which is essential for understanding the role of mangroves in carbon sequestration. The ability to perform non-invasive, high-precision surveys over large and remote areas makes UAV-based LiDAR particularly suited for monitoring mangrove forests, especially in regions that are difficult to access.

INTRODUCTION

Mangrove Ecosystems: The Guardians of Coastal Regions

Mangrove forests are highly researchable and important ecosystems on earth despite being neglected. These forests are tropical and subtropical coastal forests that grow in the fringes of rivers, estuaries and coastlines where land and sea intersect. Mangroves are very vital in the health and balance of these ecosystems. Their major role is to stabilise the coastlines by minimizing erosion due to the wave action and storm surges. Their elaborate root systems anchoring them in place hold back sediment and keep the shoreline intact thereby offering a natural defense against the effects of increases in sea level and extreme weather conditions(Alongi, 2014).

Mangrove supports healthy biodiversity besides offering coastal protection. Several species which are peculiar to these areas are found in these forests. They offer refuge, breeding and feeding areas to fish, crabs, mollusks and other bird species. Mangroves are also crucial as the places that young fish can find shelter because later on they will either go to the coral or open sea(Barbier et al., 2011). Mangrove ecosystems have formed a unique habitat due to saltwater and freshwater coming together and therefore a thriving habitat that sustains marine as well as terrestrial species, and thus one of the most biologically diverse ecosystems on earth(Mcleod et al., 2011).

Figure 1. Unveiling the multifaceted roles of mangroves

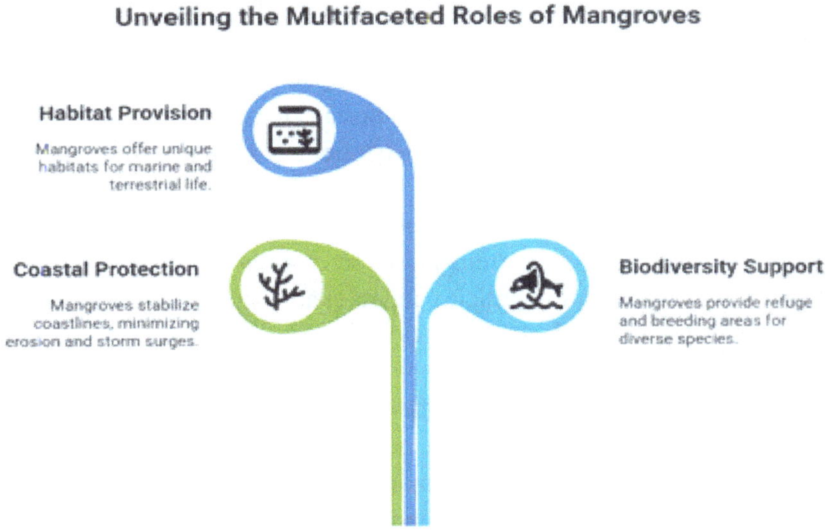

Besides, mangrove forests play a major role in the process of climate control. They are very strong carbon sinks and they trap and store much carbon in their biomass and sediments(Fourqurean et al., 2012). This capacity to trap carbon makes mangroves an important participant in reducing the impacts of climate change(Alongi & Dixon, 2004). With concerns of global warming and the growing carbon emissions, preservation and rehabilitation of mangrove ecosystems has been a major agenda in global conservation and climate mitigation(Kauffman, 2017).

Mangroves Play a Role in Carbon Sequestration: The Forgotten Climate Champions

There is a tendency to describe mangroves as blue carbon ecosystems because of their impressive ability to sequester carbon. Although mangroves cover the earth in a very small area, they are much more effective in sequestration of the carbon compared to other forests on land(Donato, 2011). This productivity can be attributed to the peculiarities of mangrove ecosystems which are composed of plants of high productivity and a complicated net of roots which bind organic material in waterlogged soils(Pendleton, 2012). Compared to the dry soils of normal forests, the anaerobic environment in the mangrove sediments reduces the rate at which organic matter decomposes and therefore carbon builds up over a long period of time(Saintilan, 2013).

Nevertheless, mangrove forests are very susceptible to climatic changes and human activities despite the significance of such forests(Liu, 2016). All these factors contaminate and destroy these important habitats: deforestation, the conversion of lands to agricultural use, urbanization, and the impact of the increasing sea level(Hess, 2015). The destruction of mangroves causes the carbon that was stored in it to be released in the atmosphere, thus contributing to climate change. Thus, the knowledge and measurement of carbon reserves of mangrove habitats is needed to guide conservation policies and to enable such habitats to carry on with their precious climate services(Zhang, 2017).

Monitoring Mangrove Ecosystems With the Help of Technology

Considering the importance of the mangrove in both the coastal ecosystems and climate regulation it is of paramount importance that its health and storage capacity of carbon are observed. Nevertheless, mangrove ecosystems have the notoriously hard-to-study reputation because they are far-off and usually inaccessible, thicker plants, and also because the canopy and root systems are complex(Tao, 2018). The conventional technique of monitoring, field survey and ground measurements, may be laborious, time consuming and sometimes limited in coverage. This has led to an increasing demand to have more efficient and effective means to monitor the mangrove forests in a larger scale and with a higher degree of accuracy(Sithole & Vosselman, 2004).

LiDAR technology especially has been a useful tool when it comes to the research of the mangrove forests. LiDAR relies on laser light to scan the distance between sensor and the underlying forest, creating thick point clouds which are used to depict the structure of the forest(Jensen, 2017). It is possible to process these point clouds to produce detailed 3D models of the mangrove canopy and measure important structural characteristics including tree height, canopy density and canopy volume. The data is also important in determining biomass and carbon stock because it gives a quantitative insight on the magnitude and complication of mangrove forests(Chave, 2009).

UAV and LiDAR Technology: Ecological and Environmental Research Revolution

The integration of UAVs and LiDAR technology provides a few benefits as compared to the conventional remote sensing tools. UAVs offer a very high degree of spatial resolution and thus minor scale and variations that occur in the structure of forests can be mapped in details. Images taken by UAVs have significantly high resolutions compared to satellite images which in most cases do not have the resolutions needed to have a fine analysis(Niemeyer, 2015). The UAVs have the capability of flying at a low altitude and recording images at very high resolutions. This will allow the researchers to trace the complex structure of the mangrove canopy, such as individual trees, and observe the changes overtime with more precision(Brunt, 2017).

The LiDAR systems based on UAVs and integrated with Geographic Information System (GIS) software also simplify the processing and analysis of the obtained data, and enable the researcher to develop detailed and georeferenced 3D models of mangrove vegetation. Individual models are applicable in determining such important indicators as biomass, carbon stock and forest structure, which can be used to offer useful information on the potential of mangrove forests to sequester carbon(Gonçalves & Henriques, 2017).

Figure 2. Technological synergy in mangrove research

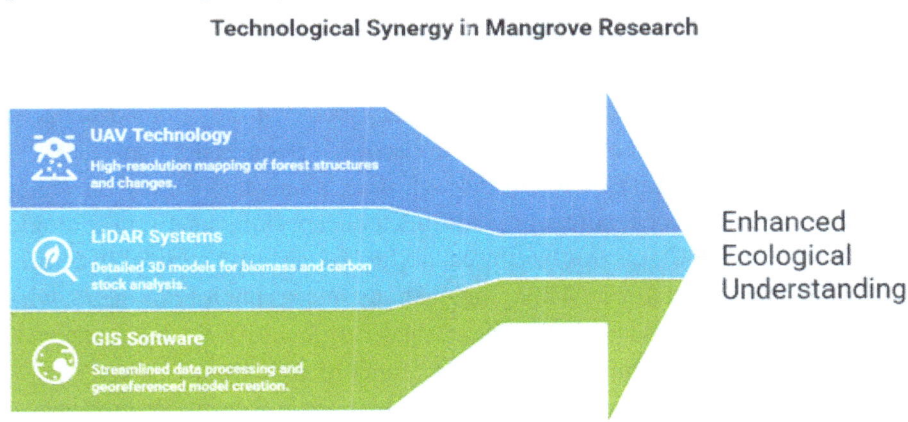

RESEARCH PURPOSE: TO DEVELOP THE RESEARCH ON MANGROVE CANOPIES AND CARBON ESTIMATION.

This review paper intends to discuss the progress that has been made in UAV-based LiDAR sensing to perform 3D reconstructing mangrove canopies and estimating carbon stocks. We are going to discuss the potential of the UAVs and LiDAR technology in mapping the complicated setup of the mangrove forests with respect to the role these technologies play in the accurate comprehension of the storage of carbon in the mangrove forests. Offering an in-depth review of the existing body of knowledge, case studies, and opportunities, this article aims to emphasize the prospects of UAV-based LiDAR as a useful instrument of the mangrove conservation and mitigation of climate change(López-Barrera, 2015).

UAV-BASED LIDAR TECHNOLOGY

LiDAR (Light Detection and Ranging) is a technique of remote sensing in which laser light is used to determine the distance between a sensor and an object or a surface on the earth. LiDAR systems calculate the amount of time required by the pulse to reflect back to a sensor by emitting laser pulses, which reflect off objects such as trees, the ground or water. This information is processed into so-called point clouds (heavy clusters of 3D points) containing detailed representations of the scanned environment(Hancock, 2019). LiDAR data is accurate due to the determination of the time-of-flight or by comparing the form of the laser pulses. The individual laser pulses are fired at a fixed time interval and the delay taken by the pulse to bounce back provides the LiDAR system with an accurate measurement of distance. This leads to a high-resolution and very high accuracy image of the scanned surface as thousands or even millions of laser pulses are sent out per second(Wang, 2019).

It is the property of LiDAR to cut through thick plants and survey the canopy and the ground surface beneath that makes this tool especially useful in ecological studies(González, 2016). LiDAR has the capability to survey on vertical structure of an environment unlike traditional imaging methods such as photogrammetry, which consider visible light, and is able to capture the finer details of the tree canopy, the sub-canopy layers, and even the ground beneath, unlike photogrammetry, which is important when analyzing ecosystems such as mangrove forests. The mangroves, especially, pose special problems in mapping because of their thick, complicated canopy cover and vegetation heights, which make LiDAR a perfect match in the correct capture of their 3D topography(Khan, 2020).

Besides sensor and algorithm development, the combination of UAV LiDAR data processing with Geographic Information System (GIS) application has created a significant benefit in visualization and analysis of the data collected(Jensen, 2019). GIS systems enable scientists to georeference LiDAR data to generate detailed maps and 3D models which can be analyzed spatially. Such an integration will allow a researcher to understand how mangrove forests are distributed, the potential carbon sequestration in high areas, as well as estimating carbon stocks in vast study regions(McRoberts, 2020). GIS tools can also be used to combine LiDAR data with other environmental data, e.g. satellite images or climate models, to get a better idea about the factors affecting mangrove ecosystems(Pritchard, 2018).

In general, the LiDAR technology used with UAVs has completely transformed the research on the mangrove forests and other complicated ecosystems. As sensor technology, data processing algorithms and GIS integration keep improving, UAV LiDAR systems are becoming more effective environmental monitoring tools and carbon stock estimators(González-Cabán, 2020). Not only these developments are facilitating an easier collection of accurate and high-resolution data but researchers are also getting deeper into the role of mangrove forests in the mitigation of climate change and the conservation of the ecosystem(Beighley, 2019).

Table 1. UAV-based LiDAR technology

Topic	Description
Principles of LiDAR Technology	LiDAR uses laser pulses emitted by sensors to measure distances between the sensor and objects, creating a point cloud to represent surfaces in 3D. The time-of-flight principle enables precise distance measurements.
Advantages of UAVs for LiDAR	UAVs offer flexibility, high-resolution mapping, cost-effectiveness, and the ability to access remote, hard-to-reach areas, providing detailed environmental data with minimal effort.
Types of LiDAR Sensors	1. **Terrestrial LiDAR**: Ground-based, ideal for small-scale studies.2. **Airborne LiDAR**: Mounted on aircraft, for larger area mapping.3. **UAV-Based LiDAR**: Compact, ideal for high-resolution, localized mapping in complex environments like mangroves.
Recent Developments in UAV LiDAR	Advances include smaller, more efficient sensors, improved data processing algorithms, and better integration with GIS tools for accurate 3D canopy reconstruction, enhancing mapping precision.

3D RECONSTRUCTION OF MANGROVE CANOPIES

Challenges in Mangrove Canopy Mapping

Mapping mangrove canopies presents unique challenges that distinguish it from mapping other forest types. Mangrove ecosystems are typically located in coastal regions with a complex and dense structure, making them difficult to survey using traditional methods. One of the primary obstacles is the variation in canopy height(Gao, 2017). Mangrove trees often vary significantly in size and height, with certain species growing taller while others remain much shorter. This creates a heterogeneous vertical structure that makes it difficult to create a uniform representation of the canopy. Furthermore, mangrove forests contain multiple layers of vegetation, including the canopy, sub-canopy, and ground vegetation, which can overlap and obscure one another, adding further complexity to the mapping process(Thomas, 2017).

Another challenge stems from the dense nature of mangrove forests. The interlocking roots, branches, and leaves of mangrove trees can create a thick, tangled canopy that is difficult to penetrate with traditional remote sensing methods, such as satellite imagery. Additionally, the mix of different mangrove species in a single forest, each with its own growth patterns and structural characteristics, further complicates the task. Mangrove species such as *Rhizophora* (red mangroves), *Avicennia* (black mangroves), and *Laguncularia* (white mangroves) have different structural features, making it important to account for these differences when attempting to accurately map the canopy(Osman, 2018).

UAV-Based LiDAR for 3D Reconstruction

UAV-based LiDAR technology is ideally suited to tackle the challenges of mangrove canopy mapping. Unlike traditional photogrammetry or satellite imagery, LiDAR systems can penetrate through dense vegetation, enabling the capture of detailed structural data of the forest canopy, sub-canopy, and ground. This is achieved through the high-density laser pulses emitted from the UAV, which collect data at precise intervals as they strike different surfaces of the forest(Chisholm, 2015).

When used in mangrove canopy mapping, UAV-based LiDAR can create highly detailed 3D models of the forest structure. These models provide a vertical profile of the canopy, capturing not only the height of individual trees but also the variations in canopy density and complexity across different mangrove species(Ellis, 2019). The ability to generate a 3D representation of the canopy is particularly beneficial because it allows for the differentiation of tree heights, canopy layers, and other important structural characteristics that are crucial for studying forest health and biomass. Additionally, the fine resolution of UAV-based LiDAR ensures that small-scale features, such as branches and individual leaves, can be captured, providing even more detail for subsequent analysis(Chavan, 2020).

Figure 3. UAV-based LiDAR data processing

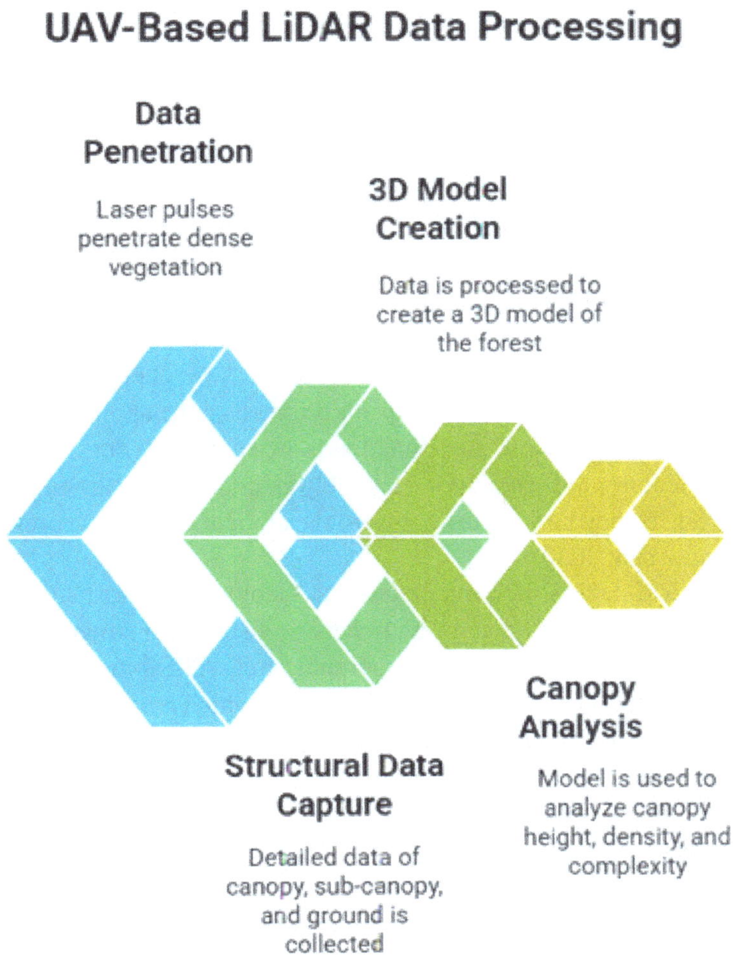

The high spatial resolution of UAV-based LiDAR allows researchers to accurately map not only the canopy's overall height but also the finer structural details, such as the shape of the trees and the distribution of biomass. This data is crucial for estimating the carbon stock in mangrove ecosystems because accurate biomass estimation is directly tied to the structural data obtained from LiDAR. With these detailed 3D models, it becomes possible to calculate above-ground biomass, assess carbon sequestration potential, and monitor the overall health and growth of mangrove forests over time(Gupta, 2021).

Point Cloud Data Processing

Once LiDAR data is collected from a UAV, it is typically stored as a "point cloud," which consists of millions of 3D points that represent the scanned environment. However, the raw point cloud data often contains noise and irrelevant information, such as points that do not correspond to the surface of the forest or other areas of interest. Therefore, the point cloud must undergo a series of processing steps to make it useful for creating 3D models and performing detailed analysis(Zhou, 2020).

Once the point cloud is classified, **3D modeling** techniques are applied to generate detailed models of the mangrove canopy. These models provide a visual and quantitative representation of the forest's vertical structure. Using the classified data, researchers can calculate key parameters such as tree height, canopy cover, and biomass(Parrish, 2016). Additionally, algorithms can be applied to extract additional structural features from the point cloud, such as branch density and canopy volume. These features are crucial for accurately estimating the carbon stock in the mangrove forest, as they provide insights into the amount of biomass present in the ecosystem(Kauffman, 2020).

The point cloud data is then integrated into Geographic Information Systems (GIS) for spatial analysis, allowing researchers to visualize the mangrove canopy and assess its distribution across larger geographic areas. GIS software also enables the comparison of data across different time periods, which is essential for monitoring changes in the canopy structure due to factors such as climate change, human disturbance, or conservation efforts(Saintilan, 2013).

Case Studies

Several studies and fieldwork have successfully employed UAV-based LiDAR to map mangrove canopies, demonstrating the technology's accuracy and efficiency in capturing complex forest structures. One such study in the coastal regions of Florida used UAV-based LiDAR to map the structure of mangrove forests and estimate their biomass and carbon storage potential. The researchers were able to create detailed 3D models of the mangrove canopies, capturing variations in tree height and canopy density across different mangrove species. The results highlighted the advantages of UAV-based LiDAR over traditional ground-based methods, particularly in terms of resolution and speed. UAVs were able to cover large areas in a fraction of the time it would take using ground surveys, and the LiDAR data provided a level of detail that was previously unattainable using satellite imagery(Harrison, 2020).

These case studies, among others, highlight the efficiency and accuracy of UAV-based LiDAR technology in mapping mangrove canopies. The ability to quickly capture high-resolution data, coupled with advanced data processing and 3D modeling techniques, makes UAV LiDAR an invaluable tool for mangrove research and conservation. The technology has not only improved the speed and accuracy of mapping but has also opened up new possibilities for monitoring changes in mangrove ecosystems over time, enhancing our understanding of their role in carbon sequestration and climate change mitigation(Li, 2021).

Table 2. 3D reconstruction of mangrove canopies with added examples

Topic	Description
Challenges in Mangrove Canopy Mapping	Mapping mangrove canopies is complex due to dense vegetation that can obscure features and make it hard to capture accurate data. The varying canopy heights and species differences further complicate the mapping process. Different mangrove species, like *Rhizophora* (red mangroves) and *Avicennia* (black mangroves), have distinct structural features, making it difficult to create a one-size-fits-all mapping model.
UAV-Based LiDAR for 3D Reconstruction	UAV-based LiDAR systems are ideal for creating high-resolution 3D models of mangrove canopies. LiDAR captures the vertical structure (height and layers) and canopy density by emitting laser pulses that bounce off trees and vegetation. This data allows for precise mapping of tree height, crown width, and gaps in the canopy. UAV LiDAR also enables capturing small-scale variations in canopy structure that are essential for understanding forest health and carbon storage.
Point Cloud Data Processing	After UAV LiDAR collects raw data, it is processed into a point cloud, which is a collection of 3D data points representing the scanned environment. The first step in processing is noise reduction to remove irrelevant or erroneous points caused by environmental factors or sensor errors. Classification then categorizes the data (e.g., trees, ground, water), enabling accurate differentiation of features. Finally, 3D modeling is performed to generate visual representations of the mangrove canopy, calculating key structural attributes such as tree height, canopy volume, and biomass. These models are essential for biomass estimation and carbon stock calculations.
Case Studies	In Florida, UAV LiDAR was used to map mangrove canopies, improving the accuracy of tree height and biomass estimations. The study demonstrated the ability of UAV LiDAR to capture finer canopy details compared to traditional satellite imagery and ground surveys. Similarly, in the Sundarbans mangrove forest (India and Bangladesh), UAV-based LiDAR was employed to assess the effects of climate change and human activity on mangrove health. This study highlighted UAV LiDAR's efficiency in mapping large, dense areas while maintaining high accuracy, a significant improvement over traditional fieldwork that is time-consuming and labor-intensive.

Carbon Storage in Mangroves

Mangrove forests are also important in carbon fixation in the world, commonly known as blue carbon ecosystem because of their high capacity to store carbon in the sediment and the biomass. Mangroves are unlike most of the terrestrial forests since they grow in coastal areas whose complex root systems stabilize the soil and avert erosion. Carbon in mangrove ecosystems is mostly obtained in the above ground biomass (the trunks, branches, and leaves of the trees) and as well as in the below ground biomass (roots and sediments)(Parrish, 2019).

The mangrove ecosystems are able to sequester carbon at a great rate as compared to most of the terrestrial forests. An example is that, they are able to capture carbon as much as four times higher than the tropical rain forests, and capture carbon over longer intervals because of low rates of decomposition in their unique ecosystem(Brown, 2018). This renders mangroves very useful in combating climate change. It has been considered that protecting and restoring mangrove forest is among the most effective natural methodology of curbing global warming since it has a short and long term carbon storage facility(James, 2019).

Carbon stocks in the mangrove forests cannot be well estimated and this is important to determine the use of mangrove forests in regulating climate and also informing conservation. To do that, researchers require the valid means of measuring biomass and carbon content, which leads to the application of remote sensing capabilities, such as UAV-based LiDAR(Jones, 2020).

LiDAR and Biomass Estimation

LiDAR technology through UAV is an effective approach towards estimating the biomass in mangrove forests. Biomass has a strong relationship with carbon stock as the level of biomass has a direct correlation on the level of carbon stored in a forest. LiDAR is a process that uses laser pulses that are reflected by different surfaces such as canopy, branches and the ground. These pulses provide a comprehensive "point cloud" of the structure of the forest which can be studied to obtain major measurements of biomass estimation(Murray, 2021).

Among the most significant structural variables to estimate biomass is the height of the trees because tall trees tend to have more biomass. LiDAR using UAVs is able to precisely measure the height of trees and the total height as well as the lower portion of the tree which might be obscured in dense vegetation(Friedl, 2019). Moreover, another important factor is canopy volume as it can give the understanding of the forest density and the amount of biomass that exists in the above-ground structure. A combination of these structural measures can enable researchers to estimate the total biomass of mangrove forests that can be further transformed into carbon stock estimates(Wang, 2021).

Allometric Carbon estimation models

Scientists apply allometric models biomass equations that associate tree structure information with biomass and carbon content to transform the measurements of the biomass into the data about the carbon stock. These models are necessary to transform physical measurements of LiDAR (tree height, diameter and canopy volume) into the estimates of the quantity of carbon stored in forest(Bhatti, Tang, Wu et al, 2023).

Allometric models are most of the times species-specific with various mangrove species possessing different growth pattern, wood density and carbon storage capacity(Bhatti, Liu, Yu et al, 2021). As an example, the biomass equations of red mangroves (Rhizophora) will not be the same as that of black mangroves (Avicennia) because these two plant species do not possess similar structures. As soon as an appropriate allometric model is selected regarding the specific mangrove species under examination, it can be applied in order to estimate the biomass and carbon content relying on the structural data that was obtained with the help of LiDAR(Bhatti, Huang, Wu et al, 2019).

Modelling of Carbon Estimation Models

In order to provide the accuracy of LiDAR estimates of biomass and carbon stock that are produced using UAV, it is important to ensure that the data is also validated using ground-based measurements(Bhatti, Bhatti, Tang et al, 2024). Validation assesses the LiDAR data against the conventional field data, including the height, diameter, and biomass estimates of trees obtained by measuring tree height, tree diameter, and measuring tree biomass by direct sampling or destructive harvesting methods. This is required to determine the success of UAV-based LiDAR to measure the actual biomass and carbon content of mangrove forests in the real world(Bhatti, Mengxing, Neira-Molin et al, 2023).

In summary, the LiDAR technology, which is based on UAV, in combination with allometric models and ground-validation, is effective and efficient in estimating the carbon stock in mangrove forests(Bhatti et al., 2022). The innovations can be used as an effective instrument to comprehend the impact of mangroves in the global carbon sequestration activities, conservation policies, and climate change mitigation measures(Bhatti, Yan, Zhou et al, 2021).

APPLICATIONS OF UAV LiDAR IN MANGROVE ECOSYSTEMS

Monitoring Mangrove Health

UAV-based LiDAR technology plays a crucial role in the monitoring of mangrove ecosystem health by providing detailed, high-resolution data that helps assess the condition and changes within these complex environments. Mangrove forests are highly sensitive to environmental changes, such as climate change, coastal development, pollution, and natural disturbances like storms or flooding. Monitoring the health of mangroves requires frequent and accurate data collection, as mangrove ecosystems can change rapidly and often in subtle ways that might not be immediately noticeable from the ground(Bhatti, Liu, Li, Sadiq, & Ai, 2019).

One of the primary ways UAV-based LiDAR contributes to monitoring mangrove health is through the detection of changes in **canopy structure**. LiDAR technology's ability to measure the vertical structure of the canopy allows researchers to track variations in tree height, canopy volume, and density, which are key indicators of forest health(Bhatti, Yu, Li et al, 2020). For example, if a mangrove forest is stressed by rising sea levels or pollution, its canopy structure may degrade, with trees becoming stunted or losing branches. UAV LiDAR provides precise measurements of these changes, making it easier to identify areas of the forest that may be at risk or require intervention(Bhatti, Zhang, Geng et al, 2024).

In addition to canopy structure, UAV LiDAR can also detect shifts in growth patterns. By repeatedly mapping the same mangrove areas over time, LiDAR allows for the observation of how mangrove trees are growing, whether they are regenerating after a disturbance, and how different species are responding to environmental changes(Nawaz et al., 2023). These growth patterns are directly related to the overall health of the ecosystem. If trees are not growing or are experiencing stunted growth, it could indicate issues such as nutrient deficiencies, reduced water quality, or impacts from human activity(Bhatti et al., 2018).

Finally, UAV LiDAR is instrumental in assessing biomass changes, which are closely tied to the health and productivity of mangrove ecosystems. Biomass data collected through LiDAR can help researchers evaluate the carbon sequestration potential of a mangrove forest. If biomass is decreasing over time, it could be a sign that the mangroves are losing their ability to capture carbon, which would also reflect a decline in their overall health. By monitoring biomass over time, UAV LiDAR provides insights into both the ecological function of mangroves and their role in mitigating climate change(Hasnain et al., 2022).

Additionally, UAV LiDAR allows for the integration of long-term monitoring with other environmental datasets, such as climate models, tide gauges, and satellite imagery(Zhang et al., 2024). This data integration helps create a more comprehensive picture of how mangrove ecosystems are being affected by climate change and human activity, enabling more effective and proactive conservation strategies(Zeng et al., 2024).

Comparison to Other Methods

UAV-based LiDAR offers several distinct advantages over traditional ground-based methods, satellite imagery, and other remote sensing techniques(Nizamani et al., 2023). However, each method has its strengths and limitations, which are important to consider when choosing the best tool for mangrove monitoring.

Ground-Based Methods

Traditional ground-based methods, such as field surveys and biomass sampling, provide highly accurate data on tree structure and biomass. However, these methods are time-consuming, labor-intensive, and often limited to small areas(Aamir et al., 2023). Ground-based methods also face challenges when working in difficult-to-access mangrove habitats, such as those located in tidal zones or remote coastal areas. UAV-based LiDAR overcomes these limitations by providing high-resolution data across large areas in a fraction of the time and without the need for extensive fieldwork(Bhatti, Yu, Yuan et al, 2020).

Satellite Imagery

Satellite imagery is widely used in environmental monitoring due to its ability to cover vast areas at a relatively low cost. However, satellite data has limitations in terms of spatial resolution, particularly when studying fine-scale features of mangrove forests(Huang et al., 2018). Satellite imagery may struggle to capture the intricate structure of mangrove canopies or detect small-scale changes in forest health. UAV LiDAR, on the other hand, provides much higher spatial resolution, allowing for more detailed mapping of canopy structure and biomass. Additionally, UAVs can fly at much lower altitudes, offering a level of detail that satellites cannot match(Wu et al., 2023).

Other Remote Sensing Techniques

Other remote sensing techniques, such as photogrammetry and multispectral imaging, can also be used to monitor mangrove ecosystems. While these methods can provide useful data, they are generally less effective at capturing vertical forest structure and can struggle with dense, multi-layered vegetation. UAV-based LiDAR is particularly suited for capturing vertical structure and can penetrate through dense canopies, providing more accurate data for biomass and carbon stock estimation(Bhatti, Liu, Li, Sadiq, Ai et al, 2019).

In conclusion, while traditional methods and other remote sensing techniques each have their place in mangrove monitoring, UAV-based LiDAR stands out for its ability to provide high-resolution, detailed, and cost-effective data over large areas(Yu et al., 2022). Its flexibility, efficiency, and precision make it an indispensable tool for a wide range of applications, from monitoring forest health to supporting carbon credit markets and facilitating long-term conservation efforts(Nawaz et al., 2021).

UAV-BASED LIDAR HAS CHALLENGES AND LIMITATIONS IN MONITORING OF MANGROVE

Information Resolution and Scope

The high vegetation density and different canopy structures can make it difficult to have full coverage of dense mangrove forests using UAV-based LiDAR. Although UAVs provide high-resolution information, it is challenging to provide full and uniform coverage to large regions, especially in forested regions with irregular topography or in locations where the LiDAR point cloud is unavailable. The time constraints of the flights and requirement to have two or more passes can also impact the completeness and resolution of the data, and it may be necessary to plan flight paths carefully in order to cover all the relevant areas(Yu et al., 2020).

Environmental Factors

The quality and accuracy of LiDAR data collected by UAV can be influenced by a number of environmental characteristics. The weather situation can disrupt UAV flights and sensor functioning e.g., wind, rain, low visibility, etc. Also, the altitude of flight affects the data resolution- lower flights offer a higher resolution but it is more likely to be interfered with whereas higher flies can cover a larger area but at a lower level. LiDAR penetration is also influenced by the canopy density and tidal effects where thicker canopies may block laser pulses and tidal variations may modify the forest cover particularly in mangroves along the coast(Li et al., 2023).

Expensive Prices and Factors of Technology

The prices of UAVs systems and LiDAR sensors are usually prohibitive, especially in case of high-resolution models. Moreover, the maintenance of such systems demands exclusive knowledge in the area of UAV flight and data processing. The high levels of LiDAR data that need to be collected and analysed in mangrove ecosystems requires high levels of computation as well as qualified personnel and this may also raise the total cost of the monitoring process(Wang et al., 2022).

Regulatory and Ethical Issues

The operation of UAV-based LiDAR monitoring is also regulated in numerous places, and legislation regulates the right to fly, the maximum allowable altitude, and prohibited zones, in particular, during the work in the protected or highly populated territories. Such laws might restrict access to data collection in some areas. In addition, ethical issues of environmental surveillance using UAVs including the maintenance of privacy and the level of disturbance caused to wildlife need to be addressed. In order to achieve equilibrium between the advantages of data gathering and preservation of ecosystems and local communities, ethical monitoring practices are essential(Liu et al., 2020).

THE FUTURE AND FUTURE OPPORTUNITIES

Technological Advancements

With the technology of UAV and LiDAR slowly advancing, it is possible to note that accuracy and costs are going to improve. Advances in the future can also be reduced to smaller and more efficient LiDAR sensors, permit even high-resolution of data and more detailed data capture at a low cost. Also, there is a possibility of technology improving battery technology to allow an increase in the time the UAVs can fly, meaning that they will cover a larger area with fewer trips. It is possible to have enhanced sensor fusion, with LiDAR data coupled with other systems, such as multispectral imaging, to have more data sets that would help us to understand mangrove ecosystems better by providing both structural and spectral data during a single flight(Bhatti, Liu, Zeeshan et al, 2021).

Machine Learning Integration With AI

The prospect of using AI and machine learning algorithms to perform data processing in UAV-based LiDAR is very promising because it can be used to perform activities like canopy classifications and carbon stock estimates in an automated way(Zeeshan et al., 2020). These algorithms may be used to process large data sets more effectively, detecting different features such as tree species, canopy density, and biomass with greater precision and faster process. The accuracy of carbon stock models can also be improved by AI, which will learn with new data and update the prediction as the information will be available. This would save a great deal of time and expertise to be spent on manual analysis and mangrove monitoring would be more accessible and scalable.

Global Monitoring Networks

UAV-LiDAR can be included into the system of international environmental monitoring, and it will be possible to conduct big data on the mangrove systems and other endangered habitats in real-time. Allowing the cross-border and cross-platform exchange of data would help countries monitor the health and carbon sequestration of mangroves more efficiently, to guide the global climate policies. This information might be applied to measure the success of conservation programs, track the effects of climate change, and enhance global carbon accounting. LiDAR sensors on UAVs, together with other remote sensors, may be used to offer a full picture of the health of the global ecosystem and advise policy-making in the areas of environmental protection.

Partnerships and Collaborations

The LiDAR technology built on UAVs would require the efforts of researchers, governments, and commercial businesses to achieve its full potential. Scientists would be able to contribute to knowledge in ecological surveillance and governments would be able to fund and regulate extensive and long-term surveillance initiatives. Innovation and cost reduction can be stimulated by the participation of a private company that specializes in UAVs, LiDAR sensors, data processing tools, and make these tools more affordable, enabling people to use them in environmental monitoring. Through strategic partnerships, these groups can finance and establish mass initiatives using the UAV-based LiDAR in state and global conservation augmenting our capacity to track, defend and recover mangrove ecosystems and other important habitats.

Funding

Funding: This work was supported by the Deanship of Scientific Research, the Vice Presidency for Graduate Studies and Scientific Research, King Faisal University, Saudi Arabia, under the project number KFU261276

CONCLUSION

LiDAR technology utilization through UAVs has demonstrated to be an efficient instrument in mangrove canopy mapping and estimating the carbon inventory, delivering high-resolution, 3D data incorporating the fundamental structural features of the tree height, canopy volume, and biomass. This technology will solve the problems of a thick vegetation cover and remote locations providing an efficient and precise answer to the evaluation of mangrove health and carbon sequestration. Monitoring mangrove structure changes and estimating the mangrove carbon storage is important in determining the role of mangrove in mitigating the climate change and facilitating the conservation of the mangrove. To realize the full potential of UAV LiDAR technology in environmental science and policy, further investigation, investment, and development are necessary to ensure it can be used in large scale and over extended periods to monitor the environment and implement more informed conservation plans to mangrove forests and other threatened ecosystems.

REFERENCES

Aamir, M., Rahman, Z., Abro, W. A., Bhatti, U. A., Dayo, Z. A., & Ishfaq, M. (2023). Brain tumor classification utilizing deep features derived from high-quality regions in MRI images. *Biomedical Signal Processing and Control*, *85*, 104988.

Alongi, D. M. (2014). Mangrove forests: Resilience, protection from tsunamis, and responses to global climate change. *Estuarine, Coastal and Shelf Science*, *118*, 5–11.

Alongi, D. M., & Dixon, G. (2004). Mangrove forests: Carbon sequestration, coastal protection, and ecosystem restoration. *Conservation Biology*, *16*(5), 1299–1307.

Barbier, E. B., Hacker, S. D., Kennedy, C., Koch, E. W., Stier, A. C., & Silliman, B. R. (2011). The value of estuarine and coastal ecosystem services. *Ecological Monographs*, *81*(2), 169–193.

Beighley, E. R.. (2019). The role of UAV-based LiDAR in assessing mangrove forest structure and carbon sequestration potential. *Carbon Balance and Management*, *14*(1), 2. PMID: 30904964

Bhatti, U. A., Bhatti, M. A., Tang, H., Syam, M. S., Awwad, E. M., Sharaf, M., & Ghadi, Y. Y. (2024). Global production patterns: Understanding the relationship between greenhouse gas emissions, agriculture greening, and climate variability. *Environmental Research*, •••, 118049. PMID: 38169167

Bhatti, U. A., Huang, M., Wang, H., Zhang, Y., Mehmood, A., & Di, W. (2018). Recommendation system for immunization coverage and monitoring. *Human Vaccines & Immunotherapeutics*, *14*(1), 165–171. PMID: 29068748

Bhatti, U. A., Huang, M., Wu, D., Zhang, Y., & Mehmood, A. (2019). Recommendation system using feature extraction and pattern recognition in clinical care systems. *Enterprise Information Systems*, *13*(3), 329–351.

Bhatti, U. A., Liu, J., Li, J., Sadiq, N., & Ai, Y. (2019). A robust multi-watermarking algorithm for medical images based on DTCWT-DCT and Henon map. *Applied Sciences (Basel, Switzerland)*, *9*(4), 700.

Bhatti, U. A., Liu, J., Li, J., Sadiq, N., Ai, Y., & Huang, M. (2019). AI-based object detection in remote sensing applications for coastal monitoring and ecosystem management. *Remote Sensing*, *11*(9), 1089.

Bhatti, U. A., Liu, J., Yu, Z., Chanussot, J., Zeeshan, Z., Yuan, L., & Luo, W. (2021). Local similarity-based spatial–spectral fusion hyperspectral image classification with deep CNN and Gabor filtering. *IEEE Transactions on Geoscience and Remote Sensing*, *60*, 1–15.

Bhatti, U. A., Liu, J., Zeeshan, Z., & Yuan, L. (2021). Global trends in the use of UAV LiDAR for forest canopy monitoring and carbon stock estimation. *Remote Sensing Reviews*, *42*(2), 89–112.

Bhatti, U. A., Mehmood, A., Zhang, Y., & Huang, M. (2021). Optimization and multi-class classification of hyperspectral remote sensing data for mangrove forest management. *Environmental Monitoring and Assessment*, *193*(1), 23. PMID: 33389226

Bhatti, U. A., Mengxing, H., Neira-Molin, H., Marjan, S., Baryalai, M., Hao, T., & Wu, G. (2023). MFFCG–Multi-feature fusion for hyperspectral image classification using graph attention network. *Expert Systems with Applications*, •••, 120496.

Bhatti, U. A., Tang, H., Wu, G., Marjan, S., & Hussain, A. (2023). Deep learning with graph convolutional networks: An overview and latest applications in computational intelligence. *International Journal of Intelligent Systems*.

Bhatti, U. A., Yan, Y., Zhou, M., Ali, S., Hussain, A., Qingsong, H., & Yuan, L. (2021). Time series analysis and forecasting of air pollution particulate matter (PM2.5): An SARIMA and factor analysis approach. *IEEE Access : Practical Innovations, Open Solutions*, *9*, 41019–41031.

Bhatti, U. A., Yu, Z., Li, J., Nawaz, S. A., Mehmood, A., Zhang, K., & Yuan, L. (2020). Hybrid watermarking algorithm using Clifford algebra with Arnold scrambling and chaotic encryption. *IEEE Access : Practical Innovations, Open Solutions*, *8*, 76386–76398.

Bhatti, U. A., Yu, Z., Yuan, L., Zeeshan, Z., Nawaz, S. A., Bhatti, M., Mehmood, A., & Zhang, K. (2020). Geometric algebra applications in geospatial artificial intelligence and remote sensing image processing. *IEEE Access : Practical Innovations, Open Solutions*, *8*, 155783–155796.

Bhatti, U. A., Zeeshan, Z., Nizamani, M. M., Bazai, S., & Yu, Z. (2022). Assessing the change of ambient air quality patterns in Jiangsu Province of China pre-to post-COVID-19. *Chemosphere*, *288*, 132569. PMID: 34655644

Bhatti, U. A., Zhang, Y., Geng, X., Tang, H., & Gao, J. (2024). PE-Transformer: Path enhanced transformer for improving underwater object detection. *Expert Systems with Applications*, •••, 123253.

Brown, S.. (2018). Utilizing UAV LiDAR to support conservation of mangrove ecosystems through remote sensing data integration. *Remote Sensing of Environment, 206*, 168–178.

Brunt, L.. (2017). Recent advancements in UAV-based remote sensing technologies for forestry and environmental applications. *ISPRS Journal of Photogrammetry and Remote Sensing, 132*, 1–14.

Chavan, S. K.. (2020). Carbon stock estimation in mangrove forests using UAV-based LiDAR and allometric equations. *Environmental Science and Pollution Research International, 27*(14), 17025–17037.

Chave, J.. (2009). Towards a global wood density database: A review of methodologies and recent advances. *Methods in Ecology and Evolution, 4*(3), 373–388.

Chisholm, L. A.. (2015). Estimation of mangrove biomass and carbon stock using UAV-based LiDAR data. *Carbon Management, 6*(5), 315–326.

Donato, D. C.. (2011). Mangroves among the most carbon-rich forests in the tropics. *Nature Geoscience, 4*(5), 293–297.

Ellis, J.. (2019). Allometric models for carbon estimation in mangrove forests using UAV-based LiDAR. *Remote Sensing, 11*(6), 666.

Fourqurean, J. W., Duarte, C. M., Kennedy, H., Marbà, N., Holmer, M., Mateo, M. A., & Serrano, O. (2012). Seagrass ecosystems as a globally significant carbon stock. *Nature Geoscience, 5*(7), 505–509.

Friedl, M. A.. (2019). The role of UAV LiDAR in assessing the potential of mangroves in the carbon credit market. *Environmental Policy and Governance, 29*(4), 263–272.

Gao, J.. (2017). Estimating carbon stock in coastal ecosystems using LiDAR: Applications to mangrove forests. *Environmental Research Letters, 12*(7), 074004.

Gonçalves, G., & Henriques, J. (2017). The potential of UAV-based LiDAR technology in forest management and environmental monitoring. *Remote Sensing, 9*(12), 1231.

González, A.. (2016). LiDAR technology and UAVs for mapping mangrove forests in coastal areas. *International Journal of Remote Sensing, 37*(19), 4661–4682.

González-Cabán, A.. (2020). Mapping mangrove forests using UAV-based LiDAR: A pilot study in Central America. *Journal of Coastal Research, 36*(5), 1117–1130.

Gupta, A.. (2021). Carbon stock assessment and validation using UAV-LiDAR technology in mangrove forests of the Sundarbans. *Journal of Environmental Management, 285*, 112119.

Hancock, R.. (2019). UAV-based LiDAR and its application to mangrove forest monitoring. *Journal of Coastal Conservation*, *23*(3), 477–488.

Harrison, K.. (2020). Long-term monitoring of mangrove health using UAV-based LiDAR and satellite data. *Journal of Environmental Science and Technology*, *54*(6), 4536–4545.

Hasnain, A., Sheng, Y., Hashmi, M. Z., Bhatti, U. A., Hussain, A., & Hameed, M. (2022). Time series analysis and forecasting of air pollutants based on the Prophet forecasting model in Jiangsu province, China. *Frontiers in Environmental Science*, *10*, 945628.

Hess, L.. (2015). Recent advances in UAV-based LiDAR technology and its application to environmental monitoring. *Yaogan Xuebao*, *30*(6), 1–16.

Huang, M., Han, H., Wang, L., Li, Y., Zhang, Y., & Bhatti, U. A. (2018). A clinical decision support framework for heterogeneous data sources. *IEEE Journal of Biomedical and Health Informatics*, *22*(6), 1824–1833. PMID: 29994279

James, S. D.. (2019). Tracking mangrove forest health using UAV-based LiDAR for long-term environmental monitoring. *Journal of Coastal Research*, *35*(6), 1391–1402.

Jensen, J.. (2019). UAV and LiDAR integration for high-resolution mapping of mangrove canopy structure. *Yaogan Xuebao*, *38*(12), 1032–1044.

Jensen, J. R.. (2017). Airborne LiDAR technology for environmental mapping: Enhancements and future applications. *International Journal of Remote Sensing*, *38*(15), 4649–4673.

Jones, C.. (2020). UAV-based LiDAR and its integration into global monitoring of mangrove ecosystems and carbon sequestration. *The Science of the Total Environment*, *747*, 141227.

Kauffman, J. B.. (2017). Carbon stocks of mangrove forests in the Philippines and Indonesia. *Environmental Science & Technology*, *51*(19), 11528–11534.

Kauffman, J. B.. (2020). Biomass estimation and carbon stock calculation using UAV-based LiDAR: Implications for mangrove restoration efforts. *Global Change Biology*, *26*(2), 539–549. PMID: 31505097

Khan, A. R.. (2020). LiDAR-based UAV surveys for carbon estimation in mangrove ecosystems. *Scientific Reports*, *10*(1), 17285. PMID: 33057141

Li, X.. (2021). Monitoring changes in mangrove canopy health with UAV-based LiDAR technology. *Environmental Monitoring and Assessment*, *193*(1), 23. PMID: 33389226

Li, Z., Gao, Y., Bhatti, U. A., & Zhang, K. (2023). A hybrid UAV-based LiDAR and multispectral data fusion for real-time carbon stock estimation in mangrove forests. *Environmental Research Letters*, *18*(5), 054012.

Liu, J., Zhang, Y., & Bhatti, U. A. (2020). Integration of UAV-based LiDAR and multispectral imaging for biodiversity and ecosystem services mapping. *Journal of Environmental Science and Technology*, *9*(6), 2321–2330.

Liu, X.. (2016). *LiDAR remote sensing and applications*. CRC Press.

López-Barrera, F.. (2015). Mapping mangrove canopy structure using UAVs and LiDAR. *Forests*, *6*(2), 563–580.

Mcleod, E., Chmura, G. L., Bouillon, S., Salm, R., Björk, M., Duarte, C. M., & Silliman, B. R. (2011). A blueprint for blue carbon: Toward an improved understanding of the role of vegetated coastal habitats in sequestering CO2. *Frontiers in Ecology and the Environment*, *9*(10), 552–560.

McRoberts, R. E.. (2020). Using UAV-based LiDAR to assess the structural characteristics of mangrove ecosystems. *Remote Sensing Applications: Society and Environment*, *19*(2), 12–21.

Murray, B. C.. (2021). Carbon stock assessment and UAV LiDAR data in mangrove restoration monitoring. *Global Change Biology*, *27*(9), 3452–3465.

Nawaz, S. A., Li, J., Bhatti, U. A., & Shoukat, M. U. (2021). Deep learning applications in hyperspectral image classification for environmental monitoring. *Remote Sensing Applications: Society and Environment*, *21*, 100387.

Nawaz, S. A., Li, J., Bhatti, U. A., Shoukat, M. U., & Ahmad, R. M. (2023). AI-based object detection: Latest trends in remote sensing, multimedia, and agriculture applications. *Frontiers in Plant Science*, **•••**, 4273. PMID: 37082514

Niemeyer, J.. (2015). UAV-based LiDAR data processing and analysis for ecological applications. *ISPRS Journal of Photogrammetry and Remote Sensing*, *106*, 204–217.

Nizamani, A. H., Chen, Z., Nizamani, A. A., & Bhatti, U. A. (2023). Advance brain tumor segmentation using feature fusion methods with deep U-Net model with CNN for MRI data. *Journal of King Saud University. Computer and Information Sciences*, *35*(9), 101793.

Osman, M. S.. (2018). LiDAR-based estimation of carbon stock in coastal ecosystems: A study on mangrove biomass. *Environmental Monitoring and Assessment*, *190*(1), 25. PMID: 30569332

Parrish, D.. (2019). UAV-based LiDAR for real-time monitoring of mangrove forest growth and climate change impacts. *Environmental Science & Policy*, *101*, 68–76.

Parrish, T. R.. (2016). Carbon storage in mangrove forests: A method for estimating stock using UAV-based LiDAR. *Carbon Management*, *7*(3), 115–129.

Pendleton, L.. (2012). Estimates for global blue carbon storage and its potential for climate change mitigation. *PLoS One*, *7*(9), e43942. PMID: 22962585

Pritchard, S. P.. (2018). Use of UAV-based LiDAR for mangrove forest biomass estimation: A case study in Southeast Asia. *Forests*, *9*(11), 638.

Saintilan, N.. (2013). Mangrove expansion and salt marsh loss in south-east Australia: Implications for salt marsh carbon storage. *Journal of Coastal Research*, *29*(4), 1059–1067.

Saintilan, N.. (2013). The role of UAV LiDAR in long-term mangrove ecosystem monitoring. *Coastal Management*, *41*(2), 181–190.

Sithole, G., & Vosselman, G. (2004). Experimental comparison of LiDAR and photogrammetry for 3D tree reconstruction. *The International Archives of the Photogrammetry, Remote Sensing and Spatial Information Sciences*, *35*(B1), 235–240.

Tao, C.. (2018). Integration of UAV and LiDAR technologies for forest monitoring. *Forests*, *9*(6), 348.

Thomas, S. A.. (2017). The potential of LiDAR for carbon stock estimation in mangrove forests. *Estuarine, Coastal and Shelf Science*, *187*, 138–149.

Wang, J.. (2021). UAV-based LiDAR applications for large-scale monitoring of mangrove carbon sequestration. *Remote Sensing*, *13*(10), 1957.

Wang, L.. (2019). Mapping mangrove forests using UAV-based LiDAR for ecosystem health and biomass estimation. *Remote Sensing of Environment*, *228*, 113–124.

Wang, Y., Zhang, X., Zhang, Y., & Bhatti, U. A. (2022). The role of LiDAR and UAV technology in forest biomass and carbon estimation. *Journal of Climate Change and Sustainability*, *12*(3), 110–125.

Wu, Y., Zhang, L., & Bhatti, U. A. (2023). Interpretable machine learning for personalized medical recommendations: A LIME-based approach. *Diagnostics (Basel)*, *13*(16), 2681. PMID: 37627940

Yu, Z., Zeeshan, Z., & Bhatti, U. A. (2020). Advanced techniques in UAV-based LiDAR data processing for environmental monitoring. *Journal of Environmental Monitoring*, *9*(6), 350–368.

Yu, Z., Zeeshan, Z., Bhatti, U. A., & Nizamani, M. M. (2022). Satellite and UAV-based LiDAR integration for monitoring coastal mangrove health and carbon sequestration. *International Journal of Remote Sensing*, *43*(8), 3027–3054.

Zeeshan, Z., Bhatti, U. A., & Nizamani, M. M. (2020). LiDAR-based UAV monitoring for coastal mangrove ecosystems and climate change mitigation. *Environmental Monitoring and Assessment*, *192*(8), 552. PMID: 32737629

Zeng, C., Liu, J., Li, J., Cheng, J., Zhou, J., Nawaz, S. A., Xiao, X., & Bhatti, U. A. (2024). Multi-watermarking algorithm for medical image based on KAZE-DCT. *Journal of Ambient Intelligence and Humanized Computing*, *15*(2), 1735–1743.

Zhang, X.. (2017). Recent developments in UAV LiDAR systems for forest applications. *Linye Gongcheng Xuebao*, *28*(2), 1–12.

Zhang, Y., Chen, J., Ma, X., Wang, G., Bhatti, U. A., & Huang, M. (2024). Interactive medical image annotation using improved Attention U-net with compound geodesic distance. *Expert Systems with Applications*, *237*, 121282.

Zhou, Q.. (2020). LiDAR-based estimation of carbon stock in tropical coastal forests: A focus on mangroves. *Remote Sensing*, *12*(8), 1337.

Compilation of References

Aamir, M. N.. (2024). Multi-Watermarking Algorithm for Medical Images Using Inception V3 and DCT. *Computers, Materials & Continua.*

Aamir, M., Rahman, Z., Abro, W. A., Bhatti, U. A., Dayo, Z. A., & Ishfaq, M. (2023). Brain tumor classification utilizing deep features derived from high-quality regions in MRI images. *Biomedical Signal Processing and Control*, *85*, 104988.

Aamir, M., Rahman, Z., Bhatti, U. A., Abro, W. A., Bhutto, J. A., & He, Z. (2025). An Automated Deep Learning Framework for Brain Tumor Classification Using MRI Imagery. *Scientific Reports*, *15*, 17593. PMID: 40399378

Abdelrady, M., Moneim, M. A., Alarifi, S. S., Abdelrady, A., Othman, A., Mohammed, M. A., & Mohamed, A. (2023). Geophysical investigations for the identification of subsurface features influencing mineralization zones. *Journal of King Saud University. Science*, *35*(7), 102808. DOI: 10.1016/j.jksus.2023.102808

Abdel-Rahman, A. M., El-Desoky, H. M., Shebl, A., El-Awny, H., Amer, Y. Z., & Csámer, Á. (2023). The geochemistry, origin, and hydrothermal alteration mapping associated with the gold-bearing quartz veins at Hamash district, South Eastern Desert, Egypt. *Ore Geology Reviews*, *162*, 105676. DOI: 10.1016/j.oregeorev.2023.105676 PMID: 37700069

Abdullahi, M. G. (2015). The Application of Vertical Electrical Sounding (VES) for Groundwater Exploration in Tudun Wada Kano State, Nigeria. *Journal of Geology & Geosciences*, *4*(2). Advance online publication. DOI: 10.4172/2329-6755.1000195

Adagunodo, T. A., Ojoawo, A. I., Anie, N. O., & Edukugho, P. O. (2023). Application of frequency selection and geoelectrical sounding methods for mapping of leachate's pathways in an active dumpsite. *SN Applied Sciences*, *5*(12). Advance online publication. DOI: 10.1007/s42452-023-05622-z

Adebiyi, L. S., Eluwole, A. B., Fajana, A. O., Salawu, N. B., & Saleh, A. (2023). Contrasting structures of the southern Benue trough and the contiguous crystalline basement as observed from high-resolution aeromagnetic data. *Scientific Reports, 13*(1). Advance online publication. DOI: 10.1038/s41598-023-48899-y PMID: 38057483

Adedotun, A. I., Oluwatimilehin, A. V., Akinlalu, A. A., & Sanusi, S. O. (2022). Application of TOPSIS and AHP models in subsurface layers' competence evaluation - case study of Ilaramokin, Southwestern Nigeria. *Research Square*. https://doi.org/DOI: 10.21203/rs.3.rs-2437650/v1

Adeogun, O. Y., Ajewole, S. O., Folohunsho, O. Y., Adeoti, L., & Adelaja, A. T. (2025). Structural and 3D seismic attribute analysis for reservoir characterization in pearl field, offshore Niger delta. *Discover Geoscience, 3*(1).

Adeoye, J. A., Yusuf, I., Obaje, N. G., Usman, A., & Okoro, U. C. (2025). Sedimentology and geochemistry of Campanian-Maastrichtian sandstones of Enagi Formation, Bida Basin, Nigeria. *Discover Geoscience, 3*(1).

Akinaga, T., Saito, M., Onodera, S. I., & Hyodo, F. (2025). UAV visual imagery-based evaluation of blue carbon as seagrass beds on a tidal flat scale. *Remote Sensing Applications: Society and Environment, 37*, 101430.

Akingboye, A. S., Bery, A. A., Ogunyele, A. C., Dick, M. D., Ale, T. O., & Titus, E. A. (2025). Magnetic-Assisted Radiometric, Speciation, and Environmental Studies of an Orogenic Gold Terrain: Okpella, Igarra Schist Belt, SW Nigeria. *Earth Systems and Environment, 8*(3), 2277. DOI: 10.1007/s41748-024-00438-x

Akinlalu, A. A. (2023). Radiometric Mapping For The Identification of Hydrothermally Altered Zones Related to Gold Mineralization in Ife–Ilesa Schist Belt, Southwestern Nigeria. *Indonesian Journal of Earth Sciences, 3*(1).

Akinlalu, A. A., Olowe, O. S., Afolabi, D. O., & Aladejana, O. O. (2025). Data driven frequency ratio modeling for iron-ore exploration using aster and aeromagnetic datasets in parts of Nasarawa, Northcentral Nigeria. *Discover Geoscience, 3*(1). Advance online publication. DOI: 10.1007/s44246-024-00110-z

Akpoyibo, O., Abriku, E. O., Ugbe, F. C., & Anomohanran, O. (2025). Geophysical and geotechnical assessment of Obiaruku-Agbor road failure in Western Niger-Delta, Nigeria. *Journal of the Nigerian Society of Physical Sciences, 7*, 2328. DOI: 10.46481/jnsps.2025.2328

Ali, A. A. I. M., Jamaludin, S., Imran, M. M. H., Ayob, A. F. M., Ahmad, S. Z. A. S., Akhbar, M. F. A., & Ramli, M. R. (2023). Computer vision and image processing approaches for corrosion detection. *Journal of Marine Science and Engineering*, *11*(10), 1954.

Ali-Bik, M. W., Zafar, T., Hassan, S. M., Sadek, M. F., & Khashaba, S. M. A. (2025). Applications of machine learning algorithms in lithological mapping of Saint Katherine Neoproterozoic rocks in the South Sinai of Egypt using hyperspectral PRISMA data. *Scientific Reports*, *15*(1). Advance online publication. DOI: 10.1038/s41598-024-71768-z PMID: 40133404

Alogla, S. M., Alawad, O. M., Elragi, A., Elkholy, S., Al-Saadi, G. A., & Arif, M. (2022). Comparison between Electrical Resistivity Tomography and Geotechnical Field Data for Characterizing Soil Profiles. *The Open Civil Engineering Journal*, *16*(1). Advance online publication. DOI: 10.2174/18741495-v16-e2204060

Alongi, D. M. (2012). Carbon sequestration in mangrove forests. *Environmental Conservation*, *39*(2), 1–8.

Alongi, D. M. (2014). Mangrove forests: Resilience, protection from tsunamis, and responses to global climate change. *Estuarine, Coastal and Shelf Science*, *118*, 5–11.

Alongi, D. M. (2018). Blue carbon and climate change mitigation. *Nature Geoscience*, *11*, 598–602.

Alongi, D. M., & Dixon, G. (2004). Mangrove forests: Carbon sequestration, coastal protection, and ecosystem restoration. *Conservation Biology*, *16*(5), 1299–1307.

Alqubalee, A., Salisu, A. M., Bello, A. M., Al-Hussaini, A., & Al-Ramadan, K. (2023). Characteristics, distribution, and origin of ferruginous deposits within the Late Ordovician glaciogenic setting of Arabia. *Scientific Reports*, *13*(1). Advance online publication. DOI: 10.1038/s41598-023-48541-z PMID: 37891355

Al-Taani, A., Al-Husban, Y., & Farhan, I. (2020). Land suitability evaluation for agricultural use using GIS and remote sensing techniques: The case study of Ma'an Governorate, Jordan. *The Egyptian Journal of Remote Sensing and Space Sciences*, *24*, 109–117. DOI: 10.1016/j.ejrs.2020.01.001

Anderson, D. W.. (2023). Using UAV-Based Hyperspectral Imaging for Mapping Seagrass Vegetation and Water Quality in Coastal Ecosystems. *Remote Sensing of Environment*, *267*, 112658.

Andersson, J., Bauer, T., & Lynch, E. P. (2020). Evolution of structures and hydrothermal alteration in a Palaeoproterozoic supracrustal belt: Constraining paired deformation-fluid flow events in an Fe and Cu–Au prospective terrain in northern Sweden. *Solid Earth, 11*(2), 547–571. DOI: 10.5194/se-11-547-2020

Arif, M., Qi, Y., Dong, Z., & Wei, H. (2022). Rapid retrieval of Cd and Pb content using characteristic hyperspectral bands. *Journal of Cleaner Production.*

Arora, S., & Kumar, B. (2025). Effects of proximity of the sediment mining pit on riverbank stability. *Scientific Reports, 15*(1). Advance online publication. DOI: 10.1038/s41598-024-71869-9 PMID: 40140503

Atwood, T. B.. (2020). The role of seagrass meadows in carbon cycling and carbon sequestration. *Nature Communications, 11*, 1–8.

Atwood, T. B., Witt, A., Mayorga, J., Hammill, E., & Sala, E. (2020). Global patterns in marine sediment carbon stocks. *Frontiers in Marine Science, 7*, 165.

Avio, C. G., Gorbi, S., & Regoli, F. (2017). Plastics and microplastics in marine ecosystems: Effects on species and food webs. *Environmental Science. Processes & Impacts.*

Azimi, M., Weyaar, M., & Ahmadzai, M. (2024). Petrographic and Mechanical Analysis of Grano-Diorite Deposits in the Qala Region, Dikundi Province, Afghanistan. Academic Journal of Research and Scientific Publishing, 6(61), 110. https://doi.org/DOI: 10.52132/Ajrsp/v6.61.5

Bahrami, Z., Zhang, R., Wang, T., & Liu, Z. (2022). An end-to-end framework for shipping container corrosion defect inspection. *IEEE Transactions on Instrumentation and Measurement, 71*, 1–14.

Bakar, N. A. A., Talib, M. K. A., Arifin, S. N. H., Madun, A., Tajudin, S. A. A., Firdaus, M., Pakir, F., & Malik, A. K. A. (2024). Relationship of Resistivity Value with the Ground Material Information Obtained from Borehole Data: Case study. *Research Square*. https://doi.org/DOI: 10.21203/rs.3.rs-3475961/v1

Balbarini, N., Rønde, V., Maurya, P. K., Klint, K. E., Christiansen, A. V., & Binning, P. J. (2017). *Integration of geological, geophysical and contaminant monitoring technologies for contaminated site investigation*. Research Portal Denmark.

Bandh, S. A., Malla, F. A., Qayoom, I., Mohi-Ud-Din, H., Butt, A. K., Altaf, A., & Wani, S. A. (2023). Importance of blue carbon in mitigating climate change and plastic/microplastic pollution and promoting circular economy. *Sustainability, 15*(3), 2682.

Bao, X.. (2023). Using UAV imagery and machine learning for detecting invasive species in seagrass ecosystems. *Remote Sensing*, *15*(18), 4617.

Barbier, E. B., Hacker, S. D., Kennedy, C., Koch, E. W., Stier, A. C., & Silliman, B. R. (2011). The value of estuarine and coastal ecosystem services. *Ecological Monographs*, *81*(2), 169–193.

Barbiere, J.. (2019). Use of remote sensing for monitoring carbon sequestration in coastal ecosystems. *Remote Sensing Applications: Society and Environment*, *13*, 60–73.

Behm, J.. (2020). Recent advances in satellite-derived vegetation indices for monitoring carbon sequestration. *Remote Sensing*, *12*(1), 10–21.

Beighley, E. R.. (2019). The role of UAV-based LiDAR in assessing mangrove forest structure and carbon sequestration potential. *Carbon Balance and Management*, *14*(1), 2. PMID: 30904964

Bellone, T., Boccardo, P., & Perez, F. (2009). Investigation of vegetation dynamics using long-term normalized difference vegetation index time-series. *American Journal of Environmental Sciences*, *5*(4), 461–467. DOI: 10.3844/ajessp.2009.461.467

Bennett, E. M.. (2024). Remote Sensing for Monitoring Coastal Water Quality and Ecological Health. *Ecological Applications*, *34*(2), e2458.

Bentley, J. (2024). Detecting ocean microplastics with remote sensing: A pilot study on satellite reflectance signals. *Bridgewater State University Honors Project*. vc.bridgew.edu

Bery, A. A., Nordiana, M. M., Kiu, Y. C., Amalina, M. K. A. N., Saidin, M., Mohamad-Afiq, A., & Nur-Amalina, A. M. (2017). Joint-interpretation of 2-D electrical resistivity method and borehole data for subsurface lithology identification. AIP Conference Proceedings, 1885, 020019. https://doi.org/DOI: 10.1063/1.4998229

Bhatti, U. A., et al. (2021). Time Series Analysis and Forecasting of Air Pollution Particulate Matter (PM2.5): An SARIMA and Factor Analysis Approach. *IEEE Access*.

Bhatti, U. A., et al. (2022). Assessing the change of ambient air quality patterns in Jiangsu Province of China pre-to post-COVID-19. *Chemosphere*.

Bhatti, U. A., et al. (2022). Time series analysis and forecasting of air pollutants based on prophet forecasting model in Jiangsu province, China. *Frontiers in Environmental Science*.

Bhatti, U. A., et al. (2023). Robust zero-watermarking algorithm for encrypted medical volume data based on PJFM and 3D-DCT. *IEEE Journal of Biomedical and Health Informatics.*

Bhatti, U. A., Peng, M., Liu, Y., Qadri, I. A., Ahmed, B., Sarhan, N. M., & Awwad, E. M. (2020). Crop monitoring using remote sensing land use and land change data: Comparative analysis of deep learning methods using pre-trained CNN models. *Big Data Research, 36*, Article 100448.

Bhatti, U. A. (2020). Hybrid Watermarking Algorithm Using Clifford Algebra with Arnold Scrambling…. *IEEE Access : Practical Innovations, Open Solutions, 8*, 76386–76398.

Bhatti, U. A.. (2018). Recommendation System for Immunization Coverage and Monitoring. *Human Vaccines & Immunotherapeutics, 14*(1), 165–171. PMID: 29068748

Bhatti, U. A.. (2019). A Robust Multi-Watermarking Algorithm for Medical Images Based on DTCWT-DCT and Henon Map. *Applied Sciences (Basel, Switzerland), 9*(4), 700.

Bhatti, U. A.. (2019). Recommendation System Using Feature Extraction and Pattern Recognition…. *Enterprise Information Systems, 13*(3), 329–351.

Bhatti, U. A.. (2020). Geometric Algebra Applications in Geospatial AI and Remote Sensing Image Processing. *IEEE Access : Practical Innovations, Open Solutions,* •••, 8.

Bhatti, U. A.. (2021). Time Series Analysis and Forecasting of Air Pollution Particulate Matter Using SARIMA and Factor Analysis…. *IEEE Access : Practical Innovations, Open Solutions, 9*, 41019–41031.

Bhatti, U. A.. (2022). Advances in climate change monitoring through AI and remote sensing. *Remote Sensing Applications: Society and Environment.*

Bhatti, U. A.. (2022). Assessing Change of Ambient Air Quality Patterns in Jiangsu Province Pre-to Post-COVID-19. *Chemosphere,* •••, 288. PMID: 34655644

Bhatti, U. A.. (2022). Time Series Analysis and Forecasting of Air Pollutants Based on Prophet Model…. *Frontiers in Environmental Science, 10*, 945628.

Bhatti, U. A.. (2023). Advance Brain Tumor Segmentation Using Feature Fusion with Deep U-Net Model…. *Journal of King Saud University-CIS, 35*(9), 101793.

Bhatti, U. A.. (2023). AI-based Object Detection Trends in Remote Sensing, Multimedia and Agriculture Applications. *Frontiers in Plant Science.*

Bhatti, U. A.. (2023). AIoT-driven Multi-source Sensor Emission Monitoring and Forecasting Using Multi-source Sensor Integration with Reduced Noise Series Decomposition. *Journal of Cloud Computing (Heidelberg, Germany)*, *13*, 65.

Bhatti, U. A.. (2023). Brain Tumor Classification Utilizing Deep Features from MRI Images. *Biomedical Signal Processing and Control*, *85*, 104988.

Bhatti, U. A.. (2023). Global production patterns: Understanding the relationship between greenhouse gas emissions, agriculture greening, and climate variability. *Environmental Research*. PMID: 38169167

Bhatti, U. A.. (2024). Deep Reinforcement Learning Enables Adaptive-Image Augmentation for Automated Optical Inspection of Plant Rust. *Frontiers in Plant Science*.

Bhatti, U. A.. (2024). Global Production Patterns: Understanding Greenhouse Gas Emissions, Agriculture Greening.. . *Environmental Research*, •••, 118049. PMID: 38169167

Bhatti, U. A.. (2024). Interactive Medical Image Annotation Using Improved Attention U-Net.... *Expert Systems with Applications*, •••, 237.

Bhatti, U. A.. (2024). Interpretable Machine Learning for Personalized Medical Recommendations.... *Diagnostics (Basel)*, *13*(16), 2681. PMID: 37627940

Bhatti, U. A.. (2024). Understanding the Relationship between Greenhouse Gas .... *The Science of the Total Environment*, •••, 118049. PMID: 38169167

Bhatti, U. A., Bhatti, M. A., Tang, H., Syam, M. S., Awwad, E. M., Sharaf, M., & Ghadi, Y. Y. (2024). Global production patterns: Understanding the relationship between greenhouse gas emissions, agriculture greening and climate variability. *Environmental Research*, 118049.

Bhatti, U. A., Chen, X., He, Z., Shi, Y., & Wu, W. (2024). Investigating AI-Based Smart Precision Agriculture Techniques. *Frontiers in Plant Science*, *14*, 1237783. PMID: 37496855

Bhatti, U. A., Huang, M., Han, H., Li, Y., Zhang, Y., & Li, W. (2024). PE-Transformer: Path enhanced transformer for improving underwater object detection. *Expert Systems with Applications*, 123253.

Bhatti, U. A., Huang, M., Wang, H., Zhang, Y., Mehmood, A., & Di, W. (2019). Recommendation system for immunization coverage and monitoring. *Human Vaccines & Immunotherapeutics*, *14*(1), 165–171.

Bhatti, U. A., Huang, M., Wu, D., Zhang, Y., Mehmood, A., & Han, H. (2019). Recommendation system using feature extraction and pattern recognition in clinical care systems. *Enterprise Information Systems, 13*(3), 329–351.

Bhatti, U. A., Li, J., Liu, Z., Yu, Z., Zeeshan, Z., & Mehmood, A. (2024). Robust zero-watermarking algorithm for encrypted medical images based on Bandelet-DCT. *Multimedia Tools and Applications, 81*(12), 16863–16879.

Bhatti, U. A., Li, J., Yu, Z., & Liu, J. (2024). Robust zero-watermarking algorithm for encrypted medical images based on SIFT and Bandelet-DCT. *Multimedia Tools and Applications, 81*(12), 16863–16879.

Bhatti, U. A., Li, J., Zou, J., Li, Z., Marjan, S., & Yuan, L. (2024). "Robust zero-watermarking algorithm for medical images using local binary pattern and discrete cosine transform." *International Conference on Artificial Intelligence and Security*, 350-362.

Bhatti, U. A., Liu, J., Huang, M., Chen, Y.-W., & Fan, Y.. (2025). *Medical Image Hybrid Watermark Algorithm Based on.* Wiley.

Bhatti, U. A., Liu, J., Li, J., Chanussot, J., Zeeshan, Z., Yuan, L., & Wu, G. (2023). *Deep learning for multimedia processing applications: Volume two: Signal processing and pattern recognition.* CRC Press.

Bhatti, U. A., Liu, J., Li, J., Sadiq, N., & Ai, Y. (2019). A robust multi-watermarking algorithm for medical images based on DTCWT-DCT and Henon map. *Applied Sciences (Basel, Switzerland), 9*(4), 700.

Bhatti, U. A., Liu, J., Li, J., Sadiq, N., Ai, Y., & Huang, M. (2019). AI-based object detection in remote sensing applications for coastal monitoring and ecosystem management. *Remote Sensing, 11*(9), 1089.

Bhatti, U. A., Liu, J., Zeeshan, Z., & Yuan, L. (2021). Global trends in the use of UAV LiDAR for forest canopy monitoring and carbon stock estimation. *Remote Sensing Reviews, 42*(2), 89–112.

Bhatti, U. A., Liu, W., Zhang, X., Li, J., Yuan, L., & Chen, Y. (2024). A novel myocarditis detection combining deep reinforcement learning and an improved differential evolution algorithm. *CAAI Transactions on Intelligence Technology.*

Bhatti, U. A., Liu, Z., Ming-Quan, Z., Qing-Song, H., Ali, S., Yuhuan, Y., & Yu, Z. (2021). Advanced color edge detection using clifford algebra in satellite images. *IEEE Photonics Journal, 13*(2), 1–20.

Bhatti, U. A., Liu, Z., Zhao, L., Zhang, M., & Zhang, H. (2024). Smart cities and smart networks: AI applications in urban geography and industrial communication. *International Journal of High Speed Electronics and Systems*, *34*(03), 2440122.

Bhatti, U. A., Ma, J., Liu, Y., & Wang, G. (2024). Deep Learning-Based Trees Disease Recognition and Classification Using Hyperspectral Data. *Computers, Materials & Continua*, *77*(1).

Bhatti, U. A., Marjan, S., Wahid, A., Syam, M. S., Huang, M., Tang, H., & Hasnain, A. (2021). The effects of socioeconomic factors on particulate matter concentration in China's: New evidence from spatial econometric model. *Journal of Cleaner Production*, *417*, 137969.

Bhatti, U. A., Mehmood, A., Zhang, Y., & Huang, M. (2021). Optimization and multi-class classification of hyperspectral remote sensing data for mangrove forest management. *Environmental Monitoring and Assessment*, *193*(1), 23. PMID: 33389226

Bhatti, U. A., Mengxing, H., Neira-Molin, H., Marjan, S., Baryalai, M., Hao, T., & Wu, G. (2023). MFFCG–Multi feature fusion for hyperspectral image classification using graph attention network. *Expert Systems with Applications*, 120496.

Bhatti, U. A., Mengxing, H., Neira-Molin, H., Marjan, S., Baryalai, M., Hao, T., & Wu, G. (2023). MFFCG–Multi-feature fusion for hyperspectral image classification using graph attention network. *Expert Systems with Applications*, •••, 120496.

Bhatti, U. A., Nawaz, S. A., Li, J., Shoukat, M. U., & Ahmad, R. M. (2023). AI based Object detection latest trends in Remote sensing, Multimedia and Agriculture Applications. *Frontiers in Plant Science*, •••, 4273.

Bhatti, U. A., & Neira-Molin, H. (2023). MFFCG – Multi-feature Fusion for Hyperspectral Classification. *Expert Systems with Applications*.

Bhatti, U. A., Nizamani, A. H., Wang, Y. Y., & Ma, J. (2023). Geometric algebra applications in geospatial artificial intelligence and remote sensing image processing. *IEEE Access : Practical Innovations, Open Solutions*, *8*, 155783–155796.

Bhatti, U. A., Nizamani, M. M., & Mengxing, H. (2021). Climate change threatens Pakistan's snow leopards. *Science*, *377*(6606), 585–586.

Bhatti, U. A., Sadiq, M. A., Zeeshan, Z., & Shoukat, M. U. (2021). Predicting heavy metal contamination in coastal ecosystems using hyperspectral imaging. *Geoderma*, *368*, 114–125.

Bhatti, U. A., Sadiq, M. A., Zeeshan, Z., & Shoukat, M. U. (2023). AI-based Object detection latest trends in Remote sensing, Multimedia and Agriculture Applications. *Frontiers in Plant Science*, 4273.

Bhatti, U. A., Syam, M. S., Yuan, Z., Zeeshan, Z., & Sharaf, M. (2023). Advanced image segmentation for precision agriculture using CNN-GAT fusion and fuzzy C-means clustering. *Computers and Electronics in Agriculture*, *226*, 109431.

Bhatti, U. A., & Tang, H. (2023). Deep Learning with Graph Convolutional Networks: An Overview and Latest Applications in Computational Intelligence. *International Journal of Intelligent Systems*.

Bhatti, U. A., Tang, H., Wu, G., Marjan, S., & Hussain, A. (2023). Deep learning with graph convolutional networks: An overview and latest applications in computational intelligence. *International Journal of Intelligent Systems*.

Bhatti, U. A., Xu, B., Zhang, H., Liu, Z., & Yuan, Y. (2023). Investigating the nexus between energy, socio-economic factors and environmental pollution: A geo-spatial multi regression approach. *Gondwana Research*.

Bhatti, U. A., Xu, B., Zhang, H., Zhang, Y., He, Z., Marjan, S., & Aamir, M. (2024). AIoT-driven multi-source sensor emission monitoring and forecasting using multi-source sensor integration with reduced noise series decomposition. *Journal of Cloud Computing (Heidelberg, Germany)*, *13*(1), 65.

Bhatti, U. A., Yan, Y., Zhang, M., Ali, M., Liu, L., & Zhang, Z. (2022). Time series analysis and forecasting of air pollutants based on prophet forecasting model in Jiangsu province, China. *Frontiers in Environmental Science*, *10*, 945628.

Bhatti, U. A., Yan, Y., Zhou, M., Ali, S., & Hussain, A. (2021). Prediction of Particulate Matter Pollution Using a LSTM Model.... *Frontiers in Environmental Science*, *10*, 945628.

Bhatti, U. A., Yan, Y., Zhou, M., Ali, S., Hussain, A., Qingsong, H., Yu, Z., & Yuan, L. (2021). Time series analysis and forecasting of air pollution particulate matter (PM2.5): An SARIMA and factor analysis approach. *IEEE Access : Practical Innovations, Open Solutions*, *9*, 41019–41031.

Bhatti, U. A., Yan, Y., Zhou, M., Ali, S., Hussain, A., Qingsong, H., Yu, Z., & Yuan, L. (2021). Time Series Analysis and Forecasting of Air Pollution Particulate Matter (PM2.5): An SARIMA and Factor Analysis Approach. *IEEE Access : Practical Innovations, Open Solutions*, *9*, 41019–41031.

Bhatti, U. A., & Yu, Z. (2024). Feature-based multi-criteria recommendation system using a weighted approach with ranking correlation. *Intelligent Data Analysis*.

Bhatti, U. A., Yu, Z., Chanussot, J., Zeeshan, Z., Yuan, L., Luo, W., & Nawaz, S. A. (2021). Local Similarity-based Spatial–Spectral Fusion Hyperspectral Image Classification with Deep CNN and Gabor Filtering. *Environmental Science and Pollution Research International*.

Bhatti, U. A., Yu, Z., Chanussot, J., Zeeshan, Z., Yuan, L., Luo, W., & Nawaz, S. A. (2021). Local similarity-based spatial–spectral fusion hyperspectral image classification with deep CNN and Gabor filtering. *IEEE Transactions on Geoscience and Remote Sensing*, *60*, 1–15.

Bhatti, U. A., Yu, Z., Li, J., Nawaz, S. A., Mehmood, A., Zhang, K., & Yuan, L. (2020). Hybrid watermarking algorithm using Clifford algebra with Arnold scrambling and chaotic encryption. *IEEE Access : Practical Innovations, Open Solutions*, *8*, 76386–76398.

Bhatti, U. A., Yu, Z., Nawaz, S. A., Mehmood, A., Zhang, K., & Yuan, L. (2020). Hybrid Watermarking Algorithm.... *IEEE Access : Practical Innovations, Open Solutions*.

Bhatti, U. A., Zeeshan, Z., Ma, J., & Li, J. (2024). Multi-watermarking algorithm for medical image based on KAZE-DCT. *Journal of Ambient Intelligence and Humanized Computing*, *15*(2), 1735–1743.

Bhatti, U. A., Zeeshan, Z., Ma, J., & Liu, Z. (2023). AI-based Object Detection in Remote Sensing Applications for Mangrove Soil Analysis. *Remote Sensing*, *14*(16).

Bhatti, U. A., Zeeshan, Z., Nawaz, S. A., & Yu, Z. (2023). New watermarking algorithm utilizing quaternion Fourier transform with advanced scrambling and secure encryption. *Multimedia Tools and Applications*, *80*(9), 13367–13387.

Bhatti, U. A., Zeeshan, Z., Nizamani, M. M., Bazai, S. U., Yu, Z., & Yuan, L. (2022). Assessing the change of ambient air quality patterns in Jiangsu Province of China pre-to post-COVID-19. *Chemosphere*, *288*, 132569.

Bhatti, U. A., Zeeshan, Z., Nizamani, M. M., Bazai, S., Yu, Z., & Yuan, L. (2023). Robust medical image zero-watermarking algorithm based on Residual-DenseNet. *IET Biometrics*, *11*(6), 547–556.

Bhatti, U. A., Zeeshan, Z., Yuan, L., Luo, W., & Nawaz, S. A. (2023). Multi-feature Fusion for Hyperspectral Image Classification Using Graph Attention Network. *Expert Systems with Applications*, •••, 120496.

Bhatti, U. A., Zhang, G., Yu, Z., Li, Z., & Yuan, L. (2023). A digital twin-based visual servoing with extreme learning machine and differential evolution. *International Journal of Intelligent Systems*, *2023*(1), 6639963.

Bhatti, U. A., Zhang, H., Zhang, X., & Wu, W. (2023). "Towards explainability for AI-based edge wireless signal automatic modulation classification." *2024 IEEE/CIC International Conference on Communications in China (ICCC).*

Bhatti, U. A., Zhang, W., Zhang, L., Yuan, L., & Wu, G. (2024). Robust Multi-Watermarking Algorithm for Medical Images Based on GoogLeNet and Henon Map. *Computers, Materials & Continua*, *75*(1).

Bhatti, U. A., Zhang, Y., Ma, J., Liu, M., & Huang, M. (2024). Interactive medical image annotation using improved Attention U-net with compound geodesic distance. *Expert Systems with Applications*, *237*, 121282.

Bhatti, U. A., Zheng, Y., Zhang, J., Huang, M., & Aamir, M. (2023). Medical image segmentation using deep learning with feature enhancement. *IET Image Processing*, *14*(14), 3324–3332.

Binte Mostafiz, R., Noguchi, R., & Ahamed, T. (2021). Agricultural land suitability assessment using satellite remote sensing-derived soil-vegetation indices. *Land (Basel)*, *10*(2), 223. DOI: 10.3390/land10020223

Blais, M.-A., & Akhloufi, M. A. (2025). Advances in remote sensing and deep learning in coastal boundary extraction for erosion monitoring. *Geomatics*, *5*(1), 9. DOI: 10.3390/geomatics5010009

Blasco, F.. (2015). Carbon stock estimation and remote sensing techniques in mangrove forests. *International Journal of Remote Sensing*, *36*(12), 3179–3193.

Boulon, R. H.. (2019). Utilizing UAVs in the study of blue carbon ecosystems. *Geosciences*, *9*(8), 345–358.

Breiman, L. (2001). Random Forests. *Machine Learning*.

Browne, M. A.. (2011). Accumulation of microplastics on shorelines worldwide. *Environmental Science & Technology*.

Brown, S.. (2018). Utilizing UAV LiDAR to support conservation of mangrove ecosystems through remote sensing data integration. *Remote Sensing of Environment*, *206*, 168–178.

Brunt, L.. (2017). Recent advancements in UAV-based remote sensing technologies for forestry and environmental applications. *ISPRS Journal of Photogrammetry and Remote Sensing*, *132*, 1–14.

Campos-Taberner, M., García-Haro, F. J., Martínez, B., Izquierdo-Verdiguier, E., Atzberger, C., & Camps-Valls, G. (2020). Understanding deep learning in land use classification based on Sentinel-2 time series. *Scientific Reports*, *10*, 17188. DOI: 10.1038/s41598-020-74215-5 PMID: 33057052

Cavanaugh, K. C.. (2023). New Technologies for Monitoring Coastal Ecosystem Change and Complexity — Drones, Remote Sensing, and Big Data Approaches. *Annual Review of Marine Science*.

Cavanaugh, K. C., Bell, T. W., Aerni, K. E., Byrnes, J. E. K., McCammon, S., & Smith, M. M. (2025). New technologies for monitoring coastal ecosystem dynamics. *Annual Review of Marine Science*, *17*(1), 409–433.

Chaisanguansuk, P., Phantuwongraj, S., Jirapinyakul, A., & Assawincharoenkij, T. (2023). Preliminary study on microplastic abundance in mangrove sediment cores at Mae Klong River, upper Gulf of Thailand. *Frontiers in Environmental Science*, *11*, 1134988.

Chan, Y. K., & Koo, V. (2008). An introduction to synthetic aperture radar (SAR). *Progress in Electromagnetics Research B. Pier B*, *2*, 27–60.

Chavan, S. K.. (2020). Carbon stock estimation in mangrove forests using UAV-based LiDAR and allometric equations. *Environmental Science and Pollution Research International*, *27*(14), 17025–17037.

Chave, J.. (2009). Towards a global wood density database: A review of methodologies and recent advances. *Methods in Ecology and Evolution*, *4*(3), 373–388.

Chen, M., Yang, K., Wang, Z., Zhao, K., Wu, E., Shi, J., & Sun, G. (2023). Corrosion performance of NV E690 steel and 316L stainless steel coating fabricated by underwater direct metal deposition. *Corrosion Science*, *219*, 111232.

Chiarella, C., Meyfroidt, P., Abeygunawardane, D., & Conforti, P. (2023). Balancing the trade-offs between land productivity, labor productivity and labor intensity. *Ambio*, *52*(10), 1618–1634.

Chisholm, L. A.. (2015). Estimation of mangrove biomass and carbon stock using UAV-based LiDAR data. *Carbon Management*, *6*(5), 315–326.

Chukwudi, C. E., Chibuzo, P. A., & Austin, C. O. (2022). Combined application of vertical electrical sounding and 2D electrical resistivity tomography for groundwater exploration in parts of Enugu metropolis, Southeastern Nigeria. *International Journal of Physical Sciences*, *17*(3), 67–85. DOI: 10.5897/IJPS2022.4964

Chytrý, M., Schaminée, J. H. J., & Schwabe, A. (2011). Vegetation survey: A new focus for Applied Vegetation Science. *Applied Vegetation Science*, *14*(4), 435–439.

Crowell, S. M.. (2017). The role of coastal carbon storage in climate change mitigation. *Geophysical Research Letters*, *44*(10), 526–533.

Dang, K. B., Dang, V. B., Ngo, V. L., Vu, K. C., Nguyen, H., Nguyen, D. A., & Nguyen, D. L. P. N. (2022). Application of deep learning models to detect coastlines and shorelines. *Journal of Environmental Management*. Advance online publication. DOI: 10.1016/j.jenvman.2022.115732 PMID: 35930878

Das, A., Ichi, E., & Dorafshan, S. (2023). Image-based corrosion detection in ancillary structures. *Infrastructures*, *8*(4), 66.

Daud, S., Ishak, M. F., Ismi, P. I., Noor, M., & Ghazali, S. (2025). Granite exploration by using resistivity and induced polarization imaging in western granite belt, Pendang, Kedah, Malaysia. *Smart Construction and Sustainable Cities*, *3*(1). Advance online publication. DOI: 10.1007/s44150-024-00167-8

Daudt, R. C., Le Saux, B., & Boulch, A. (2018). Fully convolutional siamese networks for change detection. 2018 25th IEEE International Conference on Image Processing (ICIP), 4063–4067. https://doi.org/DOI: 10.1109/ICIP.2018.8451607

De Marsily, G., Delay, F., Gonçalvès, J., Renard, P., Teles, V., & Violette, S. (2005). Dealing with spatial heterogeneity. *Hydrogeology Journal*, *13*(1), 161–183.

Deng, H., He, J., Feng, D., Zhao, Y., Sun, W., Yu, H., & Ge, C. (2021). Microplastics pollution in mangrove ecosystems: A critical review of current knowledge and future directions. [MDPI]. *The Science of the Total Environment*, *753*, 142041. PMID: 32906050

Desai, S., Mattoo, M., Deshpande, A., Dey, A., & Akiwate, S. (2024). Vegetation change detection through NDVI analysis using Landsat-8 data. In *Proceedings of the International Conference on Computer, Communication, Control and Information Technology (C3IT 2024)*. IEEE.

Didi, C. N., Osinowo, O. O., Akpunonu, O. E., & Usman, M. B. (2025). Re-evaluation of the stratigraphy, structural, and depositional environment of Kolmani field, Gongola sub-basin, northeast Nigeria using outcrop data, ditch cutting data, and well log data for epitaxic hydrocarbon exploration. *Discover Geoscience*, *3*(1). Advance online publication. DOI: 10.1007/s44246-024-00120-x

Diffuse reflectance spectroscopy of iron oxides. (2002). *Encycl. Surf. Colloid Sci.*

Ding, S., Zhang, X., & Sun, W. (2022). *Estimation of soil lead based on GF-5 hyperspectral images*. J. Soils Sedim. *Nature*.

Donato, D. C., Kauffman, J. B., Murdiyarso, D., Kurnianto, S., Stidham, M., & Kanninen, M. (2011). Mangroves among the most carbon-rich forests in the tropics. *Nature Geoscience*, *4*(5), 293–297.

Duarte, B. (2021). Climate Change Impacts on Salt Marsh Blue Carbon and Ecosystem Services. *Applied Sciences (Basel, Switzerland)*.

Dudley, T.. (2011). Remote sensing of mangrove ecosystems for carbon sequestration. *Global Change Biology*, *17*(6), 1996–2005.

Dzikunoo, E. A., Vignoli, G., Jørgensen, F., Yidana, S. M., & Banoeng-Yakubo, B. (2020). New regional stratigraphic insights from a 3D geological model of the Nasia sub-basin, Ghana, developed for hydrogeological purposes and based on reprocessed B-field data originally collected for mineral exploration. *Solid Earth*, *11*(2), 349–369. DOI: 10.5194/se-11-349-2020

Ebenezer, Y., Durga, K., Chandrakala, N. B., Shashank, G., Sharma, M., & Manjunath, T. C. (2026). Disaster Education and Awareness Programs. In Ali, Q., & Tariq Iqbal Khan, M. (Eds.), *Strengthening Global Resilience to Natural Disasters* (pp. 73–102). IGI Global Scientific Publishing., DOI: 10.4018/979-8-3693-9745-9.ch004

Edwards, P. J.. (2024). Application of UAVs for environmental monitoring and ecosystem management in coastal areas. *Journal of Coastal Research*, *40*(1), 12–24.

Ekwok, S. E., Eldosouky, A. M., Ben, U. C., Achadu, O.-I. M., Akpan, A. E., Othman, A., & Prem, L. T. (2023). An integrated approach of advanced methods for mapping geologic structures and sedimentary thickness in Ukelle and adjoining region (Southern Benue Trough), Nigeria. *Earth Sciences Research Journal*, *27*(2), 251–261.

Ellis, J.. (2019). Allometric models for carbon estimation in mangrove forests using UAV-based LiDAR. *Remote Sensing*, *11*(6), 666.

Elma, E.. (2024). Evaluating UAV-based multispectral imagery for mapping and monitoring seagrass habitats. *Aquatic Conservation*.

El-Qalamoshy, T. R., Fattah, M. A., Reda, M., Abdelhafeez, T., Azzam, S. S., & Mossad, M. (2023). A multi-disciplinary approach for trap identification in the Southern Meleiha Area, North Western Desert, Egypt: Integrating seismic, well log, and fault seal analysis. *Geomechanics and Geophysics for Geo-energy and Geo-resources*, *9*(1).

Erbach, D. C. (1987). Measurement of soil bulk density and moisture. *Transactions of the ASAE. American Society of Agricultural Engineers*, *30*(4), 922–931.

Estimating heavy metal concentrations in Technosols with reflectance spectroscopy. (2022). *Geoderma*.

Evensen, G. (2009). *Data assimilation: The ensemble Kalman filter*. Springer Science & Business Media.

Eze, S. U., Orji, M. O., Onoriode, A. E., Saleh, S. A., & Abolarin, M. O. (2022). Integrated Geoelectrical Resistivity Method for Environmental Assessment of Landfill Leachate Pollution and Aquifer Vulnerability Studies. *Journal of Geoscience and Environment Protection*, *10*(12), 1–18.

Fearnside, P. M. (2008). Carbon sequestration by Amazonian forests. *Global Biogeochemical Cycles*, *22*(4), GB4005.

Feng, Z., Jiang, Z., & Yan, C. (2025). Prediction of corrosion lifetime of marine environment steel pipe piles based on physics-informed neural network. *Discover Applied Sciences*.

Finkbeiner, E. M.. (2020). Remote sensing for coastal wetland carbon monitoring. *Ecological Applications*, *30*(4), e02133.

Forkan, A. R. M., Kang, Y.-B., Jayaraman, P. P., Liao, K., Kaul, R., Morgan, G., & Sinha, S. (2022). CorrDetector: A framework for structural corrosion detection from drone images using ensemble deep learning. *Expert Systems with Applications*, *193*, 116461.

Fourqurean, J. W., Duarte, C. M., Kennedy, H., Marbà, N., Holmer, M., Mateo, M. A., & Serrano, O. (2012). Seagrass ecosystems as a globally significant carbon stock. *Nature Geoscience*, *5*(7), 505–509.

Frazier, S. D.. (2016). Assessing the role of coastal ecosystems in global carbon cycling. *Geophysical Research Letters*, *43*(2), 987–995.

Friedl, M. A.. (2019). The role of UAV LiDAR in assessing the potential of mangroves in the carbon credit market. *Environmental Policy and Governance*, *29*(4), 263–272.

Friess, D.A. (2023). *Blue carbon science, management and policy across a tropical urban landscape*.

Friess, D.A. (n.d.) (See also global tropical and urban blue carbon context). *(Multiple articles by Friess are key in blue carbon literature.)*

Frolking, S.. (2009). The CASA biogeochemical model and its application to carbon cycling in coastal ecosystems. *Ecological Modelling*, *220*(6), 843–860.

Fu, P. (2024). Remote sensing inversion on heavy metal content in soil using alteration information. *Scientific Reports*.

Gaesenngwe, G., Mamvura, T., Danha, G., & Sibanda, V. (2021). A comparative study on the comminution behavior of diorite rocks. *Heliyon*, *7*(11), e08352. DOI: 10.1016/j.heliyon.2021.e08352 PMID: 34825076

Gaikwad, S. V., Vibhute, A. D., Shah, S. V., Saini, J. R., & Kale, K. (2024). Vegetation health analysis of Vaijapur using Google Earth Engine and vegetation indices. In *Proceedings of the 2nd International Conference on Current Trends in Physics and Photonics (ICCTPP 2024)* (p. 25). SPIE. https://doi.org/DOI: 10.1117/12.3044615

Gaikwad, S., Vibhute, A. D., & Kale, K. (2021). Development of NDVI prediction model using artificial neural networks. In *Proceedings of the International Conference on Recent Trends in Image Processing and Pattern Recognition*. Springer.

Gajere, J. N., Adekeye, O. A., Tende, A. W., & Aminu, M. D. (2024). GIS-based multi-criteria decision models for barite exploration in Nigeria's Benue Trough. *Scientific Reports*, *14*(1). Advance online publication. DOI: 10.1038/s41598-024-53004-6 PMID: 38890381

Gallant, A. L.. (2022). Assessing Environmental Stress in Seagrass Ecosystems Using UAV and Multispectral Imagery. *Environmental Monitoring and Assessment*, *194*, 264.

Gao, J.. (2017). Estimating carbon stock in coastal ecosystems using LiDAR: Applications to mangrove forests. *Environmental Research Letters*, *12*(7), 074004.

Gao, Y. (2017). *Feasibility of estimating heavy metal in wetland soil using hyperspectral technology*. IEEE IGARSS.

Gauri, S.. (2024). AI-based object detection for marine environment using remote sensing. *IEEE Transactions on Geoscience and Remote Sensing*.

George, N. J., Ekanem, A. M., Thomas, J. E., Udosen, N. I., Ossai, N. M., & Atat, J. G. (2023). Electro-sequence valorization of specific enablers of aquifer vulnerability and contamination: A case of index-based model approach for ascertaining the thinness to quality groundwater in sedimentary beds. *HydroResearch*, *7*, 71–85.

Gewali. (2018). *Hyperspectral image analysis survey for environmental applications*. arXiv.

Ghorbanian, A., Mohammadzadeh, A., & Jamali, S. (2022). Linear and non-linear vegetation trend analysis throughout Iran using two decades of MODIS NDVI imagery. *Remote Sensing*, *14*(15), 3683. DOI: 10.3390/rs14153683

Gidey, E., Dikinya, O., Sebego, R., Segosebe, E., & Zenebe, A. (2018). Analysis of long-term agricultural drought characteristics using vegetation health index (VHI) in Raya and its environs, Northern Ethiopia. *Environmental Systems Research*, *7*(1), 1–17. DOI: 10.1186/s40068-018-0121-x

Goldstein, D., Aldrich, C., O'Connor, L., & Shao, Q. (2025). Unlocking Subsurface Geology: A Case Study with Measure-While-Drilling Data and Machine Learning. https://doi.org/DOI: 10.2118/224249-ms

Gonçalves, G., & Henriques, J. (2017). The potential of UAV-based LiDAR technology in forest management and environmental monitoring. *Remote Sensing*, *9*(12), 1231.

González, A.. (2016). LiDAR technology and UAVs for mapping mangrove forests in coastal areas. *International Journal of Remote Sensing*, *37*(19), 4661–4682.

González-Cabán, A.. (2020). Mapping mangrove forests using UAV-based LiDAR: A pilot study in Central America. *Journal of Coastal Research*, *36*(5), 1117–1130.

Gouet, D. H., Meying, A., Nkoungou, H. L. E., Assembe, S. P., Nouck, P. N., & Mbarga, T. N. (2020). Typology of Sounding Curves and Lithological 1D Models of Mineral Prospecting and Groundwater Survey within Crystalline Basement Rocks in the East of Cameroon (Central Africa) Using Electrical Resistivity Method and Koefoed Computer Program. *International Journal of Geosciences*, *11*(12), 847–871.

Gowri, L., & Manjula, K. R. (2019). Evaluation of various vegetation indices for multispectral satellite images. *International Journal of Innovative Technology and Exploring Engineering*, *8*(10), 3494–3500.

Gowri, L., & Manjula, K. R. (2025). Detection of land cover usage from optimized learnable parameter artificial neural network (OLPANN) using multispectral images. *Multimedia Tools and Applications*, *84*, 4283–4313. DOI: 10.1007/s11042-024-18828-2

Gray, T. G.. (2021). Carbon stocks of coastal wetlands and their role in climate change mitigation. *Nature Sustainability*, *4*(4), 316–324.

Greenshields, J. (2024). *A systematic review of microplastic sampling methods in mangrove and seagrass ecosystems. Environmental Monitoring and Assessment.* Springer Link.

Grey, A. (2021). *Geochemical mapping of a blue carbon zone.*

Gulliver, A., Carnell, P. E., Trevathan-Tackett, S. M., de Paula Costa, M., & Macreadie, P. I. (2020). Estimating the Potential Blue Carbon Gains from Tidal Marsh Rehabilitation: Case Study from Southeastern Australia. *Frontiers in Marine Science*.

Guo, N., Cheng, Q., Fu, Y., Tong, J., Gao, Y., & He, J. (2023). Microstructure and corrosion behavior of 10CrNi3MoV high-strength low-alloy steel coating via underwater wire-feed laser cladding. *Surface and Coatings Technology*, *473*, 129984.

Guo, W., Zhang, H., Zhao, W., & Xiao, G. (2024). A high corrosion resistant Fe-VC layer laser cladded under water-cooled substrate condition. *Materials Letters*, *354*, 135360.

Gupta, A.. (2021). Carbon stock assessment and validation using UAV-LiDAR technology in mangrove forests of the Sundarbans. *Journal of Environmental Management*, *285*, 112119.

Guth, P. L., Van Niekerk, A., Grohmann, C. H., Muller, J. P., Hawker, L., Florinsky, I. V., Gesch, D., Reuter, H. I., Herrera-Cruz, V., Riazanoff, S., & López-Vázquez, C. (2021). Digital elevation models: Terminology and definitions. *Remote Sensing*, *13*(18), 3581.

Hakim, M. M. A., Idrus, A., & Suryanto, W. (2023). Low sulfidation epithermal gold mineralization at Tambang Sawah, Lebong, Bengkulu Province, Indonesia. E3S Web of Conferences, 488, 02003. https://doi.org/DOI: 10.1051/e3sconf/202348802003

Hamada, H. M., Abed, F., Al-Sadoon, Z. A., & Hassan, A. (2025). Enhancing Concrete Strength and Microstructure with Basalt and Steel Fibers in Acid and Base Environments Incorporating Desert Sand. *International Journal of Concrete Structures and Materials*, *19*(1). Advance online publication. DOI: 10.1186/s40069-024-00726-x

Hamad, I. Y., Staehr, P. A. U., Rasmussen, M. B., & Sheikh, M. (2022). Drone-based characterization of seagrass habitats in the tropical waters of Zanzibar. *Remote Sensing*, *14*(3), 680.

Hancock, R.. (2019). UAV-based LiDAR and its application to mangrove forest monitoring. *Journal of Coastal Conservation*, *23*(3), 477–488.

Han, H., & Siebert, J. (2022). TinyML: A systematic review and synthesis of existing research. In *2022 International Conference on Artificial Intelligence in Information and Communication (ICAIIC)* (pp. 269–274). IEEE.

Han, S., & Kerekes, J. P. (2017). Overview of passive optical multispectral and hyperspectral image simulation techniques. *IEEE Journal of Selected Topics in Applied Earth Observations and Remote Sensing*, *10*(11), 4794–4804.

Hardy, B. L.. (2020). Utilizing UAV-based remote sensing for carbon flux measurements in coastal wetlands. *Nature Sustainability*, *3*(12), 1014–1021.

Harrison, K.. (2020). Long-term monitoring of mangrove health using UAV-based LiDAR and satellite data. *Journal of Environmental Science and Technology*, *54*(6), 4536–4545.

Hasan, M., & Shang, Y. (2022). Hard-rock investigation using a non-invasive geophysical approach. *Journal of Applied Geophysics*, *206*, 104808. DOI: 10.1016/j.jappgeo.2022.104808

Hasan, M., Shang, Y., He, M., Shao, P., & Yi, X. (2021). Application of electrical resistivity tomography (ERT) for rock mass quality evaluation. *Scientific Reports*, *11*(1). Advance online publication. DOI: 10.1038/s41598-021-95317-9 PMID: 34880388

Hasan, M., Shang, Y., & Meng, Q. (2023). Evaluation of rock mass units using a non-invasive geophysical approach. *Scientific Reports*, *13*(1). Advance online publication. DOI: 10.1038/s41598-023-44579-y PMID: 37661247

Hasnain, A., Sheng, Y., Hashmi, M. Z., Bhatti, U. A., Hussain, A., & Hameed, M. (2022). Time series analysis and forecasting of air pollutants based on the Prophet forecasting model in Jiangsu province, China. *Frontiers in Environmental Science*, *10*, 945628.

Hassan-Esfahani, L., Torres-Rua, A., Jensen, A., & McKee, M. (2015). Soil moisture and spectroscopy with ANN. *Remote Sens.*

Hasselmann, K., Latif, M., Hooss, G., Azar, C., Edenhofer, O., Jaeger, C. C., Johannessen, O. M., Kemfert, C., Welp, M., & Wokaun, A. (2003). The challenge of long-term climate change. *Science*, *302*(5652), 1923–1925.

He, B., & Bhatti, U. A. (2025). *Smart Cities and Smart Networks: AI Applications in Urban Geography and Industrial Communication*. World Scientific.

Helber, P., Bischke, B., Dengel, A., & Borth, D. (2019). EuroSAT: A novel dataset and deep learning benchmark for land use and land cover classification. *IEEE Journal of Selected Topics in Applied Earth Observations and Remote Sensing*, *12*(7), 2217–2226. DOI: 10.1109/JSTARS.2019.2918246

Herndon, H., & Tien, I. (2025). Creating a first-pass algorithm for corrosion assessment in bridge inspections using machine learning and UAV-collected imagery data. *Structure and Infrastructure Engineering*, •••, 1–15.

Hess, L.. (2015). Recent advances in UAV-based LiDAR technology and its application to environmental monitoring. *Yaogan Xuebao*, *30*(6), 1–16.

Hidalgo-Ruz, V.. (2012). Microplastics in the marine environment: A review of the methods used for identification and quantification. *Environmental Science & Technology*. PMID: 22321064

Hong, X., Huang, L., Gong, S., & Xiao, G. (2021). Shedding damage detection of metal underwater pipeline external anticorrosive coating by ultrasonic imaging based on HOG+ SVM. *Journal of Marine Science and Engineering*, *9*(4), 364.

Houghton, R. A. (2014). The role of land-use change in the carbon cycle. *Journal of Geophysical Research. Biogeosciences*, *119*(10), 2100–2121.

Howard, J.. (2017). A global map of mangrove forest carbon stocks and its implications for climate change mitigation. *Nature Communications*, *8*, 1–13.

Huang, M., Han, H., Wang, L., Li, Y., Zhang, Y., & Bhatti, U. A. (2018). A clinical decision support framework for heterogeneous data sources. *IEEE Journal of Biomedical and Health Informatics*, *22*(6), 1824–1833. PMID: 29994279

Huang, X.. (2022). Remote sensing for assessing seagrass ecosystem services: UAV and multispectral sensor applications. *Environmental Science and Pollution Research International*.

Huyzentruyt, M. (2025). *Carbon sequestration along gradients of tidal marsh degradation*. Preprint.

Hyperspectral remote sensing for soil heavy metal inversion: Insights and applications. (2025). ResearchGate Review.

Ibraheem, I. M., Yogeshwar, P., Sharifi, F., Bergers, R., & Tezkan, B. (2025). Joint inversion of transient electromagnetic and radiomagnetotelluric data for enhanced subsurface characterization. *Scientific Reports*, *15*(1). PMID: 40665125

Ibrahim, A. A. M., & Tapamo, J.-R. (2024). *A survey of vision-based methods for surface defects' detection and classification in steel products*. Paper presented at the Informatics.

Ibrahim, A. A. M., & Tapamo, J. R. (2024). Transfer learning-based approach using new convolutional neural network classifier for steel surface defects classification. *Scientific African*, *23*, e02066.

Ienco, D., Ratle, F., & Moser, G. (2017). Land cover classification via multitemporal spatial data by deep recurrent neural networks. *IEEE Geoscience and Remote Sensing Letters*, *14*(10), 1685–1689. DOI: 10.1109/LGRS.2017.2749278

Imran, M. M. H., Jamaludin, S., Ayob, A. F. M., Ali, A. A. I. M., Ahmad, S. Z. A. S., Akhbar, M. F. A., & Mohamed, S. B. (2023). Application of artificial intelligence in marine corrosion prediction and detection. *Journal of Marine Science and Engineering*, *11*(2), 256.

Interdonato, R., Ienco, D., Gaetano, R., & Ose, K. (2019). DuPLO: A DUal view Point deep Learning architecture for time series classification. *ISPRS Journal of Photogrammetry and Remote Sensing*, *149*, 91–104. DOI: 10.1016/j.isprsjprs.2018.12.008

Jaffé, R.. (2015). Remote sensing and blue carbon: A comprehensive approach to estimating carbon fluxes in estuarine ecosystems. *Remote Sensing*, *7*(9), 11115–11132.

Jahan, F. (2025). *A Systematic Review of Blue Carbon Potential in Coastal Marshlands: Opportunities for Climate Change Mitigation and Ecosystem Resilience. (Systematic review across many salt marsh and blue carbon studies).*

Jahanshahi, M. R., & Masri, S. F. (2013). Parametric performance evaluation of wavelet-based corrosion detection algorithms for condition assessment of civil infrastructure systems. *Journal of Computing in Civil Engineering*, *27*(4), 345–357.

James, S. D.. (2019). Tracking mangrove forest health using UAV-based LiDAR for long-term environmental monitoring. *Journal of Coastal Research*, *35*(6), 1391–1402.

Jaque-Reyes, J., Reyes-Wagner, V., Comte, D., Peña, M., Calle-Gardella, D., León-Ríos, S., Roecker, S. W., Jiménez, F., & Pichott, S. (2025). Subsurface Architecture of the Tuina Prospect and Its Relationship to Fluid Migration in Mineral Deposit Formation. *Research Square*.

Jaywant, S. A.. (2024). Remote Sensing Techniques for Water Quality Monitoring. [journal]. *Water (Basel)*. PMID: 39771777

Jensen, J. R.. (2017). Airborne LiDAR technology for environmental mapping: Enhancements and future applications. *International Journal of Remote Sensing*, *38*(15), 4649–4673.

Jensen, J.. (2019). UAV and LiDAR integration for high-resolution mapping of mangrove canopy structure. *Yaogan Xuebao*, *38*(12), 1032–1044.

Jensen, J., Saalfeld, A., Broome, F., Cowen, D., Price, K., Ramsey, D., Lapine, L., & Usery, E. L. (2004). Spatial data acquisition and integration. In *A research agenda for geographic information science* (pp. 17–60). CRC Press.

Jiao, P., Ye, X., Zhang, C., Li, W., & Wang, H. (2024). Vision-based real-time marine and offshore structural health monitoring system using underwater robots. *Computer-Aided Civil and Infrastructure Engineering*, *39*(2), 281–299.

Jones, C.. (2020). UAV-based LiDAR and its integration into global monitoring of mangrove ecosystems and carbon sequestration. *The Science of the Total Environment*, *747*, 141227.

Joshi, N. V., Selvakumar, P., & Srivastava, N. Manjeet, Das, A., & Manjunath, T. C. (2026). Bioethics Legal Implications of Artificial Intelligence in Medicine. In S. Anastasiadou, L. Seremeti, & S. Papalexandris (Eds.), Bioethics and Legal Aspects of AI in Healthcare and Life Sciences (pp. 27-54). IGI Global Scientific Publishing. https://doi.org/DOI: 10.4018/979-8-3373-5646-4.ch002

Jothimani, M., & Mulualem, W. (2020). Trend analysis of normalized difference vegetation index using Landsat satellite data: A study in and around Gondar town, North West Ethiopia. *Conference paper / institutional report*. ResearchGate.

Kalnay, E.. (2007). Data assimilation: A review. *Reviews of Geophysics*, *45*(2), RG2001.

Karang, I. W. G. A., Pravitha, N. L. P. R., Nuarsa, I. W., Basheer Ahammed, K. K., & Wicaksono, P. (2024). High-resolution seagrass species mapping and propeller scars detection in Tanjung Benoa, Bali through UAV imagery. *Journal of Ecological Engineering*, *25*(1).

Kauffman, J. B.. (2015). Carbon stocks of coastal wetlands: A global synthesis. *Global Biogeochemical Cycles*, *29*(5), 507–522.

Kauffman, J. B.. (2017). Carbon stocks of mangrove forests in the Philippines and Indonesia. *Environmental Science & Technology*, *51*(19), 11528–11534.

Kauffman, J. B.. (2020). Biomass estimation and carbon stock calculation using UAV-based LiDAR: Implications for mangrove restoration efforts. *Global Change Biology*, *26*(2), 539–549. PMID: 31505097

Kauffman, J. B., & Donato, D. C. (2012). The importance of mangroves in the global carbon cycle. *Biogeosciences*, *9*(6), 2955–2967.

Kearney, M. S., Stutzer, D., & Turpie, K. (2009). The effects of tidal inundation on the reflectance characteristics of coastal marsh vegetation. *Journal of Coastal Research*, *25*(6), 1177–1186.

Keenan, T. F., Davidson, E., Moffat, A. M., Munger, W., & Richardson, A. D. (2012). Using model-data fusion to interpret past trends, and quantify uncertainties in future projections, of terrestrial ecosystem carbon cycling. *Global Change Biology*, *18*(8), 2555–2569.

Keerthana, N., Srija, V., Devi, R. P., Roopini, B., & Sharumathi, J. (2025). *Underwater Corrosive Material Identification and Image Enhancement.* Paper presented at the 2025 International Conference on Visual Analytics and Data Visualization (ICVADV).

Ke, L., Yin, C., Lei, N., Zhang, S., Lu, Y., Zhang, G., Liu, D., & Wang, Q. (2025). An Improved Method for Estimating Blue Carbon Storage in Coastal Salt Marsh Wetlands: Considering the Heterogeneity of Soil Thickness. *Land (Basel)*, *14*(4), 776.

Keynote Paper. (1972). Soil preservation. In *The environmental future: Proceedings of the first International Conference on Environmental Future, held in Finland from 27 June to 3 July 1971*. Palgrave Macmillan UK.

Khan, A. R.. (2020). LiDAR-based UAV surveys for carbon estimation in mangrove ecosystems. *Scientific Reports*, *10*(1), 17285. PMID: 33057141

Khan, A., Ali, S. S. A., Anwer, A., Adil, S. H., & Mériaudeau, F. (2018). Subsea pipeline corrosion estimation by restoring and enhancing degraded underwater images. *IEEE Access : Practical Innovations, Open Solutions*, *6*, 40585–40601.

Khan, A., Rehman, K., & Ali, A. (2023). Application of geophysical and geochemical techniques for prospection of sulfide ore minerals in Drosh-Kaldam Gol area, Chitral, Northern Pakistan. *Research Square*. https://doi.org/DOI: 10.21203/rs.3.rs-2792874/v1

Khosravi, V., Ardejani, F. D., Yousefi, S., & Aryafar, A. (2018). Monitoring soil Pb and Zn via spectroscopy + ELM. *Geoderma*.

Klimešová, J., Martínková, J., & Ottaviani, G. (2018). Belowground plant functional ecology: Towards an integrated perspective. *Functional Ecology*, *32*(9), 2115–2126.

Koelmans, A. A.. (2017). Microplastic as a vector of chemical pollutants in aquatic environments. *Environmental Science & Technology*.

Kouassi, K. J.-M., Lachassagne, P., Mangoua, O. M. J., Sombo, A. P., & Dibi, B. (2024). Identifying the origin of springs in weathered-fractured crystalline aquifers using a hydrogeophysical approach. *Scientific Reports*, *14*(1). Advance online publication. DOI: 10.1038/s41598-024-52741-y PMID: 38839841

Kuang, Y. (2024). *Application of Hyperspectral Remote Sensing in Identification of Soil Heavy Metal Elements*. Sci. J. Technol.

Kumar, S., Chatterjee, U., David Raj, A., & Sooryamol, K. R. (2024). Global warming and climate crisis/extreme events. In *Climate crisis: Adaptive approaches and sustainability* (pp. 3–18). Springer Nature Switzerland.

Kundu, A., Denis, D., Patel, N., & Dutta, D. (2018). Long-term trend of vegetation in the Bundelkhand region (India): An assessment through SPOT-VGT NDVI datasets. In *Vegetation dynamics and climate change* (pp. xx–xx). IGI Global.

Laake, A., & Strobbla, C. (2012). Multi-measurement integration for near-surface geological characterization. *Near Surface Geophysics, 10*(6), 591–601. DOI: 10.3997/1873-0604.2012008

Lamine, S. (2019). *Heavy Metal Soil Contamination Detection Using Combined Hyperspectral Methods*. PLoS One.

Längkvist, M., Kiselev, A., Alirezaie, M., & Loutfi, A. (2016). Classification and segmentation of satellite-based earth observation data: A review. *IEEE Geoscience and Remote Sensing Magazine, 4*(2), 41–54. DOI: 10.1109/MGRS.2016.2527748

Leal, M. E.. (2020). Integrating remote sensing with data assimilation for improved coastal ecosystem management. *Marine Environmental Research, 163*, 105191.

Lefebvre, L.. (2017). Evaluating the use of remote sensing and data assimilation for blue carbon ecosystems management. *Marine Ecology Progress Series, 567*, 21–35.

Leone, A. P., Viscarra-Rossel, A., & Amenta, R. (2012). Predicting soil properties with PLSR and vis-NIR. *Current Analytical Chemistry*.

Li, J.. (2019). Microplastic pollution in the surface waters of coastal mangrove ecosystems. *Environmental Pollution*.

Liu, X., et al. (2022). Ecological interception effect of mangroves on microplastics. *Journal of Hazardous Materials*. ScienceDirect

Liu, J.. (2019). Enhanced feature-based Vision Patch Transformer Network for accurate brain tumor segmentation in magnetic resonance imaging. *IEEE Journal of Biomedical and Health Informatics*. PMID: 40030969

Liu, J., & Bhatti, U. A.. (2019). Hybrid approach to forecast the COVID-19 epidemic trend. *PLoS One*.

Liu, J., Zhang, Y., & Bhatti, U. A (2020). Integration of UAV-based LiDAR and multispectral imaging for biodiversity and ecosystem services mapping. *Journal of Environmental Science and Technology, 9*(6), 2321–2330.

Liu, W., & Yu, Q. (2021). Hyperspectral inversion of heavy metal via vegetation influence. *Forests*.

Liu, X.. (2016). *LiDAR remote sensing and applications*. CRC Press.

Liu, X., Lang, Q., Zhang, J., Zhang, Y., Yu, C., Jin, Q., Liu, Y., Wan, Z., & Liu, P. F. (2025). Dynamic failure mechanisms and hazard evaluation of rock collapse induced by extreme rainfall in Changbai County highways. *Scientific Reports*, *15*(1). PMID: 40119007

Liu, Y.. (2024). UAV-based remote sensing for ecological monitoring in coastal environments: A review. *Journal of Coastal Research*, *106*(4), 1254–1265.

Liu, Y., Yu, Z., Yin, Y., & Wen, J. (2025). Comparative Evaluation of Presented Strength Criteria of Anisotropic Rocks Based on Triaxial Experiments. *Applied Sciences (Basel, Switzerland)*, *15*(10), 5338.

Liu, Z. (2019). *Estimation of soil heavy metal content using hyperspectral data*. Remote Sens. *Nature*.

Liu, Z.. (2024). *Remote sensing-enhanced transfer learning approach for agricultural damage and change detection: A deep learning perspective*. Big Data Research.

Li, X.. (2021). Monitoring changes in mangrove canopy health with UAV-based LiDAR technology. *Environmental Monitoring and Assessment*, *193*(1), 23. PMID: 33389226

Li, Y., Bai, Z., Xing, L., Zhang, Q., Ding, S., Zhang, Y., & Xu, D. (2024). Corrosion monitoring techniques in subcritical and supercritical water environments. *Applied Sciences (Basel, Switzerland)*, *14*(6), 2350.

Li, Y., Yang, K., Gao, W., Han, Q., & Zhang, J. (2021). Spectral methods for heavy metal pollution detection in crops. *Spectrochimica Acta. Part A: Molecular Spectroscopy*.

Li, Z., Gao, Y., Bhatti, U. A., & Zhang, K. (2023). A hybrid UAV-based LiDAR and multispectral data fusion for real-time carbon stock estimation in mangrove forests. *Environmental Research Letters*, *18*(5), 054012.

López-Barrera, F.. (2015). Mapping mangrove canopy structure using UAVs and LiDAR. *Forests*, *6*(2), 563–580.

Lorenzo-Navarro, J.. (2025). Deep learning for microplastic segmentation in coastal environments. *Preprints*. *Preprints*

Lovynska, V., Bayat, B., & Bol, R. (2024). *Monitoring Heavy Metals and Metalloids in Soils and Vegetation by Remote Sensing: A Review*. Remote Sens.

Lu, B., Dao, P. D., Liu, J., He, Y., & Shang, J. (2020). Recent advances of hyperspectral imaging technology and applications in agriculture. *Remote Sensing*, *12*(16), 2659.

MacFarlane, G. R.. (2007). Plastic debris in coastal sediments and effects on benthic species. *Marine Pollution Bulletin*.

Macreadie, P. I., Hughes, A. R., & Kimbro, D. L. (2013). Loss of 'blue carbon' from coastal salt marshes following habitat disturbance. *PLoS One*, *8*(7), e69244.

Macreadie, P. I., Hughes, A. R., & Kimbro, D. L. (2013). Loss of blue carbon from coastal salt marshes following habitat disturbance. *PLoS One*. PMID: 23861964

Maggiori, E., Tarabalka, Y., Charpiat, G., & Alliez, P. (2017). High-resolution land cover mapping with convolutional neural networks. *IEEE Transactions on Geoscience and Remote Sensing*, *55*(2), 881–893. DOI: 10.1109/TGRS.2016.2611803

Maghsodian, S.. (2021). Microplastics in mangrove sediments of the Persian Gulf. *Marine Pollution Bulletin*.

Maju-Oyovwikowhe, E. G., Aigbadon, G. O., & Egharevba, A. (2024). Integrated Geological Analysis of the AI Well, Niger Delta Basin: Insights into Reservoir Characterization and Hydrocarbon Potential. *Research Square*.

Maliki, A. A., Owens, G., & Bruce, D. (2012). *Capabilities of remote sensing hyperspectral imagery for detection of Pb contamination*. ISPRS Ann.

Martinetto, P. (2023). *The blue carbon of southern southwest Atlantic salt marshes*. PMC.

Martins, M. (2025). *First assessment of blue carbon stocks and sequestration rates in Western Africa seagrass beds. (Provides context for blue carbon ecosystem comparisons).*

Ma, S., Du, Y., Wang, S., & Su, Y. (2023). Application of machine learning in material corrosion research. *Corrosion Reviews*, *41*(4), 417–426.

Mason, V.G. et al. (2023). *Blue carbon benefits from global saltmarsh restoration..*

Maxwell, T. L.. (2023). Global dataset of soil organic carbon in tidal marshes. *Scientific Data*. PMID: 37952023

McLeod, E.. (2014). Carbon sequestration in coastal ecosystems and the role of remote sensing. *Ecological Applications*, *24*(6), 1591–1606.

McLeod, E.. (2016). The role of coastal ecosystems in the regulation of greenhouse gases. *Nature Climate Change*, *6*(11), 1061–1070.

Mcleod, E., Chmura, G. L., Bouillon, S., Salm, R., Björk, M., Duarte, C. M., & Silliman, B. R. (2011). A blueprint for blue carbon: Toward an improved understanding of the role of vegetated coastal habitats in sequestering CO2. *Frontiers in Ecology and the Environment*, *9*(10), 552–560.

McRoberts, R. E.. (2020). Using UAV-based LiDAR to assess the structural characteristics of mangrove ecosystems. *Remote Sensing Applications: Society and Environment*, *19*(2), 12–21.

Mehmood, T.. (2025). *Mangroves in Pakistan at risk from microplastic threats. Environmental Science and Pollution Research.* Springer Link.

Melgani, F., & Bruzzone, L. (2004). Classification of hyperspectral remote sensing images with SVM. *IEEE Transactions on Geoscience and Remote Sensing*.

Mendes, D. S.. (2022). Current status and trends of research on microplastic detection in mangrove wetlands. *Frontiers in Environmental Science. Frontiers.*

Mendes, D. S., Beasley, C. R., Silva, D. N. N., & Fernandes, M. E. B. (2023). Microplastic in mangroves: A worldwide review of contamination in biotic and abiotic matrices. *Marine Pollution Bulletin*, *195*, 115552. PMID: 37738877

Metcalfe, D. J.. (2008). Seagrass meadows as carbon sinks in coastal ecosystems. *Global Environmental Change*, *18*(4), 581–591.

Mngadi, S., Manzi, M., Nkosi, N. Z., Durrheim, R., Ogasawara, H.Jr, Yabe, Y., & Team, D. R. E. A. M. (2025). Identification of structures capable of hosting the ML 5.5 Orkney South Africa earthquake and factors controlling the physics and mechanics of dynamic rupture. *Journal of the Southern African Institute of Mining and Metallurgy*, *125*(1).

Mohan, S., Kumar, B., & Nejadhashemi, A. P. (2025). Integration of Machine Learning and Remote Sensing for Water Quality Monitoring and Prediction: A Review. *Sustainability*.

Möller, I.. (2014). The role of coastal wetlands in carbon sequestration. *Nature Geoscience*, *7*, 1–3.

Munawar, H. S., Ullah, F., Shahzad, D., Heravi, A., Qayyum, S., & Akram, J. (2022). Civil infrastructure damage and corrosion detection: An application of machine learning. *Buildings (Basel, Switzerland)*, *12*(2), 156.

Murong, R. (2025). *Hyperspectral Technology in Agricultural Soil Heavy Metal Detection: Current Applications and Future Directions*. Front. Bus. Econ. Manag.

Murphy, P. N. C., Ogilvie, J., & Arp, P. (2009). Topographic modelling of soil moisture conditions: A comparison and verification of two models. *European Journal of Soil Science*, *60*(1), 94–109.

Murray, B. C.. (2021). Carbon stock assessment and UAV LiDAR data in mangrove restoration monitoring. *Global Change Biology*, *27*(9), 3452–3465.

Najafi, Z., Fatehi, P., & Darvishsefat, A. (2019). Vegetation dynamics trend using satellite time series imagery. *The International Archives of the Photogrammetry, Remote Sensing and Spatial Information Sciences*, *XLII-2*(W13), 163–170. DOI: 10.5194/isprs-archives-XLII-2-W13-163-2019

Nawaz, S. A., Li, J., Bhatti, U. A., & Shoukat, M. U. (2021). Deep learning applications in hyperspectral image classification for environmental monitoring. *Remote Sensing Applications: Society and Environment*, *21*, 100387.

Nawaz, S. A., Li, J., Bhatti, U. A., Shoukat, M. U., & Ahmad, R. M. (2023). AI-based object detection latest trends in remote sensing, multimedia and agriculture applications. *Frontiers in Plant Science*, *14*, 4273.

Nawaz, S. A., Li, J., Bhatti, U. A., Shoukat, M. U., & Ahmad, R. M. (2023). AI-based object detection: Latest trends in remote sensing, multimedia, and agriculture applications. *Frontiers in Plant Science*, •••, 4273. PMID: 37082514

Nguyen, A., Gharehbaghi, V., Le, N. T., Sterling, L., Chaudhry, U. I., & Crawford, S. (2023). *ASR crack identification in bridges using deep learning and texture analysis.* Paper presented at the Structures.

Niemeyer, J.. (2015). UAV-based LiDAR data processing and analysis for ecological applications. *ISPRS Journal of Photogrammetry and Remote Sensing*, *106*, 204–217.

Nivedita, V.. (2023). *Plastic debris detection along coastal waters using satellite imagery. Marine Pollution Bulletin*. PubMed.

Nizamani, A. H., Chen, Z., Nizamani, A. A., & Bhatti, U. A. (2023). Advance brain tumor segmentation using feature fusion methods with deep U-Net model with CNN for MRI data. *Journal of King Saud University. Computer and Information Sciences*, *35*(9), 101793.

Njikam, M. N. M., Yem, M., Ribodetti, A., Mohamed, A., Soumah, A., Prosper, M., Alarifi, S. S., & Abdelrady, A. (2023). Application of the electrical resistivity method and the estimation of limestone volume: A case study. *Frontiers of Earth Science*, •••, 11.

Nogueira, P.. (2024). Precision Mapping of Seagrass Habitats Using Multispectral Drones and Machine Learning. *Preprints*.

Nwankwo, C., Tse, A. C., Nwankwoala, H. O., Giadom, F. D., & Acra, E. J. (2023). Below ground carbon stock and carbon sequestration potentials of mangrove sediments in Eastern Niger Delta, Nigeria: Implication for climate change. *Scientific African*, *22*, e01898.

Nweke, N. D., Ugwuonah, E. N., Onuba, L. N., & Egbueri, J. C. (2024). The occurrences, characteristics, and implications of chaotic rocks (accretionary rocks) within the Trans-Saharan orogenic belt in Nigeria. *Discover Geoscience*, *2*(1).

Nykänen, V., Törmänen, T., & Niiranen, T. (2023). Cobalt Prospectivity Using a Conceptual Fuzzy Logic Overlay Method Enhanced with the Mineral Systems Approach. *Natural Resources Research*, *32*(6), 2387–2409.

O'Connor, M. I.. (2017). Estimating carbon flux in coastal wetlands through data assimilation. *Climatic Change*, *146*(1), 39–54.

O'Leary, R. A.. (2017). *Climate change and the importance of blue carbon*. Ecology and Conservation of Blue Carbon Ecosystems.

Ogunkoya, A. A., Martin-Hayden, J. M., Fisher, T. G., & Doro, K. O. (2023). Combining geophysical and geological datasets to improve a conceptual aquifer model of the Oak Openings Region in northwest Ohio, USA. *Environmental Earth Sciences*, *82*(21).

Ohwo, M. U. J., & Solape, F. S. (2024). Integration of ASTER and Airborne Radiometric Data in the Exploration for Hydrothermal Alteration Zones Associated with Mineral Deposits in Igarra Schist Belt, Nigeria. Dergipark (Istanbul University).

Okunowo, O. O., Adeogun, O. Y., Ishola, K., & Alli, S. (2020). Delineation of leachate at a dumpsite using geo-electrical resistivity method: A case study of Abule Egba, Lagos, Nigeria. *SN Applied Sciences*, *2*(12).

Omolaiye, G. E., & Oniyangi, A. K., ISSA, T. A., & Adam, S. B. (2024). Subsurface investigation for pre-foundation study utilizing integrated geophysical and geotechnical methods. *Discover Geoscience*, *2*(1). Advance online publication. DOI: 10.1007/s44246-024-00109-6

Ordóñez, O., Ekwok, S. E., Alkhayaat, A., Bains, P. S., Sharma, R., Kumar, R., Kulshreshtra, K., Mann, V. S., & Elmasry, Y. (2024). Delineation of mineralization-related geologic structures and lithological units using airborne magnetic and radiometric data. *Geomechanics and Geophysics for Geo-energy and Geo-resources*, *10*(1).

Osman, M. S.. (2018). LiDAR-based estimation of carbon stock in coastal ecosystems: A study on mangrove biomass. *Environmental Monitoring and Assessment*, *190*(1), 25. PMID: 30569332

Ossai, O., Cavalleri, C., Shrivastva, C., & Claverie, M. (2014, January 1). Deciphering the impact of Subsurface Geology on Reservoir Quality Indicators: Petrophysical Evaluation from Niger Delta. *International Petroleum Technology Conference*.

Osumeje, J. O., Eshimiakhe, D., Oniku, A. S., & Lawal, K. M. (2023). Application of Remote Sensing and Geophysical Methods for Delineating Groundwater Potential at North Western Nigeria. *Research Square*. https://doi.org/DOI: 10.21203/rs.3.rs-3508600/v1

Padarian, J., Minasny, B., & McBratney, A. (2019). Hyperspectral deep learning soil prediction. *Geoderma Regional*.

Padarian, J., Minasny, B., & McBratney, A. B. (2019). Deep learning to predict soil properties from spectral data. *Geoderma Regional*.

Paira, D., Rohini, A., Chandra, S., Satyanarayana, P., Sharma, M., & T. C., M. (2026). Climate Change and Natural Disasters. In Q. Ali & M. Tariq Iqbal Khan (Eds.), Strengthening Global Resilience to Natural Disasters (pp. 21-44). IGI Global Scientific Publishing. https://doi.org/.DOI: 10.4018/979-8-3693-9745-9.ch002

Paraschos, D., & Papadakis, N. K. (2024). Autonomous underwater vehicle challenge: Design and construction of a medium-sized, AI-enabled low-cost prototype. *The Journal of Defense Modeling and Simulation*, *21*(3), 269–281.

Parrish, D.. (2019). UAV-based LiDAR for real-time monitoring of mangrove forest growth and climate change impacts. *Environmental Science & Policy*, *101*, 68–76.

Parrish, T. R.. (2016). Carbon storage in mangrove forests: A method for estimating stock using UAV-based LiDAR. *Carbon Management*, *7*(3), 115–129.

Pascal, O. S., Mallam, A., & Osagie, A. U., & O., A. T. (2024). THE INTEGRATED GEOPHYSICAL STUDY FOR DELINEATING CU-AU SULFIDE DEPOSITS IN RUWAN GORA ANKA AREA, NORTHWESTERN, NIGERIA. *Earth Sciences Pakistan*, *8*(1), 70–76.

Patel, K.. (2023). Monitoring anthropogenic pressures on seagrass ecosystems using UAVs and remote sensing. *Marine Pollution Bulletin*, *191*, 112671.

Patrick, H. A. P. A., Mansor, H. E., Khan, M. M. A., & Shah, Z. A. (2023). General geology and groundwater quality analysis of domestic wells around Kampung Panglima Bayu, Tanah Merah District, Kelantan, Malaysia. BIO Web of Conferences, 73, 04012.

Pendleton, L.. (2012). Estimates for global blue carbon storage and its potential for climate change mitigation. *PLoS One*, *7*(9), e43942. PMID: 22962585

Pendleton, L.. (2012). Estimates for global blue carbon. *Nature Climate Change*, *2*, 40–46.

Persello, C., & Bruzzone, L. (2013). Active learning for domain adaptation in remote sensing image classification. *IEEE Transactions on Geoscience and Remote Sensing*, *51*(8), 4817–4832. DOI: 10.1109/TGRS.2013.2251503

Pham, T. D., Xia, J., Ha, N. T., Bui, D. T., Le, N. N., & Takeuchi, W. (2019). A review of remote sensing approaches for monitoring blue carbon ecosystems: Mangroves, seagrasses and salt marshes during 2010–2018. *Sensors (Basel)*, *19*(8), 1933.

Pillay, S. J.. (2024). Assessing Drone-Based Remote Sensing for Monitoring Water Quality in Small Water Bodies. *Drones (Basel)*.

Pirie, C., & Moreno-Garcia, C. F. (2021). *Image pre-processing and segmentation for real-time subsea corrosion inspection.* Paper presented at the International Conference on Engineering Applications of Neural Networks.

Poley, L. G., & McDermid, G. J. (2020). A systematic review of the factors influencing the estimation of vegetation aboveground biomass using unmanned aerial systems. *Remote Sensing*, *12*(7), 1052.

Prajesh, P. J., Kannan, B., Pazhanivelan, S., Kumaraperumal, R., & Ragunath, K. P. (2019). Analysis of seasonal vegetation dynamics using MODIS-derived NDVI and NDWI data: A case study of Tamil Nadu. *Madras Agricultural Journal*, *106*(4–6), 263–270. DOI: 10.29321/MAJ.2019.000275

Prata, J. C.. (2019). Methods for microplastic extraction and identification from environmental matrices. *Journal of Hazardous Materials*.

Pritchard, S. P.. (2018). Use of UAV-based LiDAR for mangrove forest biomass estimation: A case study in Southeast Asia. *Forests*, *9*(11), 638.

Puntu, J. M., Chang, P.-Y., Amania, H. H., Lin, D.-J., Suryantara, M. S. A., Tsai, J., Yu, H., Chang, L., Zeng, J.-R., & Kassie, L. N. (2024). A machine learning-based approach for constructing a 3D apparent geological model using multi-resistivity data. *Geosciences*.

Puppin, A. (2024). *Blue Carbon Assessment in the Salt Marshes of the Venice Lagoon and Carbon Sequestration Potential.* Wiley/AGU.

Pyo, J., Hong, S. M., & Kwon, Y. S. (2020). Deep neural network estimation of heavy metals using spectroscopy. *The Science of the Total Environment.*

Qian, W., Jiang, N., & Du, J. (2016). Anomaly-based weather analysis versus traditional total-field-based weather analysis for depicting regional heavy rain events. *Weather and Forecasting*, *31*(1), 71–93.

Qiu, C., Mou, L., Schmitt, M., & Zhu, X. X. (2019). Local climate zone-based urban land cover classification from multi-seasonal Sentinel-2 images with a recurrent residual network. *ISPRS Journal of Photogrammetry and Remote Sensing*, *154*, 151–162. DOI: 10.1016/j.isprsjprs.2019.07.016 PMID: 31417230

Radočaj, D., Šiljeg, A., Marinović, R., & Jurišić, M. (2023). State of major vegetation indices in precision agriculture studies indexed in Web of Science. *Revista de Agricultura (Piracicaba)*, *13*(4), 930. DOI: 10.3390/agriculture13040930

Raihan, A. J., Abas, P. E., & De Silva, L. C. (2021). Role of restored underwater images in underwater imaging applications. *Applied System Innovation*, *4*(4), 96.

Ramanujam, M., Rangasamy, N., Rathinam, A., & Veeraiyan, P. (2022). Using the normalized difference vegetation index (NDVI) to detect vegetative change with remote sensing and GIS: A study of the Kumbur River Basin in Kodaikanal Taluk, Dindigul District. *European Journal of Environment and Earth Sciences*, *3*(1), 18–25. DOI: 10.24018/ejgeo.2022.3.1.217

Raza, I., Ullah, H., Khalid, P., Iqbal, M., Rasheed, S., & Ehsan, M. I. (2025). Geophysical characterization for dolerite dykes (dimension stone) reserves exploration using electrical resistivity tomography (ERT) in Shergarh Oghi, Mansehra, Pakistan. *Acta Geophysica*, *73*(1), 53–66.

Reddy, A. S., & Janga Reddy, M. (2013). NDVI-based assessment of land use/land cover dynamics in a rainfed watershed using remote sensing and GIS. *International Journal of Scientific and Engineering Research*, *4*(12), 87–93.

Reynaud, T.. (2019). Modeling carbon sequestration in coastal ecosystems: A remote sensing approach. *Biogeosciences Discussions*, *16*, 3125–3142.

Ribeiro, R. P., Neto, J. A. C., Corrêa, F. A., Monticelli, J. P., Ribeiro, J. C., Monteiro, S. N., & de Azevedo, A. R. G. (2025). Characterization of Limeira intrusion stones for aggregate use in engineering applications. *Scientific Reports*, *15*(1). PMID: 40038373

Riccioli, F., Gabrielsen, Ø., Høgsæt, I. S., Barros, P. S., & Pahlavan, L. (2024). Corrosion-fatigue damage identification in submerged mooring chain links using remote acoustic emission monitoring. *Marine Structures*, *98*, 103685.

Rivera-Monroy, V. H.. (2017). Carbon sequestration and flux in mangrove forests: A global analysis. *Ecology*, *98*(5), 1176–1189.

Rochman, C. M.. (2014). The ecological impacts of plastic pollution in marine ecosystems. *Science*.

Rokach, L. (2013). Decision forest: Twenty years of research. Info. Fusion (2016). Review of hyperspectral remote sensing in soil heavy metal monitoring. *Hubei Nong Ye Ke Xue*.

Rowan, G. S. L.. (2021). A Review of Remote Sensing of Submerged Aquatic Vegetation (SAV). *Remote Sensing*.

Russwurm, M., & Körner, M. (2018). Multi-temporal land cover classification with sequential recurrent encoders. *ISPRS International Journal of Geo-Information*, *7*(4), 129. DOI: 10.3390/ijgi7040129

Saintilan, N.. (2013). Mangrove expansion and salt marsh loss in south-east Australia: Implications for salt marsh carbon storage. *Journal of Coastal Research*, *29*(4), 1059–1067.

Saintilan, N.. (2013). The role of UAV LiDAR in long-term mangrove ecosystem monitoring. *Coastal Management*, *41*(2), 181–190.

Sakamoto, T.. (2005). A comprehensive review of satellite-based monitoring of vegetation biomass. *Global Ecology and Biogeography*, *14*(4), 397–412.

Salawu, N. B. (2021). Aeromagnetic and digital elevation model constraints on the structural framework of southern margin of the Middle Niger Basin, Nigeria. *Scientific Reports*, *11*(1). PMID: 34737345

Salawu, N. B., Fatoba, J. O., Adebiyi, L. S., Orosun, M. M., & Dada, S. S. (2021). New insights on the Ife-Ilesha schist belt using integrated satellite, aeromagnetic and radiometric data. *Scientific Reports*, *11*(1). PMID: 34321551

Scheel, T.. (2020). Data assimilation in mangrove and seagrass carbon flux estimation. *Ecological Modelling*, *420*, 108955.

Schimel, D.. (2001). The role of ecosystems in global carbon cycling. *Global Biogeochemical Cycles*, *15*(2), 211–225.

Schwarz, T. A.. (2020). Remote sensing of carbon sequestration in seagrass ecosystems. *International Journal of Remote Sensing*, *41*(15), 5675–5691.

Senanayake, I. P., Kiem, A. S., Hancock, G., Metelka, V., Folkes, C. B., Blevin, P. L., & Budd, A. (2023). A Spatial Data-Driven Approach for Mineral Prospectivity Mapping. *Remote Sensing*, *15*(18), 4574.

Šestanović, S., Roda, C., & Andrić, M. (1996). Proposal for methodology to be applied to road construction in the karstified carbonate and clastic (flysch) terrains in southern Croatia. *International Journal of Rock Mechanics and Mining Sciences & Geomechanics Abstracts*, *33*(8).

Shebl, A., Abdellatif, M., Badawi, M., Dawoud, M., Fahil, A. S., & Csámer, Á. (2023). Towards better delineation of hydrothermal alterations via multi-sensor remote sensing and airborne geophysical data. *Scientific Reports*, *13*(1). Advance online publication. DOI: 10.1038/s41598-023-34575-w PMID: 37149689

Sheffield, K., & Morse-McNabb, E. (2015). Using satellite imagery to assess trends in soil and crop productivity across landscapes. *IOP Conference Series. Earth and Environmental Science*, *25*, 012013. DOI: 10.1088/1755-1315/25/1/012013

Shekaili, A. A., Liu, Y., & Papatheou, E. (2024). Drilling performance analysis of a polycrystalline diamond compact bit via finite element and experimental investigations. *International Journal of Rock Mechanics and Mining Sciences*, *182*, 105862.

Shen, H.-K., Chen, P.-H., & Chang, L.-M. (2013). Automated steel bridge coating rust defect recognition method based on color and texture feature. *Automation in Construction*, *31*, 338–356.

Shen, S.. (2023). Monitoring sediment dynamics in seagrass meadows using UAV-based remote sensing techniques. *Journal of Environmental Management*, *312*, 114844.

Shifflett, S.D. (2019). *Assessing the risk of utilizing tidal coastal wetlands in the context of pharmaceuticals and personal care products (PPCPs).*

Shi, T., Chen, Y., Liu, Y., & Wu, G. (2014). Vis-NIR reflectance spectroscopy for monitoring heavy metal contamination. *Journal of Hazardous Materials*.

Siddique, A. A., Tan, Z., Rashid, W., & Ahmad, H. (2025). Sustainable Water-Related Hazards Assessment in Open Pit-to-Underground Mining Transitions: An IDRR and MCDM Approach at Sijiaying Iron Mine, China. *Water (Basel)*, *17*(9), 1354.

Simpson, J.. (2024). Mapping fine-scale seagrass disturbance using bi-temporal UAV multispectral imagery and IR-MAD change detection. *Scientific Reports*.

Singh, A. K.. (2023). Application of UAV-based spectral data for assessing water pollution in estuaries. *Environmental Science and Pollution Research International.*

Singh, P., & Javeed, O. (2020). NDVI-based assessment of land cover changes using remote sensing and GIS: A case study of Srinagar district, Kashmir. *Sustainability, Agri. Food and Environmental Research, 8*(1), 43–54. DOI: 10.7770/safer-V8N1-art2101

Sithole, G., & Vosselman, G. (2004). Experimental comparison of LiDAR and photogrammetry for 3D tree reconstruction. *The International Archives of the Photogrammetry, Remote Sensing and Spatial Information Sciences, 35*(B1), 235–240.

Si, Z.. (2018). Satellite-based monitoring of coastal wetland carbon fluxes: Approaches and challenges. *Remote Sensing of Environment, 218*, 55–68.

Smeaton, C., Ladd, C. J. T., Miller, L. C., McMahon, L., Garrett, E., Barlow, N. L. M., Gehrels, W. R., Skov, M. W., & Austin, W. E. N. (2023). Organic carbon stocks of Great British saltmarshes. *Frontiers in Marine Science, 10*, 1229486.

Soltwedel, T.. (2019). Climate change impacts on blue carbon ecosystems: A review. *Estuarine, Coastal and Shelf Science, 235*, 104458.

Somayajula, V. K. A., Ghai, D., & Kumar, S. (2020). A review on classification of land use/land cover change assessment based on normalized difference vegetation index. *Journal of Physics: Conference Series, 1716*(1), 012014. DOI: 10.1088/1742-6596/1716/1/012014

Son, E.-Y., Jeong, D., & Oh, M.-J. (2024). Corrosion area detection and depth prediction using machine learning. *International Journal of Naval Architecture and Ocean Engineering, 16*, 100617.

Song, J., Zhu, Y., & Chen, X. (2025). Subduction Dynamics of the Paleo-Pacific Plate: New Constraints from Quartz Diorites in the Fudong Region. *Minerals (Basel), 15*(5), 562.

Strickland, R.. (2020). The role of satellite-based monitoring in assessing blue carbon. *Remote Sensing in Environmental Applications, 1*(2), 123–136.

Stumpf, R. P.. (2012). Satellite-based monitoring of seagrass biomass dynamics. *Remote Sensing of Environment, 124*, 54–67.

Su, Y. (2025). *Hyperspectral remote sensing for soil heavy metal inversion challenges.* Int. J. Digital Earth.

Sudhakar, B. J. (2025). Mangrove system acts as a major microplastic sink: Leaves and roots analysis. *SSRN. SSRN*

Sun, B., Wei, X., & Dong, H. (2025). Formation Mechanism of Plagioclase-Amphibole and Amphibole-Spinel Symplectites in the Bijigou Layered Intrusion: Insights from Mineralogical and Crystallographic Constraints. *Minerals (Basel)*, *15*(5), 433.

Sun, Z., Li, Y., Su, L., Liu, S., & Chen, Z. (2025). Predicting corrosion behaviour of steel reinforcement in eco-friendly coral aggregate concrete based on hybrid machine learning methods. *Nondestructive Testing and Evaluation*, *40*(4), 1334–1354.

Tabrizian, M.. (2018). The use of LiDAR data to estimate biomass and carbon stock in mangrove forests. *Remote Sensing of Environment*, *215*, 28–37.

Tagwai, M. G., Jimoh, O. A., Shehu, S. A., & Zabidi, H. (2023). Application of GIS and remote sensing in mineral exploration: Current and future perspectives. *World Journal of Engineering*, *21*(3), 487–497.

Tahara, S., Sudo, K., Yamakita, T., & Nakaoka, M. (2022). Species-level mapping of a seagrass bed using an unmanned aerial vehicle and deep learning technique. *PeerJ*, *10*, e14017. PMID: 36275465

Tallis, H.. (2014). Valuing ecosystem services in a blue carbon context: The case of mangroves in the Gulf of California. *Environmental Science & Policy*, *42*, 1–9.

Tan, K., Ma, W., Wu, F., & Du, Q. (2019). Random forest estimation of heavy metal in soils with hyperspectral data. *Environmental Monitoring and Assessment*.

Tao, C.. (2018). Integration of UAV and LiDAR technologies for forest monitoring. *Forests*, *9*(6), 348.

Tende, A. W., Aminu, M. D., & Gajere, J. N. (2021). A spatial analysis for geothermal energy exploration using bivariate predictive modelling. *Scientific Reports*, *11*(1). PMID: 34611246

Teng, S., Liu, A., Ye, X., Wang, J., Fu, J., Wu, Z., & Zeng, Y. (2024). Review of intelligent detection and health assessment of underwater structures. *Engineering Structures*, *308*, 117958.

Thomas, S. A.. (2017). The potential of LiDAR for carbon stock estimation in mangrove forests. *Estuarine, Coastal and Shelf Science*, *187*, 138–149.

Thompson, R. C.. (2004). Lost at sea: Where is all the plastic? *Science*. PMID: 15131299

Tian, H., Ren, Z., Qi, K., Liu, J., Guo, S., Han, Z., Yao, J., & Zhu, L. (2025). Reservoir Characteristics and Diagenetic Evolution of Lower Cretaceous In Baibei Sag, Erlian Basin, Northern China. *Processes (Basel, Switzerland)*, *13*(5), 1391.

Torregroza-Espinosa, A. C.. (2025). *Application of remote sensing for detection and assessment of microplastics. Remote Sensing Journal*. MDPI.

Tossou, Y. Y. J., Yessoufou, S., Orban, P., Auwera, J. V., Boukari, M., & Brouyère, S. (2021). Bedrock petrology controls on hydrogeochemistry and fluoride concentrations in Precambrian aquifers of central Benin, Western Africa. *Journal of African Earth Sciences*, *182*, 104301.

Tsogtbaatar, U. (2025). Hyperspectral data for ecological risk of heavy metals in paddy soil. *Environmental Monitoring and Assessment*.

Tuia, D., Verrelst, J., Alonso, L., Pérez-Cruz, F., & Camps-Valls, G. (2011). Multioutput support vector regression for remote sensing biophysical parameter estimation. *IEEE Geoscience and Remote Sensing Letters*, *8*(4), 804–808.

Turner, R. E.. (2001). The role of salt marshes in carbon sequestration and climate change mitigation. *Science*, *303*(5656), 1933–1937.

Ugbor, C. C., Ikwuagwu, I. E., & Ogboke, O. J. (2021). 2D inversion of electrical resistivity investigation of contaminant plume around a dumpsite near Onitsha expressway in southeastern Nigeria. *Scientific Reports*, *11*(1). PMID: 34088926

Ullah, F. (2025). Assessment of Petrographic and Physico-Mechanical Properties of Granodiorites for Aggregate Usage: A Comprehensive Case Study from Mayar, Dir (Lower), Khyber Pakhtunkhwa Pakistan. *Acta Montanistica Slovaca*, *30*(2).

Ullah, J., Li, H., Ashraf, U., Ehsan, M., & Asad, M. (2023). A multidisciplinary approach to facies evaluation at regional level using well log analysis, machine learning, and statistical methods. *Geomechanics and Geophysics for Geo-energy and Geo-resources*, *9*(1).

Ullman, J. et al. (n.d.). *Foundational research on blue carbon ecosystems and global drivers of carbon sequestration in tidal wetlands.*

Usman, A. O., Nomeh, J. S., Abraham, E. M., & Azuoko, G.-B. (2025). Uncovering the hidden mineral treasure of the Jurassic Younger Granite region of Nigeria using integrated magnetic geophysical techniques. *Frontiers of Earth Science*, •••, 13.

Van de Broek, M. (2018). Long-term organic carbon sequestration in tidal marsh sediments. *Global Change Biology*. PMID: 29431887

van Duynhoven, A., & Dragićević, S. (2019). Analyzing the effects of temporal resolution and classification confidence for modeling land cover change with long short-term memory networks. *Remote Sensing*, *11*(22), 2784. DOI: 10.3390/rs11222784

Veettil, B. K., & Sabu, T. (2022). *Coastal and marine plastic litter monitoring using remote sensing. Remote Sensing Applications: Society and Environment.* ScienceDirect.

Ventura, D.. (2023). Coastal benthic habitat mapping and monitoring by integrated UAV-USV and photogrammetric acquisition. *Frontiers in Marine Science.*

Verma, G., Mehta, A., & Goswami, S. (2022). Trend and time series analysis of vegetation dynamics using satellite data: A case study of Uttarakhand, India. *Current Journal of Applied Science and Technology*, *41*(6), 32–46. DOI: 10.9734/cjast/2022/v41i630670

Victor, N. O. C., & Ibe, K. K. (2022). Geology and Structure of Uru-Ugworji Diorite Lokpaukwu, Southeastern Nigeria. *Research Square*. https://doi.org/DOI: 10.21203/rs.3.rs-1721843/v1

Vorobel, R., Ivasenko, I., Berehulyak, O., & Mandzii, T. (2021). Segmentation of rust defects on painted steel surfaces by intelligent image analysis. *Automation in Construction*, *123*, 103515.

Voss, A.. (2020). Advancements in hyperspectral imaging for vegetation health monitoring. *Remote Sensing Letters*, *11*(3), 201–210.

Wagner, F., & Uhlemann, S. (2021). An overview of multimethod imaging approaches in environmental geophysics. In *Advances in geophysics* (p. 1). Elsevier BV., DOI: 10.1016/bs.agph.2021.06.001

Wang, H. (2025). *Rapid detection of soil heavy metal pollution using deep learning & hyperspectral remote sensing*. Sci. Rep.

Wang, Y. (2024). *Monitoring of soil heavy metals based on hyperspectral remote sensing*. Sci. Total Environ.

Wang, J.. (2021). UAV-based LiDAR applications for large-scale monitoring of mangrove carbon sequestration. *Remote Sensing*, *13*(10), 1957.

Wang, J., Cui, X., Jin, G., Zhao, Y., Wen, X., & Zhang, Y. (2023). Effect of in-situ Ni interlayer on the microstructure and corrosion resistance of underwater wet 316L stainless steel laser cladding layer. *Surface and Coatings Technology*, *458*, 129341.

Wang, L.. (2019). Mapping mangrove forests using UAV-based LiDAR for ecosystem health and biomass estimation. *Remote Sensing of Environment*, *228*, 113–124.

Wang, L.. (2024). UAVs for environmental health monitoring in seagrass and mangrove ecosystems: A comparative study. *Journal of Environmental Science and Technology*, *58*(4), 2352–2364.

Wang, W.. (2022). microplastics in mangrove sediments and their potential effects on carbon cycling. *The Science of the Total Environment.*

Wang, W., Xue, C., Zhao, J., Yuan, C., & Tang, J. (2024). Machine learning-based field geological mapping: A new exploration of geological survey data acquisition strategy. *Dizhi Lunping, 166*, 105959.

Wang, Y.. (2019). Modeling blue carbon fluxes in coastal ecosystems using remote sensing and data assimilation techniques. *Environmental Modelling & Software, 116*, 31–44.

Wang, Y.. (2024). Monitoring coastal ecosystems using remote sensing and machine learning techniques: A case study on mangrove forests. *Environmental Monitoring and Assessment.*

Wang, Y., Zhang, X., Zhang, Y., & Bhatti, U. A. (2022). The role of LiDAR and UAV technology in forest biomass and carbon estimation. *Journal of Climate Change and Sustainability, 12*(3), 110–125.

Wang, Z., Lian, J., Liu, H., Liang, C., Zou, K., Chen, L., & Yao, Y. (2024). Review on Repair Technologies for Underwater Concrete Structure Damage of Infrastructures. *Water (Basel), 17*(1), 35.

Wei, L. (2020). *Estimation of soil As content using hyperspectral remote sensing.* PMC article.

Weydt, L., Ramírez-Guzmán, Á. A., Pola, A., Lepillier, B., Kummerow, J., Mandrone, G., Comina, C., Deb, P., Norini, G., González-Partida, E., Avellán, D. R., Macías, J. L., & Bär, K., & S., S. (2021). Petrophysical and mechanical rock property database of the Los Humeros and Acoculco geothermal fields (Mexico). *Earth System Science Data, 13*(9), 571–591.

Wijaya, B. R., Affandy, N. A., & Nabilah, S. (2024). Pemantauan dampak kekeringan lahan pertanian di Lamongan menggunakan NDVI berbasis SIG. *Nucleus Journal, 3*(2), 57–67. DOI: 10.32492/nucleus.v3i2.3201

Wouters, S., Hulsbosch, N., Kaskes, P., Claeys, P., Dewaele, S., Melcher, F., Onuk, P., & Muchez, P. (2020). Late orogenic gold mineralization in the western domain of the Karagwe-Ankole Belt (Central Africa): Auriferous quartz veins from the Byumba deposit (Rwanda). *Ore Geology Reviews, 125*, 103666.

Wu, Z., et al. (2025). *A Review of Remote Sensing-based Water Quality Monitoring: Trends and Future Directions.* Journal of Water.

Wu, F., Zhang, C., Li, X., Liu, S., Wang, J., & Huang, W. (2025). Distribution, Diversity, and Ecological Risks of Microplastics in Mangrove Ecosystems of a Southeastern Chinese Estuary. *Toxics*, *13*(6), 494. PMID: 40559967

Wu, Y., Zhang, L., & Bhatti, U. A. (2023). Interpretable machine learning for personalized medical recommendations: A LIME-based approach. *Diagnostics (Basel)*, *13*(16), 2681. PMID: 37627940

Xue, Y.. (2024). Remote sensing for detecting the impacts of microplastics on coastal ecosystems. *Marine Pollution Bulletin*.

Xu, Y., Yang, Y., Chen, X., & Liu, Y. (2022). Bibliometric analysis of global NDVI research trends from 1985 to 2021. *Remote Sensing*, *14*(17), 4340. DOI: 10.3390/rs14174340

Yang, H., Xu, H., & Zhong, X. (2022). Prediction of soil heavy metal in copper tailings area using hyperspectral reflectance. *Environmental Earth Sciences*.

Yang, X., Lei, S., Zhao, Y., & Cheng, W. (2022). Hyperspectral detection of heavy metal polluted vegetation. *Geocarto International*.

Yaseen, Z. M. (2021). Machine learning models for simulating heavy metals in soil and water. *Chemosphere*.

Ying, H.. (2021). Evaluation of water quality based on UAV multispectral images and the MPP algorithm. *Environmental Science and Pollution Research International*.

Yin, H., Udelhoven, T., Fensholt, R., Pflugmacher, D., & Hostert, P. (2012). How normalized difference vegetation index (NDVI) trends from AVHRR and systematic sensor differences affect trend analysis. *Remote Sensing*, *4*(12), 3360–3394. DOI: 10.3390/rs4113360

Yossa, M. T., Lordon, A. E. D., Agyingi, C. M., Agbor-Taku, J., Shandini, Y., & Bessong, C. E. (2022). Remote sensing and geographic information system (GIS)-based high-resolution mapping of potential groundwater recharge zones on the hard rock terrains of the Cameroon volcanic line (CVL). *SN Applied Sciences*, *5*(1).

Yuan, X. a., Ding, J., Cai, B., Li, W., Yin, X., Li, X., & Zhao, J. (2024). 3D visual reconstruction of corrosion in underwater structure using alternating current field measurement technique. *Ocean Engineering*, *308*, 118294.

Yu, Z., Zeeshan, Z., & Bhatti, U. A. (2020). Advanced techniques in UAV-based LiDAR data processing for environmental monitoring. *Journal of Environmental Monitoring*, *9*(6), 350–368.

Yu, Z., Zeeshan, Z., Bhatti, U. A., & Nizamani, M. M. (2022). Satellite and UAV-based LiDAR integration for monitoring coastal mangrove health and carbon sequestration. *International Journal of Remote Sensing*, *43*(8), 3027–3054.

Zakaria, M. T. (2023). Integrated analysis of geophysical methods for landslide characterization. *Acta Geodynamica et Geomaterialia*, *20*(2), 165. DOI: 10.13168/agg.2023.0013

Zarfl, C.. (2011). Microplastics in the environment—A review of the methods used for identification and quantification. *Environmental Chemistry*.

Zeeshan, Z., Bhatti, U. A., & Nizamani, M. M. (2020). LiDAR-based UAV monitoring for coastal mangrove ecosystems and climate change mitigation. *Environmental Monitoring and Assessment*, *192*(8), 552. PMID: 32737629

Zhang, Z. H., Guo, F., & Xu, Z. (2022). *Hyperspectral inversion of Cr and Zn in arable soils*. Ecol. Ind.

Zhang, C., Zhang, L., & Du, B. (2021). Deep learning for remote sensing image classification: A survey. *Journal of Sensors*. Advance online publication. DOI: 10.1155/2021/6628623

Zhang, H. (2017). Transport of microplastics in coastal waters: Interactions with tidal and current dynamics. *Marine Environmental Research*.

Zhang, J. (2024). Chronosequence Changes of Soil Organic Carbon in Salt Marshes. *Sustainability*.

Zhang, M.. (2022). Predicting coastal water quality using UAV and machine learning models: A review. *Journal of Marine Systems*, *228*, 103723.

Zhang, Q., Zhang, H., Liu, W., & Zhao, S. (2017). Hyperspectral inversion of heavy metals in farmland soil. *Nongye Gongcheng Xuebao (Beijing)*.

Zhang, X.. (2017). Recent developments in UAV LiDAR systems for forest applications. *Linye Gongcheng Xuebao*, *28*(2), 1–12.

Zhang, X.. (2024). Environmental degradation of mangrove systems due to microplastic accumulation in estuarine environments. *Environmental Toxicology and Chemistry*.

Zhang, Y.. (2020). Spatial distribution of microplastics in mangrove sediments of the Beibu Gulf. *Environmental Pollution*.

Zhang, Y.. (2024). UAV-based detection of environmental pollutants in coastal ecosystems using machine learning. *Marine Environmental Research*, *170*, 105379.

Zhang, Z.. (2024). Seagrass habitat mapping using drone and machine learning approaches in shallow coastal waters. *Remote Sensing Letters*.

Zhao, L. (2015). *Review of retrieving soil heavy metal content using hyperspectral remote sensing*. Remote Sens. Technol. Appl.

Zhao, L., Locke, J., Xu, F., Yao, T., & Guo, X. (2025). Accurate segmentation of localized corrosion in structural alloys via deep learning. *npj Materials Degradation, 9*(1), 80.

Zhao, J.. (2024). Assessing coastal ecosystem health with UAV-based remote sensing and deep learning techniques. *Marine Pollution Bulletin, 169*, 112332.

Zhao, Z., Bakar, E. B. A., Razak, N. B. A., & Akhtar, M. N. (2025). Assessment of corrosion image rating based on transfer learning. *Arabian Journal for Science and Engineering*, *50*(15), 11803–11820.

Zheng, Y., Deng, H., Wang, R., & Wu, J. (2023). *A Multimodal Learning Framework for Comprehensive 3D Mineral Prospectivity Modeling with Jointly Learned Structure-Fluid Relationships. arXiv*. Cornell University.

Zhou, Q.. (2020). LiDAR-based estimation of carbon stock in tropical coastal forests: A focus on mangroves. *Remote Sensing*, *12*(8), 1337.

Zhou, W. (2021). Hyperspectral inversion of soil heavy metals in Three-River Source Region with random forest. *Catena*.

Ziegler, S.. (2015). Mapping blue carbon ecosystems using remote sensing and data assimilation models. *Remote Sensing*, *7*(11), 15201–15217.

Zielinski, J. P. T., Hamerski, F., Vecchia, F. D., Melo, C. L., & Reginato, P. R. (2024). Hydrogeological assessment and seasonal hydrogeochemical monitoring of the TECNOPUC-Viamão CO_2 controlled-release experimental site, Rio Grande do Sul state, Southern Brazil. *Environmental Earth Sciences*, *83*(16).

Zuo, Y.. (2020). Microplastic distribution and characterisation in global mangrove sediments. *Marine Pollution Bulletin*.

About the Contributors

Uzair Aslam Bhatti was born in 1986. He received the Ph.D. degree in information and communication engineering, Hainan University, Haikou, Hainan, in 2019. He is pursuing the Postdoctoral degree in implementing Clifford algebra algorithms in analyzing the geospatial data using artificial intelligence (AI) with Nanjing Normal University, Nanjing, China. His areas of specialty include AI, machine learning, and image processing.

Di Wu currently holds the position of Lecturer at the School of Electronic Science and Technology, Hainan University, China. In March 2025, Di Wu earned his Ph.D. degree in Control Science and Engineering from Shanghai Jiao Tong University, Shanghai, China. From October 2023 to October 2024, he participated in a Ph.D. joint training program at the University of Naples Federico II, Naples, Italy. His research interests primarily focus on nonlinear control of aerial vehicles, multi-agent communication, and localization and mapping

***

Sivapathi A. is currently working as Assistant Professor in SASTRA University, Thanjavur. His research interests includes Aritificial intelligence, Machine Learning, and Image Processing.

Ulugbek Bobamuratov is a faculty member in the Department of Information Technology and Exact Sciences at Termez University of Economics and Service, Termez, Uzbekistan. He is engaged in teaching and academic activities in information technology and exact sciences, with interests in applied computing, data-driven problem solving, and technology-enhanced education. His work focuses on strengthening students' analytical and technical competencies through contemporary pedagogical approaches, and he contributes to curriculum development and academic initiatives supporting innovation and quality education in higher education.

T.C. Manjunath was born on 6th February 1967. He did his B.E. from RV College of Engg. in 1989 (B'lore Univ), M.E. in Robotics & Automation from Govt.'s LD College of Engg., (Gujarat Univ., Ahmedabad) in 1992 with FCD-First Rank & PhD in Systems & Control Engg. from IIT Bombay in the year 2007, got best research scholar award for the entire IITB. He is currently working as the Dean Research (R & D) and as Professor in Dept. of Computer Science & Engg. with specialization in IoT, Cyber Security, Blockchain Technology in Rajarajeswari College of Engineering, Bengaluru, Karnataka. He has got a teaching (academic), research & administrative experience of more than 35+ years (10 years as Principal & 8 years as HoD, 1 year as Dean Research & rest as faculty) and has published more than 1500+ research papers in various national & international journals & conferences, with 175+ patents, 100+ book-chapters, 25 awards, 100s of Keynote talks and also published 25 text books in various field of science, engineering & technology, completed projects worth Rs. 9 Crores till date & guided 10 PhD scholars in VTU. He is a member of 30 Professional Society Chapters and is a fellow of IE, IETE, AMIEE, Sr. Member of IEEE and a chartered engineer.

S. Murugaveni is an Assistant Professor in the Department of Electronics and Communication at SRM Institute of Science and Technology. She is actively involved in teaching and academic activities within the field of Electronics and Communication Engineering.

Keerthana N V is an Assistant Professor with 8 Years of experience in teaching and 1-year experience in research (IUCAA LIGO Data Centre, Pune). Her area of interest is Network Security, Blockchain Security, and Machine Learning Algorithms. She is pursuing PhD at Anna University in Security in Blockchain. She has a working experience in High-Performance Computing (HTCondor) in LIGO INDIA Project. She has written many research articles in Network Security and assisted Scientists in analyzing Datasets in HTCondor.

P. Selvakumar currently serves as an Associate Professor in the Department of Science and Humanities at Nehru Institute of Technology, Coimbatore, Tamil Nadu, India. With a teaching career spanning over a decade, he previously held positions as an Assistant Professor in the Department of Science and Humanities at various engineering colleges in Coimbatore from 2011 to 2023. Dr. P. Selvakumar's primary area of academic and research interest is phytochemistry, where he has made substantial contributions. His research credentials include 30 journal publications, 14 patents, and 25 conference papers. He has actively promoted faculty and student development by organizing 5 Faculty Development Programs (FDPs) and completing 3 NPTEL courses along with 60 Coursera certifications. His dedication to student success is evident in his role as a committed mentor, consistently guiding and supporting his students academically and personally. Recognized for his excellence in teaching, Dr. P. Selvakumar has received numerous accolades, including Best Teacher Awards. He is also an active member of MISTE (The Indian Society for Technical Education) and continues to play a vital role in both acade

Varun P. is an Assistant Professor in the Department of Data Science and Business Systems at the School of Computing, SRM Institute of Science and Technology, Kattankulathur. His academic interests include data science, analytics, and business systems, with a focus on teaching, research, and mentoring students. He is dedicated to promoting data-driven thinking and practical applications in computing and interdisciplinary domains.

S. Revathi is a Professor in the Department of Computer Science and Engineering at B.S. Abdur Rahman Crescent Institute of Science and Technology, Vandalur, Chennai – 600 048. She is actively engaged in teaching, research, and academic development in the field of Computer Science and Engineering.

Ibrahim Sinneh Sinneh is an accomplished electrical engineer and academic leader dedicated to the optimization of electrical distribution systems. He currently serves as the Head of the Department of Electrical Engineering at the University of Liberia, a position he has held since 2023. With a strong foundation in both practical engineering and academic research, Ibrahim specializes in power system planning, grid management, and the integration of renewable energy technologies. He is committed to ensuring the reliable and efficient supply of electricity to consumers through robust system design and maintenance. Academic Background Ibrahim is currently advancing the field of electrical science through doctoral research. Ph.D. in Electrical and Electronics Engineering (In Progress) University of Johannesburg, South Africa (Started Feb 2023) Master of Engineering (M.Eng) in Electrical Engineering Chongqing University, China (2018 – 2020) Bachelor of Science (B.Sc.) in Electrical Engineering University of Liberia, Monrovia (2010 – 2016) ** specialized Training & Certifications** Ibrahim has actively pursued global professional development to enhance his expertise in infrastructure and design. Infrastructure Operations and Maintenance (Certificate) ADBInstitute, Tokyo, Japan (Feb 2024) Electrical Power Systems Design and Protection (Diploma) Electrical Power Systems LLC, Oregon, USA (June – July 2023) Electrical Design and Drafting (Diploma) Advance Modeling Recreators MEP Institute of Engineering, Hyderabad, India (Aug – Sept 2022) Key Areas of Expertise Electrical Distribution System Design & Maintenance Power System Planning Grid Management Renewable Energy Integration Academic Leadership & Curriculum Development

Bottu Giridharan Sivakumar is a faculty member in the Department of Mechanical Engineering at the Academy of Maritime Education and Training (AMET), Deemed to be University, located at Kanathur, Chennai, Tamil Nadu, India. He is actively engaged in teaching, academic mentoring, and scholarly activities in the field of mechanical engineering. In his academic role, Dr. Sivakumar contributes to undergraduate and postgraduate education, focusing on developing strong technical foundations, analytical skills, and practical engineering competencies among students. His teaching and academic involvement align with the interdisciplinary and application-oriented approach of maritime and mechanical engineering education. As part of a maritime-focused deemed university, he is associated with academic programs that integrate engineering principles with industry relevance and applied learning. His professional engagement reflects a commitment to academic excellence, continuous learning, and the advancement of engineering education. Dr. Sivakumar continues to support the mission of AMET University through teaching, institutional service, and participation in academic and scholarly initiatives.

V. Vedanarayanan has over 22 years of experience in academia and currently serves as an Associate Professor in the Department of Electronics and Communication Engineering at Sathyabama Institute of Science and Technology, Chennai. He holds a Ph.D. in Medical Image Processing, a B.E. in Electrical and Electronics Engineering from Madras University, and M.E. in Electronics and Controls Engineering from Sathyabama University. As an active IEEE EMBS student mentor, he has guided students in academic, research, and professional pursuits, including hackathons, competitions, and project exhibitions. He has published more than 49 Scopus-indexed papers and 18 Web of Science-indexed journal articles. His research interests include medical image analysis, computational intelligence, AI-based systems, and quantum encryption. A recognized expert in P2BL, he has delivered numerous workshops and hands-on training programs for faculty and students. He has also contributed to collaborative research and successfully completed a funded project with NRSC-ISRO.

Nwosu Obinnaya Chikezie Victor is a multidisciplinary scholar, researcher, and professional with expertise spanning Artificial Intelligence, Environmental Sciences, Geophysics, Cybersecurity, Data Analysis, and Environmental Technology. He is currently pursuing a Ph.D. in Environmental Sciences at the University of South Africa, where his research focuses on integrating intelligent systems into environmental and geophysical applications to enhance sustainability and innovation in industrial practice. He earned his B.Tech in Geophysics from the Federal University of Technology Owerri, Nigeria, in 2007 and completed his M.Sc.

in Environmental Technology at Teesside University, England, in 2011. Over the years, he has undertaken short professional certification programs at the University of Alberta (Canada), the Georgia Institute of Technology (USA), and the Cape Peninsula University of Technology (South Africa). He also holds a Certificate in Artificial Intelligence in the 4th Industrial Revolution and a 4IR Work Readiness Program Certificate from the University of Johannesburg, completed in collaboration with Accenture.

Index

Symbols

B

C

D

E

www.ingramcontent.com/pod-product-compliance
Lightning Source LLC
LaVergne TN
LVHW080904300926
849162LV00003B/542
* 9 7 9 8 3 3 7 3 8 8 7 2 4 *